THz and Security Applications

NATO Science for Peace and Security Series

This Series presents the results of scientific meetings supported under the NATO Programme: Science for Peace and Security (SPS).

The NATO SPS Programme supports meetings in the following Key Priority areas: (1) Defence Against Terrorism; (2) Countering other Threats to Security and (3) NATO, Partner and Mediterranean Dialogue Country Priorities. The types of meeting supported are generally "Advanced Study Institutes" and "Advanced Research Workshops". The NATO SPS Series collects together the results of these meetings. The meetings are co-organized by scientists from NATO countries and scientists from NATO's "Partner" or "Mediterranean Dialogue" countries. The observations and recommendations made at the meetings, as well as the contents of the volumes in the Series, reflect those of participants and contributors only; they should not necessarily be regarded as reflecting NATO views or policy.

Advanced Study Institutes (ASI) are high-level tutorial courses to convey the latest developments in a subject to an advanced-level audience

Advanced Research Workshops (ARW) are expert meetings where an intense but informal exchange of views at the frontiers of a subject aims at identifying directions for future action

Following a transformation of the programme in 2006 the Series has been re-named and re-organised. Recent volumes on topics not related to security, which result from meetings supported under the programme earlier, may be found in the NATO Science Series.

The Series is published by IOS Press, Amsterdam, and Springer, Dordrecht, in conjunction with the NATO Emerging Security Challenges Division.

Sub-Series

A.	Chemistry and Biology	Springer
B.	Physics and Biophysics	Springer
C.	Environmental Security	Springer
D.	Information and Communication Security	IOS Press
E.	Human and Societal Dynamics	IOS Press

http://www.nato.int/science
http://www.springer.com
http://www.iospress.nl

Series B: Physics and Biophysics

THz and Security Applications

Detectors, Sources and Associated Electronics for THz Applications

edited by

Carlo Corsi
C.R.E.O. Centro Ricerche Elettro Ottiche
L'Aquila, Italy

and

Fedir Sizov
Institute of Semiconductor Physics
National Academy of Sciences of Ukraine
Kiev, Ukraine

Springer

Published in Cooperation with NATO Emerging Security Challenges Division

Proceedings of the NATO Advanced Research Workshop on
THz and Security Applications- Detectors, Sources and Associated Electronics for THz Applications
Kiev, Ukraine
26–29 May 2013

Library of Congress Control Number: 2014938207

ISBN 978-94-017-8830-4 (PB)
ISBN 978-94-017-8827-4 (HB)
ISBN 978-94-017-8828-1 (e-book)
DOI 10.1007/978-94-017-8828-1

Published by Springer,
P.O. Box 17, 3300 AA Dordrecht, The Netherlands.

www.springer.com

Printed on acid-free paper

General Introduction

These proceedings have been written in an attempt to communicate the major purpose of the NATO Advanced Research Workshop (ARW), 2013, that is, to bring to light the possibilities of performance, based on the actual level, of the ever-promising THz (terahertz) technology, a kind of *Araba Fenice*, not yet known to most technical operators, especially its appeal in security applications.

To achieve this, the ARW has invited highly experienced scientists with expertise in THz science and technology and its application areas.

We begin with the consideration that the risk of mass murder due to terroristic attacks is on the rise, thus posing a threat for security in the civil and military world. To counter this problem, we look at one of the most appealing, newly emerging, technologies that is based on the THz detection of explosives and other forms of threats. However, operational difficulties (both for THz sensors and sources), especially regarding size, complexity of use, overall cost, and the need of very low temperatures for sensors, strongly limit the application of this technology. To find solutions to these and related issues, we invited expert scientists to present review papers on the most advanced sensors and sources based on THz technology, especially for security system applications.

The ARW has been conferred the major task of describing the most advanced technologies, in terms of identifying their operational strengths and weaknesses, forecasting the best technological solutions to overcome the actual operational limits, and suggesting to the NATO SPS (Science for Peace and Security) Division the most reliable ways to proceed for future developments.

To achieve a broad evaluation of the above aspects, a questionnaire on various key points with regard to the actual performance and possible future developments in the field of THz science, technology, and applications has been discussed.

Preface

Background and Justification

Compact detectors, both uncooled and cooled but not below liquid nitrogen temperatures, powerful sources and associated electronics are the key points that define the ultimate performance of THz imaging systems, both passive and active. Modeling and simulation of these components (including, e.g., matrix arrays and antenna simulation and optimization) being an important aspect made it essential also to include in the meeting specialists in this area. Further progress in physics-based modeling and simulation is important for developing better and cost-effective THz vision systems. The development of high-performance parts of vision systems, as well as of the systems as a whole, and their modeling and simulation will continue to be of interest to NATO for several reasons: First, these parts are important system components for, for example, surveillance and nonintrusive border security applications, standoff detection of explosives, and other military and security applications, as they qualify the system parameters. Second, development in the fields mentioned is likely to promote the design of more advanced THz vision systems. The development of THz imaging systems to provide the detectors, sources, and associated electronics capabilities is complex and costly. The design and manufacturing of advanced detectors, sources, and associated electronics, especially uncooled or slightly cooled parts, must be aimed toward reducing the cost and extending application capabilities of THz vision systems.

The objective of the meeting has been to review the current state of the art in uncooled or slightly cooled THz detectors, sources, and associated electronics for THz vision systems. The purpose has also been to provide a clear view on the current technologies and the required advances to achieve more efficient systems with the aim of establishing the baseline of current uncertainty estimates in physics-based modeling and identifying key areas requiring further research and development.

Topics have addressed detectors and sources physics and detector design and technology-based manufacturing and have included questions concerning associated electronics design as well as modeling and simulation of electronic circuits and antennas. The proceedings have provided deep analyses and discussions based on the experience acquired over the past years in developing advanced detectors, sources, and associated electronics.

Scientific Codirectors:
Prof. Carlo Corsi (Leader Country: Italy)
Prof. Fedir Sizov (Hosting Country: Ukraine)
Prof. Antony Rogalski (Cooperating Country: Poland)

List of Participants

Speakers

1. Carlo Corsi (co-director), *C.R.E.O. Centro Ricerche Elettro Ottiche – L'Aquila, Italy*
2. Antony Rogalski (co-director), *Institute of Applied Physics, Military University of Technology, 2 Kaliskiego St.,00-908 Warsaw, Poland*
3. Fedir Sizov (co-director), *Institute of Semiconductor Physics, Kiev, Ukraine*
4. Mykola Sochinskyi, *C.R.E.O. Madrid, Spain*
5. Manijeh Razeghi, *Center for Quantum Devices, Department of Electrical Engineering and Computer Science, Northwestern University, Illinois, USA*
6. Gregory Goltsman, *Moscow State Pedagogical University, Moscow, Russia*
7. Miriam Vitiello, *NEST, CNR Istituto Nanoscienze – Scuola Normale Superiore, Pisa, Italy*
8. Vincenzo Spagnolo, *Dipartimento Interateneo Fisica, Università Politecnico di Bari, Italy*
9. Hartmut Roskos, *Physikalisches Institut J. W. Goethe University of Frankfurt am Main-Germany*
10. Woichiech Knap, *Lab. Charles Coulomb University & TERALAB Montpellier2 & CNRS, France*
11. Norbert Palka, *Institute of Optoelectronics, Military University of Technology, Warsaw, Poland*
12. Alexander Shkurinov, *Moscow State University, Moscow, Russia*
13. Oleg Mitrofanov, *Electronic and Electrical Engineering, University College London, UK*
14. Francesco Palma, *NTT – New Tera Technology, 10126, Torino, Italy*
15. Sandro Mengali, *C.R.E.O. L'Aquila, Italy*
16. Ashok Vaseashta, *Institute of Advanced Sciences Convergence and Int.al Lean Water Institute NUARI Herndon, VA 20171, USA*

17. Arvydas Lisauskas, *Physikalisches Institut J. W. Goethe University of Frankfurt am Main-Germany*
18. Stepan Lucyszyn, *Centre THz Science and Engineering, Imperial College London, UK*
19. Olga Kosareva, *Moscow State University, Moscow, Russia*
20. Anton Somov, *Institute of Radioastronomy National Academy of Sciences of Kharkov, Ukraine*

Participants

From Ukraine

1. S. Maltsev, Vice Director, Scientific Industrial Enterprise *NDI "ORION", Kiev*
2. N. Boltovets, Dpt. Head, Scientific Industrial Enterprise *NDI "Orion", Kiev*
3. V. Shulga, Vice Director, Institute of Radioastronomy, *Khar'kov*
4. D. Vavriv, Institute of Radioastronomy, *Khar'kov*
5. E. Melezhyk, Institute of Semiconductor Physics, Kiev
6. Z. Tsybrii, Institute of Semiconductor Physics, Kiev
7. V. Gaivoronskii, Institute of Physics, Kiev
8. Oleksandr Belyaev, Institute of Semiconductor Physics, Kiev
9. Aleksandr Golenkov, Institute of Semiconductor Physics, Kiev
10. V. Zabudsky, Institute of Semiconductor Physics, Kiev
11. M. Sakhno, Institute of Semiconductor Physics, Kiev
12. K. Svezhentcova, Institute of Semiconductor Physics, Kiev
13. V. Dobrovolski, Institute of Semiconductor Physics, Kiev
14. V. Dykhanovskii, National Academy of Sciences, Kiev

From other Countries

15. Dimitry But, Lab Charles Coulomb Universitè & TERALAB, Montpellier 2 &CNRS, France

Auditors

1. Zinaida Kopylova, National Academy of Sciences, *Kiev, Ukraine*
2. Slava Khrustaleva, National Academy of Sciences, *Kiev, Ukraine*
3. Raisa Konakova, Institute of Semiconductor Physics, *Kiev, Ukraine*
4. Kostetskii Volodymyr, National Academy of Sciences, *Kiev, Ukraine*
5. Volodymyr Reva, Institute of Semiconductor Physics, *Kiev, Ukraine*
6. Joanna Gumenjuk-Sychevska, Institute of Semiconductor Physics, *Kiev, Ukraine*
7. Natalia Kuhtaruk, Institute of Semiconductor Physics, *Kiev, Ukraine*

8. Svetlana Bunchuk, Institute of Semiconductor Physics, *Kiev, Ukraine*
9. Anna Evmenova, Institute of Semiconductor Physics, *Kiev, Ukraine*
10. Olga Bunchuk, Institute of Semiconductor Physics, *Kiev, Ukraine*
11. Kathryn Andreeva, Institute of Semiconductor Physics, *Kiev, Ukraine*
12. Angelika Luchenko, Institute of Semiconductor Physics, *Kiev, Ukraine*
13. Ruslana Udovytska, Institute of Semiconductor Physics, *Kiev, Ukraine*
14. Anna Shevchik-Shekera, Institute of Semiconductor Physics, *Kiev, Ukraine*
15. Vyacheslav Kochelap, Institute of Semiconductor Physics, *Kiev, Ukraine*
16. Boris Glavin, Institute of Semiconductor Physics, Kiev, *Ukraine*
17. Rada Savkina, Institute of Semiconductor Physics, *Kiev, Ukraine*
18. Mikle Vinoslavskii, Institute of Physics, *Kiev, Ukraine*
19. Lyudmila Karachevtseva, Institute of Semiconductor Physics, *Kiev Ukraine*,
20. V. Tetyrkin, Institute of Semiconductor Physics, *Kiev, Ukraine*
21. G. Beketov, Institute of Semiconductor Physics, *Kiev, Ukraine*
22. V. Lozovski, National University, Kiev, Ukraine
23. M. Dmitruk, Institute of Semiconductor Physics, *Kiev, Ukraine*
24. V. Babentsov, Institute of Semiconductor Physics, *Kiev, Ukraine*

Questions and Answers of the Round Table

WHAT ARE THE STRENGTHS AND WEAKNESSES OF THz TECHNOLOGY, BASED ON YOUR LAB'S EXPERIENCE?

The most advanced labs attending the ARW answered this question with regard to two main THz technologies: detectors and sources.

THz detectors developed on the basis of field-effect transistors (FETs), using standard foundry technology of silicon as well as of GaN, with integrated silicon readouts, thus making possible on-chip signal processing and real-time focal plane image acquisition (A. Rogalski, W. Knap, and F. Sizov), were underscored (H. Roskos). Mercury cadmium tellurude (MCT) THz (bolometer-type) detectors that allow multielement array implementation with integrated silicon readouts, thus making possible on-chip signal processing and real-time focal plane image acquisition, were underlined as short-term, high-performance detectors to be achieved. A strong possibility of growth for room-temperature microbolometers, like VaO_x, was underlined (C. Corsi). Fast THz detectors recently approaching the issue with novel electronic and photonic concepts, state-of-the-art ($\sim 6 \times 10^{-11}$ W/$\sqrt{}$Hz) noise equivalent power (NEP) at room temperature, high speed, and cutoff frequencies up to $\sim$10 THz have been reached with 1D nanowire Si FET detectors. Promising results with room-temperature graphene field effect transistor (FET), already exploitable for practical real-time imaging and flexible arrays, have also been achieved (M. Vitiello). New techniques like the quartz enhanced photoacoustic spectroscopy (QEPAS) technique in the THz range, where no optical detection is required, would allow to avoid the use of low-noise but expensive, bulky, and cryogenic bolometers (R. Spagnolo).

All experts agreed with H. Roskos on the main tasks to be achieved that are

(i) To be able to develop cameras for real-time or near-to-real-time applications
(ii) To link THz technology with mainstream microelectronic technology and to take advantage of its many benefits: unit size reduction and fabrication reproducibility, high production yield, and mainly cost scalability (the near-exponential reduction of unit price with increasing production volume).

Most of the experts underlined the importance of experimental research and acknowledgment theory of THz detectors, electro magnetic (EM) modeling of the entire detector system, determination of the part of the detector system that contributes most to signal losses, and development of THz detector parameters measurement systems. All experts emphasized that there have emerged several cycles of upswings and downswings of the industrial acceptance of THz technology in the last years, but until now, it has not been a big success story. In fact, high costs and complexity remain to be important issues.

(F. Sizov)

Key items in our Labs, experience are the following devices and technologies. Detector's physical and technological parameters research, readout circuits design for THz multielement detectors. Conventional technologies (Si CMOS) of multielement detectors with integrated silicon readouts, making possible on-chip signal processing and real-time focal plane image acquisition. MCT THz detector (bolometer-type) technology allowing multielement array implementation.

WHAT IS THE STRONG POINT OF THE TECHNOLOGY DEVELOPED IN YOUR LAB?

(N. Palka)

In my presentation, I showed two technologies:

(a) The developed real-time processing and fusion software of THz images that provides better visibility and perception of the detected items.
(b) The p-spectrum method for the detection of covered materials in the reflection configuration. An advantage of the proposed method is that spectral features of materials are clearly visible even through two to three layers of covering.

(H. Roskos)

We develop THz detectors on the basis of field-effect transistors, using exclusively standard foundry technology of silicon as well as that of GaN. The reason to select foundry technology for our work was twofold: (i) to be able to develop cameras for real-time or near-to-real-time applications and (ii) to link THz technology with mainstream microelectronic technology and to take advantage of its many benefits: unit size reduction and fabrication reproducibility, high production yield, and mainly cost scalability (the near-exponential reduction of unit price with increasing production volume). We have seen several cycles of upswings and downswings of the industrial acceptance of THz technology, but until now, it has not been a big success story. We believe that high costs and complexity remain to be important issues. Conventional waveguide-based THz systems – with their serial and manual production processes – as well as optoelectronic THz approaches continue to be expensive technologies even with increased production volumes. Nevertheless, they retain their importance for ultrahigh-sensitivity applications (such as security screening at a distance) in the case of waveguide technologies and for broadband-spectroscopy applications in the case of optoelectronics. But many applications do not need the performance characteristics of these technologies. The transistor approach helps to make the transition from waveguide technology to planar technology, which is better suited for parallel production processes. In addition, transistors have the unique characteristic of allowing to create focal plane array chips for cameras, exploiting the camera technology developed for the VIS (visible), NIR (near-infrared), and IR (infrared) to the fullest.

An added advantage is that the transistors are sensitive devices and are efficient up to very high frequencies.

"From my group, there was also a presentation on standoff security imaging systems. This work was entirely separate from that of transistor-based detectors and, in my opinion, altogether speaks of coherent multiemitter/multidetector systems with synthetic image reconstruction. If applications allow for a wide field of view (such as static assemblies at fixed points of investigation in halls), then one can work with a good spatial resolution in the 100 GHz regime where sources and detectors are quite mature and impressively inexpensive" (H. Roskos).

Concerning THz sources, based on the impressive developments achieved in the last years (M. Razeghi and M. Vitiello) through groundbreaking research on both the fundamental and applied aspects of THz QCL (quantum cascade laser) design and implementation, we are actually developing the next generation of QCL technology that is defined by, and can be employed by, the European Photonics platform. High-power (100 mW), broadly tunable (10 % emission frequency), high-directional ($<3^{\circ}$ divergence) THz QCL sources have been recently demonstrated in the most advanced research labs. Great improvements in QCL emitted power are also expected in a short time, although working operation at room temperature might need longer times to be achieved (M. Razeghi, M. Vitiello).

The lack of comparably compact, efficient, long-lifetime emitters at higher frequencies with enough power to drive active cameras has been underlined. This quite excludes optoelectronic systems with photomixers. While microelectronic sources at hundreds of GHz will soon be available, there are highly powerful oscillators at tens of GHz, especially those made using the GaN material system, which are tremendously useful. Frequency multiplication coupled with re-amplification allows to reach an ever larger part of the sub-THz region. Although the progress in Si and InP technologies is impressive, sources of 1 THz and above continue to be lacking (C. Corsi, W. Knap, H. Roskos, F. Sizov, M. Vitiello). There is an urgent need to introduce QCLs that can reach an operating temperature equal to the room temperature, or to introduce entirely new concepts with this capacity which does not currently seem possible; however, it may be achieved through vacuum electronics.

(M. Vitiello)

THz quantum cascade lasers can be used for various technological applications, such as homeland security, pollution monitoring, process control, countermeasures, and clinical diagnostics. In this respect, they also have a huge potential for improvements in the quality of life. Through groundbreaking research on both the fundamental and applied aspects of THz QCL design and implementation, we are actually developing the next generation of QCL technology that is defined by, and can be employed by, the European Photonics platform. High-power (100 mW), broadly tunable (10 % emission frequency), high-directional (<3° divergence) THz QCL sources have been recently demonstrated in our research labs. THz imaging systems are actively investigated for many appealing applications in diverse fields. Acquisition time is often crucial, and so detector technologies with high-speed multipixel arrays are highly desired. Commercial THz cameras are based on microbolometer sensors, offering moderate sensitivity and speed. Alternatively, electronic devices based on rectification of the incoming radiation hold great promise; performances, though, decrease rapidly above a few hundred GHz. We made a breakthrough innovation: fast THz detectors recently approaching the issue with novel electronic and photonic concepts, state-of-the-art (6×10^{-11} W/$\sqrt{}$Hz) NEPs at room temperature, high speed, and cutoff frequencies up to 10 THz have been reached with 1D nanowire Si FET detectors. Promising results with room-temperature graphene FETs, already exploitable for practical real-time imaging and flexible arrays, have also been achieved.

Applications

(C. Corsi, W. Knap, N. Palka, H. Roskos, F. Sizov, V. Spagnolo, A. Vaseashta, M. Vitiello)

Numerous applications for THz technology, besides the major one for security, are foreseen by all the participants, especially in environmental monitoring, medical sciences, and quality and process controls. THz sensing techniques are very promising in the area of quality and process controls, which makes this an important topic of discussion as in nondestructive evaluation of materials, for example, composite materials, plastic-based structures (especially novel materials and structures: metamaterials, liquid crystals, etc.). The distinct features of THz waves, such as their sensitivity to water, could be used to control food and agricultural products. For example, damage to fruits can be evaluated and the water content in vegetables monitored. Within the industrial and agro-food sectors, compact and high-speed THz cameras capable of providing instant quality-control information of products on conveyer belts are needed. The nature of the imaged finished products does not always allow harmful radiation like X-rays to be used. THz beams can easily pass through packing materials while being harmless to humans, animals, and food.

(C. Corsi)

THz imaging and, even more, THz spectroscopy will be key, invaluable diagnostics tools for specific cancers (skin cancer, breast cancer, prostate cancer, etc.):

(N. Palka)

1. Nondestructive evaluation of materials, for example, composite materials and plastic-based structures
2. Novel materials and structures, for example, metamaterials and liquid crystals.

(H. Roskos)

Short term: Definitively nondestructive testing. This applies to the transistor-based detection per se as well as to synthetic-aperture imaging with any type of emitter and detector.

Middle term for transistor-based detection: Gas sensing, chemical- and biosensor development in a broader sense; food inspection, and communications applications.

Long term for transistor-based detection: THz microscopy with detector chips, medical imaging, and cameras for robotics. THz technology will play an integral role in multisensor applications.

(F. Sizov)

Customs and postal packages control, food production quality control, and for any application where low cost and uncooled operation of THz are the main factors

(M. Vitiello)

(a) Sensing (atmospheric sensing and also explosive diagnostics)
(b) Cultural heritage
(c) Biology (DNA chips)
(d) Medicine (skin cancer diagnostics)
(e) Astronomy
(f) Process control
(g) Tomography
(h) Defense
(i) Wireless communication
(j) Biochemical analysis

THz imaging systems are actively investigated for many appealing applications in diverse fields. Acquisition time is often crucial, and so detector technologies with high-speed multipixel arrays are highly desired. Commercial THz cameras are based on microbolometer sensors, offering moderate sensitivity and speed. Alternatively, electronic devices based on rectification of the incoming radiation hold great promise; performances, though, decrease rapidly above a few hundred GHz. We made a breakthrough innovation: fast THz detectors recently approaching the issue with novel electronic and photonic concepts, state-of-the-art (6×10^{-11} W/$\sqrt{}$Hz) NEPs at room temperature, high speed, and cutoff frequencies up to 10 THz have been reached with 1D nanowire Si FET detectors. Promising results with room-temperature graphene FETs, already exploitable for practical real-time imaging and flexible arrays, have also been achieved.

Important THz QCL applications will be developed in various fields, such as homeland security, pollution monitoring, process control, countermeasures, and clinical diagnostics. In this respect, they also have a huge potential for improvements in the quality of life. Through groundbreaking research on both the fundamental and applied aspects of THz QCL design and implementation, we are actually developing the next generation of QCL technology that is defined by, and can be employed by, the European Photonics platform. High-power (100 mW), broadly tunable (10 % emission frequency), high-directional (<3° divergence) THz QCL sources have been recently demonstrated in my research group.

DO YOU SEE ANY WEAK POINT IN YOUR R&D WORK ESPECIALLY FOR SHORT- TO MEDIUM-TERM APPLICATIONS?

(N. Palka)

Imaging: Visibility of the hidden items depends on many factors, such as the size of the item, its temperature, attenuation of the clothes, and the number of worn clothes. It is also difficult to develop automatic recognition of the detected items due to low sensitivity and resolution of the THz camera.

P-spectrum: Disadvantages are associated with specular type of reflection and sensitivity to water vapor influence – at least now. The p-spectra are distorted by the covering layers, but we cannot expect that in the THz range we can see the material through ten layers of clothes. Due to the particular spectrum of the incident pulse, the method is limited to about 1.8 THz and only detection of RDX-based (Ciclonite or T4) explosives is possible at the present time. But shorter pulses should provide better sensitivity for higher frequencies and allow to measure also HMX and pentaerythritol tetranitrate (PETN) spectra. The method is limited to short-range applications. I still believe that standoff detection of explosives is possible but is associated with many problems, such as small contrast of explosives, water vapor influence, shape of the sample, coverings, lack of powerful tunable sources, and sensitive detectors.

(H. Roskos)

Yes, the lack of comparably compact, efficient, long-lifetime emitters at higher frequencies. They must have enough power to drive active cameras. This quite excludes optoelectronic systems with photomixers. While microelectronic sources at hundreds of GHz will soon be available, there are highly powerful oscillators at tens of GHz, especially those made using the GaN material system, which are tremendously useful. Frequency multiplication coupled with re-amplification allows to reach an ever larger part of the sub-THz region. Although the progress in Si and InP technologies is impressive, sources of 1 THz and above continue to be lacking. There is an urgent need to introduce QCLs that can reach an operating temperature equal to the room temperature, or to introduce entirely new concepts with this capacity which does not currently seem possible; however, it may be achieved through vacuum electronics.

(F. Sizov)

Insufficiency of modern equipment. To realize the active vision systems, powerful THz sources are needed. FET is expensive for prototyping. R&D manufacturing circle is long (this is the consequence of CMOS usage). Conventional complementary metal-oxide semiconductors (CMOS) technology is optimized for digital applications. Common design methods should be modified taking into account the effectiveness of signal power input in detecting elements, and FET technology should be developed for high-frequency issues (gate resistance, parasitic effects, and antenna impedance matching with FET).

(M. Vitiello)

Reliable THz QCL devices operating at room temperature are still missing.

LEVEL OF COMPETITIVENESS OF THz IN COMPARISON WITH WELL-ESTABLISHED TECHNOLOGIES LIKE INFRARED FOR SPECTROSCOPY AND MONITORING IN WARNING AND SURVEILLANCE, AND MICROWAVES IN MONITORING/STANDOFF SENSORS? WHAT ABOUT COST?

(N. Palka)

I think that due to high costs, THz technologies can find real application only in areas where other above-mentioned technologies cannot be used. The mm-wave/THz scanners and cameras are prominent technologies in this area, but their widespread use is limited mostly by their costs.

THz detectors of explosives and drugs, due to serious limitations associated with attenuation of water and covering materials, are restricted now to postal scanners and other short-range applications.

(H. Roskos)

Certainly, IR is much more advanced technologically, and it has an edge over gas sensing, and over chemical sensing in a broader sense. It has the advantage of being able to probe the rather narrow-band vibrational modes instead of the broadband rotational ones in the THz range. Also, it is an excellent thermal sensing material. But THz has also some unique properties: It can pass through many container materials, and it has proven its worth in the diagnosis of microcrystalline materials (powders), which abound in pharmaceuticals and explosives, via the phonon response of crystal lattices. Fully developed IR systems with calibrated performance are not so cheap that there is no room for THz systems. We have contact with a company which uses Fourier transform IR spectroscopy for trace gas detection, and a similar system for the THz regime could costwise well be made with the new generation of optoelectronic THz measurement systems.

Microwaves: Conventional waveguide-based sources and detectors for hundreds of GHz provided by Virginia Diodes or Radiometer Physics cost tens of thousands of euros per single device. No big price difference to optoelectronic systems. This price level clearly is a showstopper.

(V. Spagnolo)

Most THz applications are still at lab level, and it will take several years for them to be able to compete with MID-IR (mid-infrared) technologies. On the other hand, for some applications (such as selected gas sensing), the THz technique is much more promising compared to MID-IR. The need to avoid the use of cryogenic systems will become crucial in terms of cost reduction.

(F. Sizov)

Still not advanced enough to be competitive. In some applications concerning, for example, detection and identification of certain gases and substances, the THz technique is much more promising compared to IR. As regards THz imaging passive systems, they are short range (atmosphere absorption) and too expensive due to the use of low-temperature detectors. For THz active real-time imaging systems, the available powerful THz sources are not suitable and need to be further developed.

(M. Vitiello)

Spectroscopy: Many chemical species have very strong characteristic rotational and vibrational absorption lines in the THz range. In particular, for gases the typical absorption strengths are 10^3–10^6 stronger than in the microwave region. *Atmospheric science* is another area of interest for THz spectroscopy. The process of thermal emission from gases in the stratosphere and upper troposphere in the form of water, oxygen, chlorine, and nitrogen compounds is used for the study of chemical processes related to ozone depletion, pollution monitoring, and global warming. Hazardous materials and composites can be easily detected through THz technology, but it takes some time to develop cost-effective systems.

The development of imaging using THz radiation is of interest specifically because many materials that are transparent in the THz frequencies are opaque at visible frequencies and vice versa. THz rays are nonionizing and provide considerable advantages for many appealing imaging applications when compared to sub-THz radiation, mainly the possibility of detecting substance-specific spectroscopic features and sub-mm diffraction-limited lateral resolution. Most of the interest in THz imaging resides in the peculiar ability of THz rays to deeply penetrate into nonpolar and nonmetallic materials such as paper, plastic, clothes, wood, and ceramics that are usually opaque at optical wavelengths. Such materials are frequently used for packaging and coating of many objects of common use, making THz imaging systems ideal tools for noninvasive quality testing and for inspecting sealed containers. Noninvasive medical imaging of teeth or subdermal melanoma has been demonstrated with compact, low-cost THz systems. Monitoring of water levels in plants, fat content in packaged meats, and manufacturing defects in automotive dashboards and high-voltage cables has all been performed. Due to its nondestructive nature, this THz method of testing is actually of peculiar interest for many industrial applications, like high-speed data acquisition, online software analysis, and online monitoring.

Pharmaceutical science is another field which benefits from the new possibilities created by the large number of substances having spectral fingerprints in the THz region. Innovative spectroscopic characterization of materials, opportunities to prevent and detect counterfeit products, and precise control of tablet coatings and densities of active substances have been made possible thanks to THz technology, along with biosensing, stretching, and twisting modes in DNA; plasma diagnostics in fusion experiments; wireless communication; and high-speed signal processing. Nowadays, also promising security applications are realized by means of THz radiation – radar modeling, baggage screening, checkpoint people screening for hidden explosives, etc. – although with *still higher costs*.

WHICH ARE THE MOST URGENT NEEDS FOR RESEARCH AND DEVELOPMENT IN SPECIFIC FIELDS?

(N. Palka)

Spectroscopy: Sensitive, broadband (0.3–3 THz), room-temperature (or above 77 K), high-dynamic-range, and reasonably priced detectors. Widely tunable (0.3–3 THz), narrow-band, powerful, room-temperature (or above 77 K), and reasonably priced sources. Broadband (0.3–3 THz), room-temperature, and reasonably priced sources.

Imaging: Long-range (>25 m) THz cameras with good performance (resolution, sensitivity) and reasonable price.

(H. Roskos)

Definitively emitter development (see point 2 above), but new ideas are missing. So, it is not about lack of money but lack of promising concepts.

(F. Sizov)

THz detectors: Experimental research and acknowledgment theory of THz detectors, EM modeling of the entire detector system, determination of the part of the detector system that contributes most to signal losses, and development of THz detector parameters measurement systems.

(V. Spagnolo)

Development of laser sources and detectors operating at temperatures above 250 K, possibly room temperature.

(M. Vitiello)

Compact room-temperature systems with integrated high-power sources, low NEP detectors, and low-loss waveguides available for practical applications:

- Creation of an extended database for THz materials
- Development of THz standards and measurement systems
- Functional THz devices and materials

WHAT IS THE INNOVATION CYCLE OF THZ TECHNOLOGY?

(N. Palka)

I see the future of the photonic approach in the development of widely tunable, narrow-band, and powerful sources based on nonlinear crystals optical parameter oscillator (OPO) and quantum cascade laser (QCL) for the range of about 1–3 THz.

Electronic approach:

- Development of cheap, sensitive, high-resolution Si-based matrices for imaging
- Development of cheap and efficient Schottky diode-based frequency multipliers up to 2 THz

(H. Roskos)

It is the same as with all fundamentally new physical technologies. It takes two generations (40–50 years): one generation of physicists to develop the core technology and one of engineers, chemists, biologists, etc., to develop applications and bring them to the market outside of research. If the history of THz has started with Auston switches in 1984, then nearly 30 years have already passed. We should be in the second generation by now if we assume a time period of 20–25 years per generation. So there's good hope for broad penetration into the nonscientific market within the next 10–15 years.

(F. Sizov)

Concerning large-scale THz arrays and demonstration of at least two color active vision systems within 3–5 years.

(M. Vitiello)

- Room-temperature, high-power semiconductor-based THz microsources
- High-speed, low-noise, high-dynamic-range THz nano detectors implemented in array configurations
- THz fibers, waveguides, and metamaterials
- THz emission microscopes (dynamic response of materials and devices)

WHAT WILL BE THE ROLE OF NEW TECHNOLOGIES (NANO, PICO, AND ORGANIC ELECTRONICS) IN THZ SCIENCE?

(H. Roskos)

Nano: Is important (isn't a transistor with a gate length of 90 nm or less a nanodevice, and isn't also a quantum-well structure of a QCL or a THz sub-band modulator a nanodevice?). Plasmonics will play an ever more important role, but I guess a lot can happen at the micron scale.

Pico: Picosecond and sub-picosecond THz pulses, generated by femtosecond laser pulses via photomixers, are extremely important for THz spectroscopy because of their large bandwidths, and they are the basis of the new field of time-resolved nonlinear THz science. I don't know whether they will be important for practical ranging or tomographic imaging applications, because frequency-agile CW radiation can do the same as long as the measurement remains in the linear regime.

Organic: Can't see the breakthrough character. Conventional organics has low charge-carrier mobility, which is a big problem. I am pessimistic about emitters and general detectors based on graphene or CNTs. There are good opportunities for passive components (switchable filters, polarizers, etc.) and for specialized sensors, though.

(F. Sizov)

Nano: High impact. The new generation of nanoelectronics can be used for generation/detection in the THz range (e.g., better FET with lower parasitics). Systems based on graphene or on quantum-well structures of zero-gap MCT seem to operate at THz speeds due to high electron mobility and quick energy relaxation times of picosecond range. Nanotechnology should be adjustable for producing multielement THz imaging arrays. Organic electronics does not seem to be very useful as it is much slower than, for example, silicon electronics. Its usage for high-speed applications seems questionable.

(V. Spagnolo)

I believe that the main role will be played by nanostructure-based detectors and laser sources.

(M. Vitiello)

The development of novel and even more advanced nanotechnological procedures is one of the key aspects for the progressive innovation of THz technology, specifically THz quantum cascade laser sources with high-temperature operation, high optical power, and collimated beam profiles; THz photonic crystal emitters; THz plasmonic lenses; THz room-temperature plasma wave nanodetectors; THz modulators; THz waveguides and filters; and THz polarizers.

The last question was about the creation of a NETWORK for INTEGRATION of a COMMON TECHNOLOGY/SCIENCE to SUPPORT MULTI-DISCIPLINARY CROSS-COUNTRY COORDINATION.

(H. Roskos)

This is definitively important because THz devices will be an integral part of multisensor systems, in security applications, nondestructive testing, and perhaps even in the medical field, where there are possibly opportunities for cancer and diabetes detection.

(F. Sizov)

It is strategically important the collaboration in THz spectroscopy, near-field and "ordinary" vision, between NATO and Partner Countries Representatives in research and in making the prototypes as THz field is a multidisciplinary science and technique.

(V. Spagnolo)

I believe that this is a crucial point for the development of the European THz community.

(M. Vitiello)

Creation of a collaborative research program across a broad range of disciplines, including advanced nanofabrication, high-resolution and high-sensitivity spectroscopy, optical metrology, holographic imaging, nonlinear optics, and advanced solid-state theory.

SECURITY APPLICATIONS EXTENDED TO QUALITY CONTROL IN AGRO FOOD

(H. Roskos)

This is an interesting aspect. Not much developed yet. The imager technology, as developed, for example, by Rohde & Schwarz, could possibly be adapted to imaging of packaged food.

(F. Sizov)

Check if common quality deviations of food can be seen in the THz range. In this case, the security system can be used without modification.

(M. Vitiello)

The distinct features of THz waves, such as their sensitivity to water, could be used to control food and agricultural products. For example, damage to fruits can be evaluated and the water content in vegetables monitored.

Within the industrial and agro food sectors, compact and high-speed THz cameras capable of providing instant quality-control information of products on conveyer belts are needed. The nature of the imaged finished products does not always allow harmful radiation like X-rays to be used. THz beams can easily pass through packing materials while being harmless to humans, animals, plants, and food.

ELECTRONICS AND PHOTONICS APPROACHES: WHAT IS THEIR FUTURE IN THz DEVELOPMENT AND WHERE AND WHEN?

(N. Palka)

I see the future of the photonic approach in the development of widely tunable, narrow-band, and powerful sources based on nonlinear crystals (OPO) and QCLs for the range of about 1–3 THz.

Electronic approach:

- Development of cheap, sensitive, high-resolution Si-based matrices for imaging
- Development of cheap and efficient Schottky diode-based frequency multipliers up to 2 THz

(H. Roskos)

See my response to points (1) and (4), where I gave my personal opinion. Both approaches have their role now and in the future. THz mass markets, if they exist (medical? probably nondestructive testing and sensors with a multitude of specific systems), will hardly be optoelectronic. The costs are too high. Especially considering chemical sensors: The competition of cheap electronic solutions is extremely strong.

(F. Sizov)

The electronic approach seems appropriate and will dominate in the near future for operation in the spectral range up to 1 THz or a little lower. The photonic approach is appropriate, for example, for generation at a frequency range higher than 1 THz for QCLs, though they need low-temperature operation as well as photonic THz detectors in this range. As regards THz detectors, the rectification and thermal (including free-carrier heating bolometer) detectors can dominate in both these spectral regions.

(V. Spagnolo)

I believe that electronics will dominate THz applications up to a frequency range of 1 THz, while above 1 THz photonics will be the winning technology.

(M. Vitiello)

Photonics has led the way to the realization of many important THz devices like quantum cascade lasers (QCLs) and uni-traveling-carrier (UTC-PD) photodiodes. The QCL is the most powerful, compact, and reliable CW THz source dominating the 1.2–5 THz range, already exportable in compact Stirling cryocoolers for practical security and in sensing and imaging applications. UTC-PD produces sub-THz waves particularly suitable for wireless communication.

Over the last two decades, photonic technologies for the THz region of the electromagnetic spectrum have experienced an extraordinary breakthrough, thanks to the progresses made in generation, manipulation, and detection of radiation. Many novel techniques have been pioneered, motivated in part by the vast range of practical applications such as high-speed communications, molecular spectroscopy, imaging in security controls, and medical diagnostics.

Continuous advances in the performance of high-frequency electronics have also benefited THz technology. Solid-state electronic devices such as resonant-tunneling diodes and THz single-photon detectors as well as conventional electronic devices such as Schottky diodes, field-effect transistors, and high electron mobility transistors have shown impressive performance as THz and sub-THz detectors.

WHICH ARE THE DISTINGUISHED PROPERTIES BASED ON SPECTROSCOPIC CHARACTERIZATION, AND HOW CAN THESE PROPERTIES BE ENHANCED OR MATCHED TO THz?

(H. Roskos)

For THz sensors, enhancement of the sensitivity will be important. I assume that metamaterials will play a very important role here and in the development of molecular docking sites directly on the THz sensor elements – again, an interdisciplinary task for point (8).

(F. Sizov)

In THz technology, a lot of gaseous substances have their own signatures. Thus, THz spectroscopy can be very selective. But still, at the moment THz spectroscopy is a relatively unexplored area because of the lack of broadband and efficient sources, measurements, and systems. Time-domain spectroscopic systems seem to be complicated and not enough cost effective to be widely used in spectroscopic applications.

(V. Spagnolo)

The main feature of spectroscopic systems is the selectivity, and on these terms a resonant condition is required between the employed laser source and the substance to be detected. In the THz range, many molecules and compounds present their main absorption band. Thus, the use of THz spectroscopic systems may enhance the detectivity of such substances to unsurpassed levels, with respect to visible, infrared, and microwave technologies.

(M. Vitiello)

THz spectroscopy provides a very peculiar tool due to the ability to observe intermolecular vibrations in chemicals and organic molecules where the intramolecular mode appears in the infrared. Via intermolecular vibrations, it is therefore possible to elucidate the dynamics of large biomolecules, thus increasing the knowledge of the human body.

- The study of proteins, DNA, biomolecules, and various cancers is of great importance. The absorption of THz waves is highly sensitive to polar molecules (like water), and unusual reflection from cancer tissues with different hydration levels can be detected.
- THz time-domain spectroscopy systems are largely used in many industrial applications spanning from silicon solar cells to nanocomposites. THz radiation is able to evaluate peculiar semiconductor material properties like mobility, carrier density, plasma oscillations, and doping level. Similar evaluation can be applied to superconductors, manganites, ferromagnetic, multiferroics.
- Many complex organic molecules have characteristic absorption bands in the THz range, while metal objects can also be easily identified due to their high reflectivity; these properties explain the high interest in THz technology

for security applications: weapons, explosives, and illegal drugs can be easily identified even when concealed underneath typical wrapping and packaging materials.

- The spectroscopy is also widely used to investigate and localize fault points or to inspect foam insulation used for space shuttles.
- THz sensitivity to water could be used to control food and agricultural products. For example, damage to fruits can be evaluated and the water content in vegetables monitored.
- THz environmental monitoring of most atmospheric compounds: (OH, H2O, HCl, O3, ClO HOCl, BrO, HNO3, N2O, CO, HCN, CH3CN), ozone chemistry to understand global warming, understand how atmospheric composition affects climate, and study pollution in the upper troposphere is to be considered an important application.

Meeting Summary Report

The NATO ARW meeting "THz and Security Applications" ("Detectors, Sources and Associated Electronics for THz Applications") held in Kiev on 27–29 May 2013 evidenced the great interest in the subject by all the participating Countries. The high-level participation by the USA and Western and Eastern European Countries, besides Ukraine, was emphasized by the presence of important and skilled participants from these countries.

The general organization was of a high level, and the local organization, by the Institute of Semiconductor Physics, Ukraine Academy of Sciences, in Kiev, was excellent. A friendly atmosphere prevailed throughout the conference fed by the overall success of the meeting.

A very important conclusion of the meeting was about making a proposal for the creation of an International Network of Centers of Excellence in THz science, mostly present at the ARW, organized possibly with the support of the NATO SPS, that, linked by the new relevant Internet-based tools, could support synergy between NATO R&D centers and Eastern Europe and CIS Countries of the scientific world.

Acknowledgments

The Institute of Semiconductor Physics, Academy of Sciences, Kiev, Ukraine, on behalf of all the participants who attended the NATO Advanced Research Workshop (ARW), wishes to express its deep gratitude to the NATO Science for Peace and Security Division for the financial support and program suggestion and cooperation in its organization.

In addition, thanks are due to Prof, Fedir Sizov and all the staff of the Institute of Semiconductor Physics for their invaluable engagement and support in the meeting organization and to all Ukrainian friends for a friendly and pleasant atmosphere.

Contents

Chapter 1
TeraHertz: Quasi-optics or Submillimeter Waves? History, Actual Limits and Future Developments for Security Systems

Carlo Corsi

Abstract The risk of mass murder due to terroristic attacks is more and more a very dangerous problem to be countered for security in civil and military world. One the most appealing technologies recently emerging is based on THz detection of explosives and other forms of threats, although, up to now, problems in operational devices (both for sensors and sources) are strongly limiting the application of this technology especially considering size, complexity in its use, and overall the cost and the need of low temperature for sensors. A Review Paper on THz most advanced Sensors and Sources Technologies is presented especially for Security Systems applications, possibly in unattended devices at room temperature. Keeping the Rule that "Future is depending from Present and Present has been originated by Past" the history of terahertz science, for learning by historical lessons, is presented and discussed identifying winning solutions and disappointing delusions either technological and/or operational. A Review what terahertz waves are, how they are generated, transmitted and detected and an evaluation for a common technological culture and discussion about what can be expected for the future developments, is outlined.

1.1 Foreword

Terahertz (THz) region of electro-magnetic (E.M.) spectrum has been and is still considered as the "Black Hole", that is the un-known region of electro-magnetic science, even if it has been lightened by explosion of scientific and technological knowledge since the end of the Nineteenth Century thanks to the geniality of few outstanding Scientists as it is reported in the following Chapters.

C. Corsi (✉)
C.R.E.O. Centro Ricerche Elettro-Ottiche, Località Boschetto SS17, 67100 L'Aquila, Italy
e-mail: corsi@romaricerche.it

C. Corsi and F. Sizov (eds.), *THz and Security Applications*, NATO Science for Peace and Security Series B: Physics and Biophysics, DOI 10.1007/978-94-017-8828-1_1,

The reasons for this knowledge gap are various, going from basic technological weakness, like the use of reliable, high sensitivity (possibly room temperature) Detectors and even more, still to recently, the scarcity of reliable, sufficient power (tuneable) Sources.

To this lack of devices, it is to be added that THz Science needs a deep and wide knowledge in so many scientific and technological items (Devices and Systems) that very few Labs and only highly selected Scientists can afford this activity with successful results. These reasons are also justifying why THz are nowadays not widespread in their use although the explosion of applications requirements especially in Security and quite recently in Biomedicine: in fact, while the science-technology knowledge of some Centres of Excellence are demonstrating to have achieved outstanding results in both applications, there are still delays in a widespread use of THz technology based on reliable, simple use Sensors and Sources produced on large scale and purchasable at acceptable prices.

Complying with this Foreword, we shall evidence the basic lessons learnt by historical highlights in THz Science, trying to forecast future developments mainly based on the outstanding results achieved by Centres of Excellence participating to the ARW Meeting.

So, in the ARW discussion steps, after having introduced the history of Terahertz science for learning by historical lessons, we shall review what terahertz waves are, how they are generated, transmitted and detected, having a look for the base of a more diffused technological culture. Then we shall discuss what can be expected for the future developments. Taking in mind that our main considerations should concern the identification and definition of actual problems and then the effort of finding possible solutions originated among expert Scientist. The main task to be achieved is "to identify and propose possible solutions for existing or expected problems and not to sponsor solutions in search of problems".

1.2 Terahertz: A Key Technology for Security – Historical Introduction

Terahertz, as mentioned, is the Science concerning the E.M. spectral band between Infrared and Microwaves and therefore most of the considerations relative to THz are originated by these two Sciences, that is Infrared and Microwaves.

Infrared as part of E.M. spectrum was discovered by Sir William Herschel as a form of light (or radiation) beyond red light. These "calorific" rays renamed infrared rays (infra in Latin means "below") were mainly devoted to Thermal Measurement and for a long time the major advances were due to infrared thermal imaging based on radiometric measurements [1–4].

The link between Infrared and thermal measurements was so strong that Infrared was for a long time synonymous of Thermography. The first device for Radiometry was developed by G. Nobili in 1829 with the fabrication of the first thermocouple, allowing the multielements thermopile development in 1833 by Macedonio Melloni.

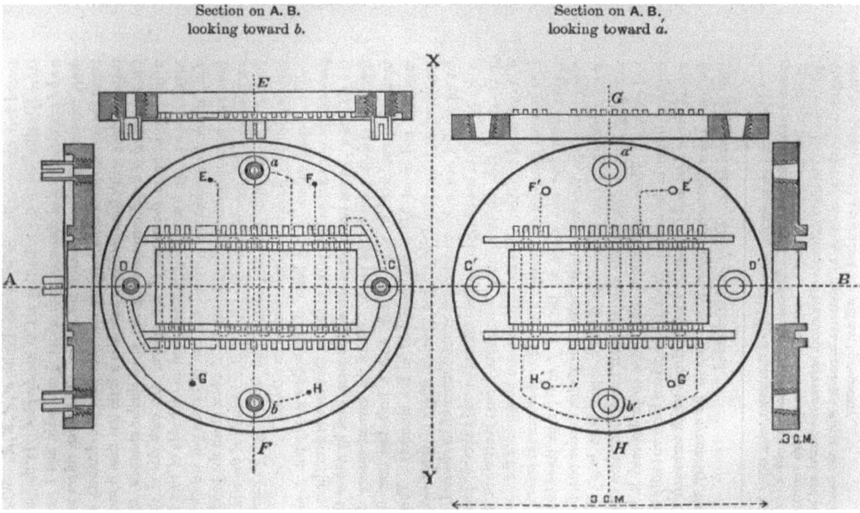

Fig. 1.1 Langley's "bolometer" composed of two sets of thin platinum strips [5]

With this device (First Radiometer) Melloni was able to show that a person 10 m away could be detected by focusing his or her thermal energy on the thermopile. Samuel Pierpont Langley, an American astrophysicist and astronomer, a pioneer in the measurement of the solar constant and the study of the infrared portion of the solar spectrum, in the late 1870s, developed the "bolometer", a device capable of accurately measuring thermal radiation. Langley's bolometer was so sensitive that it could detect the thermal radiation from a cow from 400 m away (Fig. 1.1) [5].

Probably the best example of Security application was developed in 1913 by L. Bellingham who patented "an infrared eye" capable of detecting "icebergs" (an answer to the Titanic tragedy, April 1912).

> *This invention relates to apparatus for detecting the presence at a distance of icebergs, steamships and the like, and, in particular, to arrangements on a ship of a radiation-receiver sensitive to radiations of long infra-red wave-lengths, such as a thermopile or bolometer for example, ... The radiation-receiver which I employ is hereinafter termed an infra-red eye* [6] (Fig. 1.2).

On the other side of the E.M. spectrum, that is "Microwaves", similar events were coming out, that is, as early as 1886, German physicist Heinrich Hertz showed that radio waves could be reflected from solid objects and in 1895, Alexander Popov, a physics instructor at the Imperial Russian Navy, developed an apparatus using a "coherer" tube for detecting distant lightning strikes. The next year, he added a "spark-gap transmitter". In 1897, while testing this equipment for communicating between two ships in the Baltic Sea, he took note of an "interference beat" caused by the passage of a third vessel. In his report, Popov wrote that this phenomenon might be used for detecting objects, but he did nothing more with this observation [7, 8].

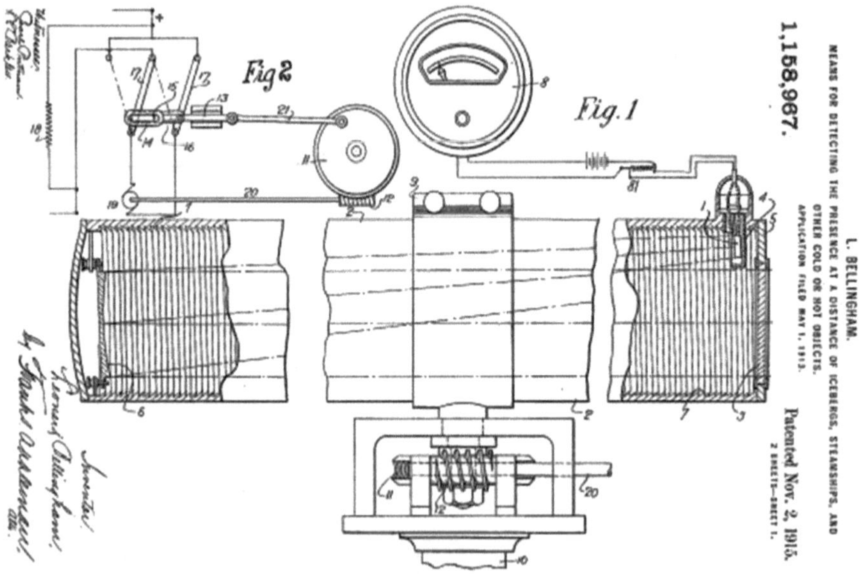

Fig. 1.2 Bellingham's Means for detecting the presence at a distance of icebergs, steamships, and other cold or hot objects [6]

The German inventor Christian Hülsmeyer was the first to use radio waves to detect "the presence of distant metallic objects". In 1904 he demonstrated the feasibility of detecting a ship in dense fog, but not its distance from the transmitter. He obtained a patent for his detection device in April 1904 and he also got a British patent for a full system, that he called a "telemobiloscope". In August 1917 Nikola Tesla outlined a concept for primitive radar-like units [7, 8].

He stated "*... by their [standing electromagnetic waves] use we may produce at will, from a sending station, an electrical effect in any particular region of the globe; [with which] we may determine the relative position or course of a moving object, such as a vessel at sea, the distance traversed by the same, or its speed*" [9].

In the same period of time, around 1900, the Italian Scientist, Nobel Prize (1909) Guglielmo Marconi, was doing first experiments with E.M. waves transmission, applying most of the above researches to the development of E.M. radio-waves transmission [10].

Even more important for Terahertz Science and Technology was the work of an Indian Scientist, J.C. Bose, who developed first THz experiments with new genial Sensors, Polarizer and Sources [11, 12].

The origins of experimental THz developments can in fact be attributed to J. Chandra Bose, defined a Noble Man without Nobel. In fact, in 1894–1896, he developed a semiconductor receiver for detection of millimetre waves (12–60 GHz) (Figs. 1.3, 1.4, and 1.5) [13].

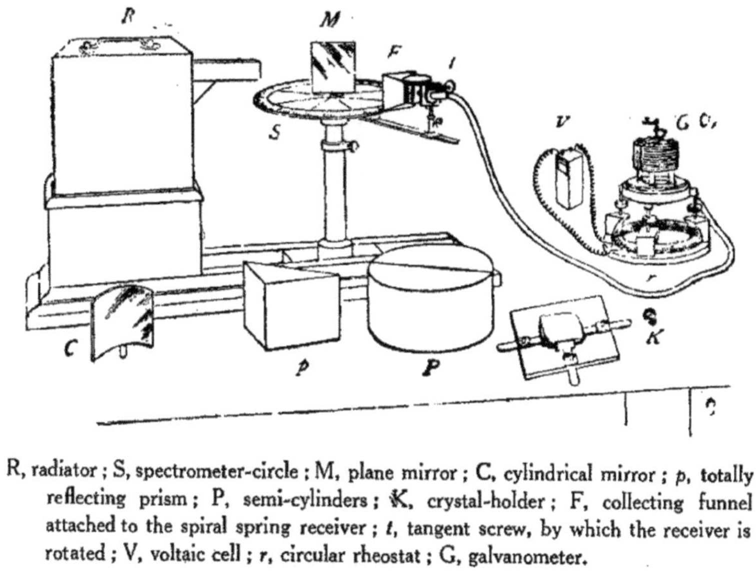

Fig. 1.3 Bose's apparatus demonstrated to the Royal Institution in London in 1897. Note the waveguide radiator on the transmitter at *left*, and that the "collecting funnel" (*F*) is in fact a pyramidal electromagnetic horn antenna, first used by Bose

Fig. 1.4 (**a**) A complete setup showing the transmitting antenna at the *left* and the receiving antenna at *right*. Note the adjustment screw on *top* of the receiving antenna, which is used to adjust the pressure of the point-contact detector. A rotating table (the "spectrometer circle") is present in the centre of the device, on which various microwave components (prisms, lenses, grids, polarisers etc.) may be mounted for study. A dual-prism attenuator is shown in this picture. (**b**) The Coherer receiver

Moreover Bose's U.S. 1904 patent, filed in 1901, was invention of galena crystal detector, a point-contact semiconductor rectifier, later used as a receiver for demodulating continuous wave radio signals, first patent for a semiconductor device in the world. Among his various pioneering solid-state semiconductor receivers are the spiral spring coherer, galena receiver and iron mercury coherer detector with a telephone link that Bose disclosed in 1897 at the Royal Society. In this

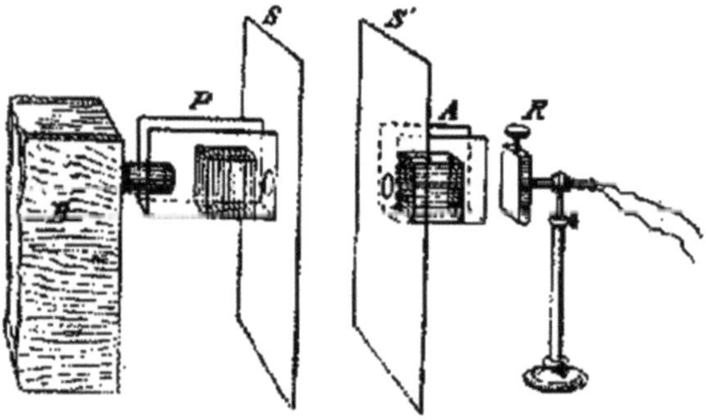

Polarisation apparatus. B, the radiating box ; P, the polariser ; A, the analyser ; S, S', the screens ; R, the receiver.

Fig. 1.5 Bose's diagram of his polarization apparatus. Note the spiral spring receiver (*Cat Whisker*) 'R' to the *right*. One of Bose's polarizer s was a cut off metal plate grating, consisting of a book (Bradshaw's Railway Timetable) with sheets of tin foil interleaved in the pages. Figure 1.6

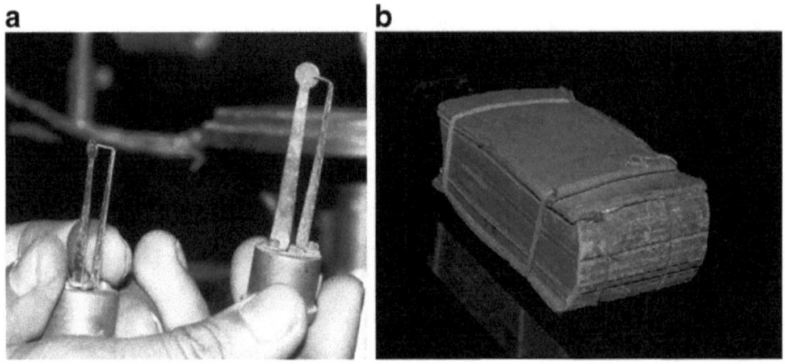

Fig. 1.6 Metal point contact "Cat Whiskers" Diode/THz Polarizer

device, thousands of steel springs (2 mm in diameter and 1 cm in length) were placed side by side in a single layer in the rectangular depression of a square piece of ebonite. The contacts between the springs acted as semiconductor junctions because the fine oxide layer on the spring acted as semiconductor. When electric radiation was absorbed by the sensitive contacts, there was a sudden decrease of the resistance and the galvanometer was deflected. This detector has been called Metal Semiconductor Metal (MSM) detector and described as a "space irradiated multicontact semiconductor (using the natural oxide of the springs)" (Figs. 1.6, 1.7, and 1.8) [20].

It was the Galena detector [14] which he invented some time during 1894–1898 [15, 19], and demonstrated in Royal Institution Discourse in 1900 [20]. In this device, a pair of point contacts (cat whiskers), in this case of galena, was connected in series with a voltage source and a galvanometer. This device could detect any kind of radiation, "Hertzian wave, light waves, and other radiation". He called his galena point contact detector an artificial retina (because by suitable arrangement it could

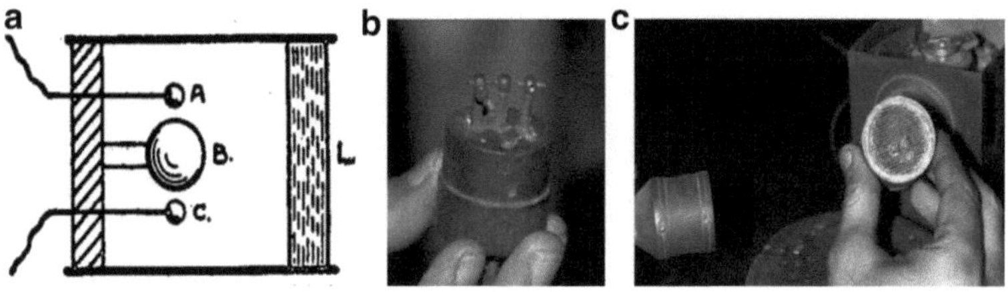

Fig. 1.7 (**a**) Photograph of one of his radiating antennas; part of the spark oscillations are generated inside the over-mode circular waveguide. A polarizing grid is built into the antenna, clearly visible at the radiating end of the waveguide. (**b**) A close-up of the dual spark gaps used for the transmitter; the sparks are generated between the two outer spheres and the inner sphere. (**c**) A transmitting antenna (*left*) and the receiver (*right*)

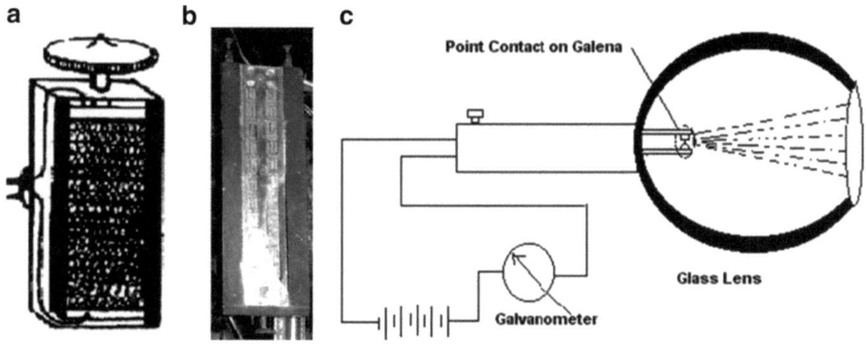

Fig. 1.8 (**a**) Bose's r spiral spring coherer [11]; (**b**) Photograph of Bose's spiral spring coherer [11]; (**c**) Functional diagram of Coherer detector [18]

be made to detect only light wave), a universal radiometer or a tejometer (in Sanskrit "tej" means radiation). According to [17, 20] this device acted as a point contact detector for millimetre waves and as photoconductive detector for light waves and millimetre waves. He intended this device for reception of "signals in wireless or other telegraphy".

From his patent application [18], *"A coherer or detector of electrical disturbances, Hertzian waves, light waves or other radiations, comprising contacting pieces of sensitive substances having a characteristic curve(giving the relation between an increasing electromotive force and the resultant current passing through the sensitive substance), which is not straight but is either convex or concave to the axis of electromotive force and in which the return curve with a decreasing electromotive force when taken slowly, approximately coincides with the former curve, in combination with means for adjusting the force of contact between said contacting pieces."*

The main difference between Bose's detector and early 1900 detectors is that Bose's detectors worked with millimetre waves and his galena detector, first use of a semiconductor crystal as a detector of radio waves directly illuminated by the

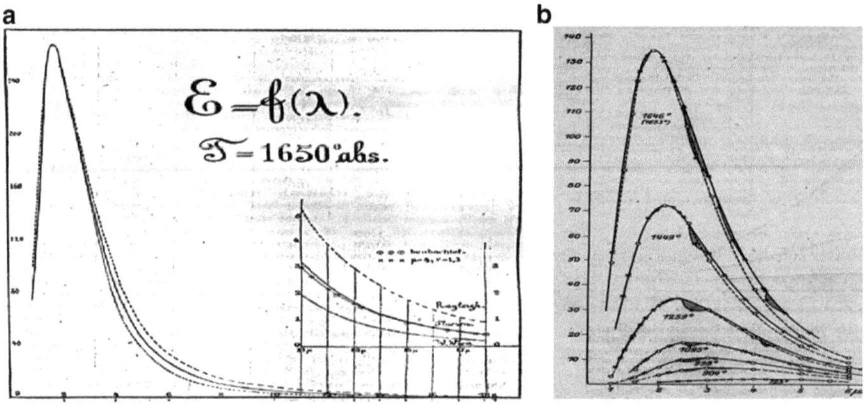

Fig. 1.9 The PTR measurements of radiation spectrum of a black body at different temperatures with the significant derivations from Wien's theoretical radiation law [19, 20]

waves. The review papers [11–14] show that semiconductor diode detectors or the crystal detector for wireless waves originated from Bose's work in the pre 1900 era it is Bose's galena detector which is the forerunner of crystal detectors.

1.2.1 Historical Introduction: The History of THz-Rays – Origins

The concept of "Terahertz waves" as part of the electromagnetic spectrum between electronic and optical regions has been studied for over a century, starting from the famous paper of H. Rubens and E.F. Nichols in which they stated) [21]: Since we have become accustomed to think of waves of electrical energy and light waves as forming component parts of a common spectrum, the attempt has often been made to extend our knowledge over the wide region which was separated the two phenomena, and to bring them closer together, either by cutting down the wave length of electrical oscillations – which, through the remarkable experiments of Righi and later of Lebedev have been reduced to 1/100 of the shortest waves measured by Hertz – or by the discovery and measurement of longer heat waves. "The great difficulty which accompanies the study of infra-red waves of great wave length is that these rays form only the minimum part of the total energy emitted by the flame or incandescent body used as a source. Consequently, if their properties are to be studied, these waves must be separated from others which completely overlap and conceal them." [21].

The study of terahertz of H. Rubens, E. F. Nichols and E. Pringsheim had a great impact on Planck's Radiation Law (Fig. 1.9) [22].

After Rubens' death in 1922, Planck wrote "Without the intervention of Rubens the formulation of the radiation law, and consequently the foundation of quantum theory, would have taken place in a totally different manner, and perhaps not at all in Germany".

1.2.2 Historical Introduction: The Concept of "Terahertz" Dates Back a Century

Few years later, on April 29 1924, Nichols and Tear were presenting the first experimental work with identification and measurement of submillimeter waves [22]. Later, this paper was being orally presented by E. Fox Nichols at the meeting of the National Academy of Sciences, when his death occurred on the platform of the Auditorium and was concerning as first item: "Extension of the short electric-wave spectrum to the wave-length of 0.22 mm", by improving methods of producing and measuring short electric waves. The source was a modified Hertzian oscillator consisting of a doublet of small tungsten cylinders sealed into the ends of glass tube sand the receiver was an adaptation of the Nichols radiometer. The receiving elements, either fine platinum wires or thin platinum films on mica, were warmed by the incident radiation, and the temperature rise was measured by means of the radiometer effect (Langley's Bolometer). Wave-length measurements were made with an interferometer of the Fabry and Perot type. In such a way "Fundamental wave-lengths of small Hertzian oscillators" were found to be approximately $4 \div 5$ times the total length of the doublet. The cylindrical electrodes of the smallest oscillator used were each 0.1 mm long and 0.1 mm in diameter, and the corresponding measured wave-length were 0.9 mm.

1.2.3 Historical Introduction: Recent History

Coming to closer times in the mid-1970s, when the term "Terahertz" has been introduced gaining popularity especially among spectroscopists, (The first occurrence of the term terahertz is attributed to Fleming in 1974, a year earlier, Kerecman applied terahertz to the frequency coverage of point contact diode detectors and Ashley and Palka used the designation to refer to the resonant frequency of a water laser) [23–26].

Later, advanced techniques in optical rectification and photoconduction had been developed making possible to produce THz radiation directly using multimode lasers and FELs. In 1989, a terahertz time-domain spectroscopy (THz-TDS) was developed allowing THz-ray generation and detection. Today, the term "Terahertz" is applied to sub-millimetre/far infrared E.M. radiation that fills the wavelength range between $1,000$ and $100\,\mu$m (300 GHz–3 THz).

1.2.4 Historical Introduction: Actual State and Future

The success of the diffusion of the "Terahertz" term is due to the explosion of applications in recent times which pushed the development of improved performances of THz Sources and Sensors. Terahertz science has become increasingly popular due largely to the advent of time-domain spectroscopy (TDS) with ultra short-pulse laser sources, which make it possible to carry out time-resolved "Far-Infrared" (FIR) studies and to explore spectroscopy and imaging applications in the submillimetre wavelength region especially in Security Systems. Moreover, new advances in different technologies for the development of acceptable or even high, performances Sources and Sensors have made possible to use the previously unused terahertz frequency bands for spectroscopy analysis and even for imaging systems. Nowadays, applications include detection of biological and chemical hazardous agents, explosive detection, applications in building and airport security weapons concealed underneath clothing (airports), environmental monitoring and control, product inspection (industry), spectroscopy (chemistry, astronomy and space research), material characterization (physics) to biomedicine diagnostics (recently have been developed new detection systems for cancer pathologies) and wide bands secure point to point TLC and high definition SAR. Specific THz transmission properties have open new applications: in fact Terahertz waves can penetrate through materials opaque to other parts of the E.M. spectrum. Many non-metallic or non-polar materials (plastics, glasses, woods, cardboards, earth/soil fabric, ceramics) are transparent to terahertz to some degree, therefore hidden or buried layers can be detected in structures containing layers, packaging or clothing [23, 27, 29]. Thanks to the property that terahertz radiation can penetrate many clothing and packing materials with modest attenuation, it makes it a potentially powerful technique in security screening. Moreover, many chemical substances and explosive materials exhibit characteristic spectral responses at terahertz frequencies that can, under certain conditions, identify these objects. Terahertz techniques therefore combine safe-to-use high-resolution imaging and sample identification through spectroscopy, even when samples are hidden under clothes or in packaging/mail materials. Terahertz imaging-systems might be even passive, that is detecting the terahertz part of the thermal blackbody radiation naturally emitted, for example, the human body naturally can radiate about 1 W of power in the THz band. Due to the need of very high sensitive detectors, generally cooled down to very low temperature, active systems are nowadays implemented thanks to the availability of sufficient powerful terahertz source for illuminating the object being imaged [39–43].

1.3 Key Devices: Sources, Filters, Sensors

THz systems and applications as all the E.M. waves are based on three main, key devices: all of them objects of great efforts for the development of reliable and good performances, but really too highly costly, complex and specialized for being used on large scale.

Recently, great improvements in reliable, room temperatures Sources and Sensors are really promising a better future for THz applications. In particular, keeping the distinction from electronics and optoelectronics cultures, for sources, we can distinguish among:

– **Thermal:**

• Lamps/Black Bodies (Globar),

– **Electrical**:

• Gunn Diodes/Mixers,
• Backward Wave Oscillators,

– **Optical/Laser Based:**

• CO_2 Pumped Gas Laser,
• Optical Parametric Oscillator,
• Heterodyne C.W. Photo-mixing,
• Terahertz Pulse Methods.

Looking at THz Sources' history, ignoring Thermal Sources because of their low energy below the far infrared especially in narrow bands for spectroscopy studies (excluding natural high intensity sources, like the Sun for atmosphere-space studies) first Electrical Sources can be attributed to the great Scientists mentioned above, that is J. Chandra Bose and E. F. Nichols and J. D. Tear with their "spark gaps" generating "Short Electric Waves". While for Electrical/Electronic waves the origin can be attributed to Townes' Maser who created a "molecular generator" for the terahertz band (leading to the laser). Historically, the word "Maser" is derived from the original upper-case acronym MASER, which stands for "Microwave Amplification by Stimulated Emission of Radiation" (Fig. 1.10) [28].

As Electrical Sources historically, thanks to Radar technologies, good Sources were Backward-Wave Oscillators (a vacuum tube that is used to generate microwaves up to the terahertz range), although are presenting problems of free electron travels and physical space, with some tuning capability, but limited in range (even if high power output is possible, high voltages are required and some operational difficulties are present). In addition to THz electronics technologies relying on the frequency multiplication mostly using Schottky diodes, Gunn diodes, and IMPATT diodes (see Fig. 1.12) and THz transistor technology has emerged, since the device feature sizes have shrunk to the point, where ballistic mode of electron transport [16] becomes important or even dominant. Hetero-structure Bipolar Transistors (HBTs) and High Electron Mobility Transistors (HEMTs) are capable of operation in the sub-terahertz region, and recently have reached cut off frequencies approaching 1 THz [29].

Different history can be found in Optoelectronics Sources, in fact, after the Free Electron Laser (FEL) with high power, large bandwidth, but with limited use due to the great dimensions, complexity and cost, and the CO_2, pumped Molecular Laser Narrow line with tuneable width and the Difference Frequency Mixing of

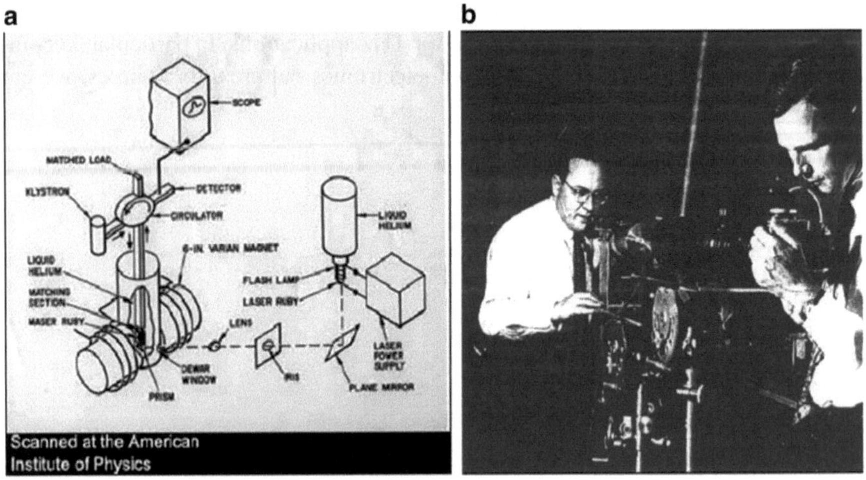

Fig. 1.10 Charles Townes and his colleagues were the first to build a "Maser," which operated in the microwave frequency range. It was the precursor of the laser. Few years later the Maser was extended to Optical and Infrared Wavelengths Regions [34, 44]

two stabilised CO_2, the emerging Solid State Laser Technology is appearing to be quite promising in various technologies. After p-Germanium laser, $10 \div 100$ μW power with $1.5 \div 5$ THz tuning range, but with cryogenic needs and requiring a superconducting magnet, difference frequency generation using two lasers, C.W. or pulsed, and optical beating in a non-linear material/device such as Photoconductive, *GaP*, *GaSe*, allows tuning of terahertz frequencies generation in the range 0.1–5 THz. Last the new innovative technology of Quantum Cascade Laser (QCLs) which is showing to be quite promising. In fact, although the need of a Complex structure to enable population inversion in electronic band-gap, QCLs supply high output powers, even in C.W. operation, and intermediate (even close up to room temperature) operation, at frequencies from 1 THz up, with tuneable laser narrow line [32].

Resuming the main characteristics of Optoelectronics THZ Sources:

– **Free Electron Laser (FEL):**

- high power, large bandwidth,
- Great Dimensions, complexity and cost.

– CO_2 **Pumped Molecular Laser:**

- Narrow line Width. Tuneable!

– CO_2 **Difference Frequency Mixing:**

- Mix two stabilised CO_2 with RF source.
- Needs care stabilization circuit Laser.

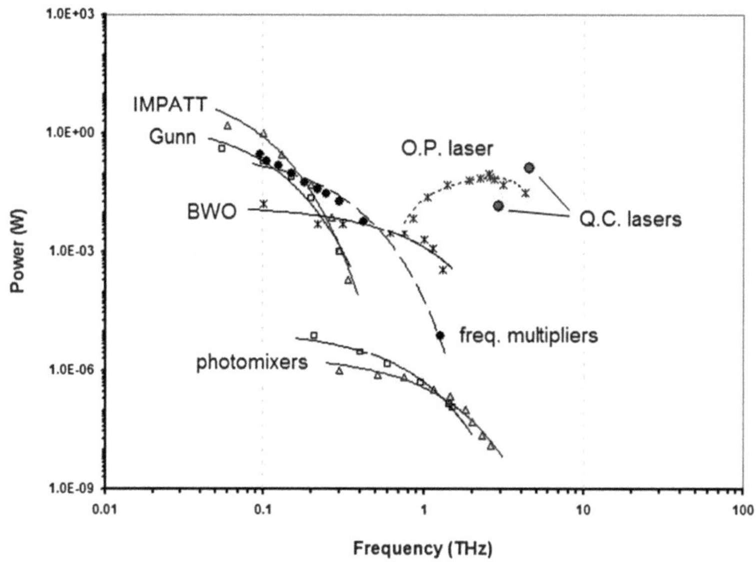

Fig. 1.11 Power output versus frequency of available Continuous Wave *CW* Terahertz and Sub-terahertz sources (Source: Michael Shur Terahertz Electronics CS MANTECH Conference, April 14–17, 2008, Chicago, IL, USA)

– **p-Germanium laser:**

- $10 - 100\ \mu W$ power 1.5–5 THz tuning range,
- cryogenic needs requiring a superconducting magnet.

– **Gunn diodes:**

- High power but limited frequencies available,
- Multiplier can help extend band,
- Heat and output power is a problem.

– **Quantum Cascade Laser (QCLs)** [35]:

- Complex structures enable population inversion,
- High output powers, C.W. operation,
- Intermediate temperature operation,
- From 1 THz up, narrow line. Tunability?

Key Performance to be achieved: Tuneable Room Temperature Sources [33]. In Figs. 1.11 and 1.12 is reported the power output versus frequency of the most relevant solid state THz Sources [29].

Although many of the THz Sources have recently achieved great improvement in their performances, they are still showing some limitations. In fact many are limited in output power or require extensive cryogenic cooling, some are inherently

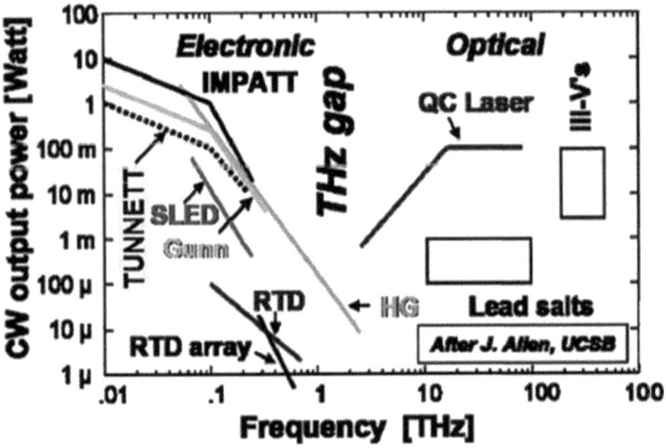

Fig. 1.12 Available solid-state sources. Note: *CW* continuous wave, *TUNNETT* tunnel injection transit time, *SLED* super luminescent diode light source, *RTD* resonant tunnel diode, *HG* harmonic generation, *QC laser* quantum cascade laser, *III-Vs* compounds from the periodic chart columns III–V (e.g., GaAs), *IMPATT* impact ionization avalanche transit time and gunn (Source: Heribert Eisele, University of Leeds, Leeds, UK)

limited to pulsed operation and others are limited by their size, complexity or cost. For these reasons efforts for developing the "Perfect THz Source" is going on with this distinction between the two main techniques. In the microwave band side, acceptable performances have been achieved by solid-state electronic devices, such as transistors, Gunn oscillators and Schottky diode multipliers, with the limit of rolls off owing to both transit-time and resistance – capacitance effects and of the available power generated above 1 THz which is generally limited to milliwatt level. In the Photonic side terahertz sources were mainly showing limits due to the lack of appropriate materials with sufficiently small bandgaps: for example the longest wavelength lead salt laser diodes do not extend below 15 THz. Despite these limitations, a wide variety of techniques have been developed to generate radiation above 1 THz, many, for example, based on down-conversion from the visible regime by using nonlinear or photoconductive effects, or by multiplication up from the millimeter-wave regime, or through direct generation, such as with optically pumped molecular gas lasers or free-electron lasers [27, 33].

A great change is now foreseeable in Solid State lasers due to the advent of Quantum Cascade Laser QCL, which is really changing the forecasting of future successful applications: particularly the extension to THz wavelengths is a real breakthrough that allows to forecast advanced spectroscopy applications in Security and Biomedicine [27]. Since the first QC laser was demonstrated in the near-middle IR Region at Bell Labs in 1994, Quantum Cascade Lasers have become the dominant mid-infrared sources, with spectral coverage from wavelengths of Near IR up to Far IR [48–50]. In October of 2001 the first QC laser with a photon energy

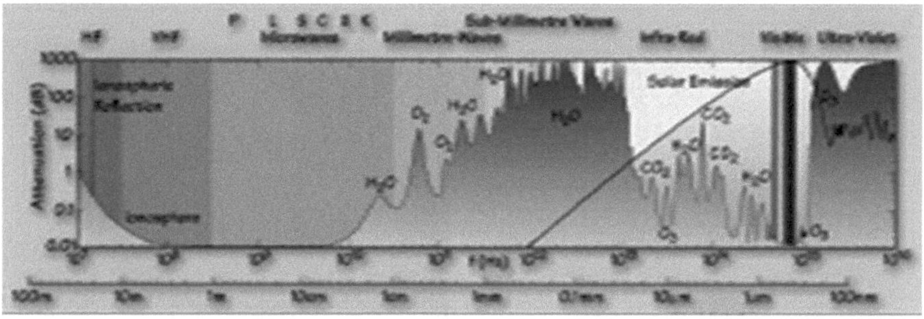

Fig. 1.13 Atmospheric absorption effects (Source: Mark J. Rosker and H. Bruce Wallace, "Imaging through the atmosphere at terahertz frequencies," IEEE MTT-S International Symposium, June 2007)

less than the semiconductor optical phonon energy was demonstrated at 4.4 THz ($\lambda = 67$ μm) at Scuola Normale Superiore in Pisa, Italy [35].

Although these first devices lased only in pulsed mode with peak powers of a few milliwatts and ceased lasing above temperatures of several tens of Kelvin, intensive research over the past few years has produced rapid improvements: nowadays quantum-cascade lasers can deliver hundreds milliwatt or more in continuous wave Recently at the Institute of Photonics – University of Technology Wien they have reached one Watt by using more semiconductor layers: a higher number of layers means that the electron changes its energy states when it passes through the structure, and therefore the number of emitted photons increases. The realization of such multi-layer structures by merging two separate quantum cascade lasers in a so-called bonding process has been achieved allowing one watt of radiation, that is quite an important step for THz applications in a variety of technological fields [45–47, 51].

1.3.1 Filters

Considering the Filters Block and their applications, mainly in spectroscopy, it is important to take vision of atmospheric transmission in the THz Band where a strong absorption, especially due to water vapour, is present (Fig. 1.13).

First examples of Filters for Selective transmission were mainly based on electronics culture with the development of metal mesh filters, which have proven in the 1970s to be the best choice for band selection in FIR/submillimetre wave instrumentation. Their performances are based on the modelling that the optical transmission properties of metal meshes can be accurately modelled by considering single mesh as a lumped circuit element in a free space transmission line. Importantly, the lumped element impedance for metallic structure is only dependent on the

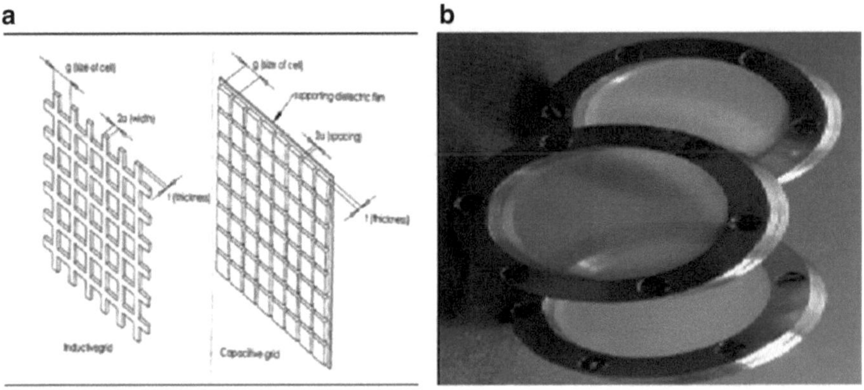

Fig. 1.14 Quasi-optics multilayer filters with selective transmission capacitive and inductive grids used in metal-mesh filters [30]

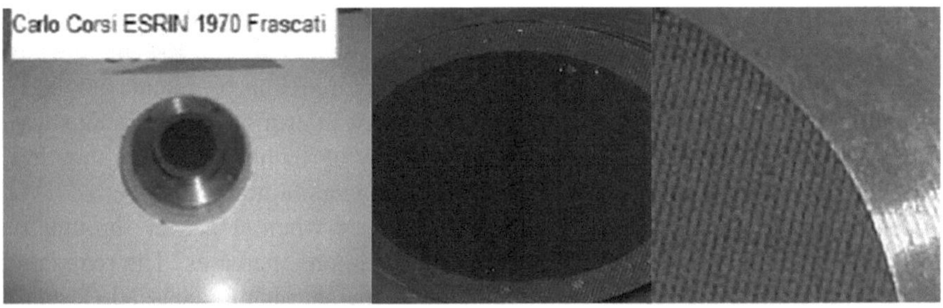

Fig. 1.15 Quasi-optics multilayer filters with selective transmission capacitive and inductive grids used in solid state metal-mesh filters [31]

geometric properties of the metallic pattern. This analogy between the geometrical parameters and the optical response is a key parameter as it allows designing the required optical properties before manufacturing [30]. Examples of Multilayer Mesh Metallic Filters for Selective Transmission with capacitive and inductive grids structure have been realized at ESRIN Frascati Rome in 1970 [31].

Interference THz Multilayer Filters were realized by depositing thin films of gold layers on thin Mylar films (few microns thick) with ten soft stainless steel spacers for waves tuning. The Project was developed by evaluating the equivalence with a tuning filter network based on Capacitors (Squares) and Inductances (Crosses) with a side dimension of 25 μm (as Detector was used an InSb Liquid Helium Sensor Bolometer), Figs. 1.14 and 1.15.

The advent of QCLs is strongly reducing the need of pass band filters in Spectroscopy applications.

1.3.2 *Terahertz Detectors*

The last block of THz technology is concerning the Detectors, key devices that till recent times (before the development of reliable and powerful Sources) were really the key devices for any application. Also in this case it is possible to arrange a rough table of the main characteristics of different technologies:

Thermal:

- Low Temperature Bolometer•••Golay Cell•••Pyroelectric device.

Electrical:

- Photo-acoustic•••Diode•••Optical/Laser

Based Terahertz Pulse Methods. In this technique based on Electro-Optic (EO) Coherent Detection, there is normally a pump-probe method and requires the laser pulse used in optical rectification generation to sample the detected signal. Terahertz co-propagates with optical beam through a non-linear crystal and encodes a polarisation change on the optical beam proportional to the terahertz field strength achieving excellent bandwidth and sensitivity [36].

- Bolometers, are Thermal device Frequency insensitive: operating at 4 K show excellent sensitivity,
- Golay Cell, a Thermal device working at room temperature with acceptable sensitivity, but slow, vibration sensitive, easily damaged and suffering of ageing and linearity limits.
- Photo-acoustic sensors are based on closed air-cell and a pressure transducer which detects terahertz radiation. Calibration can be provided by ohmic heating of a thin metal film within the cell with good sensitivity and easiness to operate.
- Pyroelectric are thermal detectors which convert dielectric constant changes into electrical signal, with limited sensitivity but easy to use.
- The new FPAs Microbolometers, show reliable, easy working, with acceptable sensitivity performances [2, 3, 37–39, 52].

In particular, recently good performances have been obtained with uncooled Microbolometer technology which has enabled: low cost, high number pixels arrays. one of the most reliable and high performances has been obtained using Vanadium Oxide (VO_x) thin film material as thermally isolated microbridge with high Temperature Coefficient of Resistance (TCR). These microbolometres show sensitivity from infrared up to submillimetre wavelengths with good dynamic range and enough sensitivity (D* up to 10^{10} cm·Hz $^{1/2}$/W).

Moreover microbolometres show reduced system complexity, low cost, no cooler or scanner with high reliability, lightweight and outstanding value of completely integrability with silicon microcircuit (Fig. 1.16a) [37].

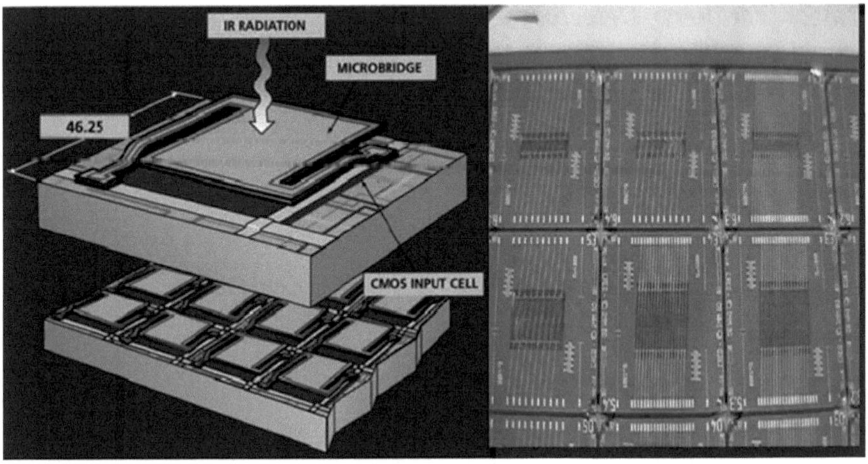

Fig. 1.16 Microbolometers: (**a**) Scheme of microbolometres [37], (**b**) CREO linear arrays of large area microbolometres (active area 280 × 5,000 μm) with high signal to noise [38]

Choice of large area (out of mainstream) is justified by applications for which high *S/N* is more important than short time constant (Fig. 1.16b) [38].

Recently Detection of terahertz radiation from quantum cascade laser has been demonstrated using VO_x microbolometer FPAs [52].

1.4 THz Systems: Security Applications

Recent terroristic events have led to great changes to the techniques employed in security screening, in fact after the discovery of hidden shoe bombing, it is common the request for shoes to be removed and X-rayed at airport checkpoints. There is therefore an increasing focus on new technologies that can be applied to security screening, either to simplify or to render more reliable the check-in process. Terahertz technology is a promising, emerging candidate [39–42].

The delay and resistance in THz use up to now are to be imputed to the so-called "Terahertz Gap", mainly due to the fact that until recently there was a lack of high power, reliable sources and high sensitivity, low cost detectors operating between 100 GHz (3 mm) and 10 THz (30 μm), that is between the millimetre -microwave part of the E.M. spectrum and the far-infrared. Recently, most of these limits are overcome and on the microwave side, new sources (recently solid state) are capable of generating enough power radiation efficiently produced at frequencies above 100 GHz, whereas on photonic side new solid state laser sources, especially QCL (Quantum Cascade Laser), limited up to now by thermal effects in their performances have been developed with performances that allow to foresee generation and detection of THz radiation truly commercially viable in short terms. Indeed, room-temperature lasers and even within the range of thermoelectric coolers at about 240 K, would facilitate the use in many applications, particularly those where portability and/or power efficiency are important (e.g. hand-held sensors for detection of explosives or bio-agents and environmental gas control). It seems probable that continuing advances with existing designs and new epitaxial growth can allow operation over 240 K and surely with much greater emitted power.

Fast and effective detection of explosives is a key security issue against possible terrorist attacks. Especially at public places like airports, railway or coach stations efficient detection systems are needed. Usual methods at airport security checks are wiping carry-on baggage/laptop computers etc. to collect samples which are analyzed subsequently [43] by e.g.

- gas chromatography mass spectroscopy (GC-MS);
- gas chromatography chemo luminescence (GC-CL)
- ion mobility spectrometers (IMS).

Great effort are expected in near future for Standoff detection showing hidden explosives and other items on personnel especially with selectivity and specificity capable of detecting differences in the spectra of two or more species. In general as for any Surveillance, Detection and Alarm Systems for detection of explosives as well as chemical warfare agents, very good sensitivity, very low false alarm rate and specificity are required as a must.

Recently terroristic events, based on new types of threats and explosives, have pushed towards the developments of new techniques of detection and alarm employing different parts of the electromagnetic spectrum, particularly extending from infrared to terahertz radiation, that is the E.M. band between infrared and microwave. Generally in new developments there is a priority to the developing of imaging systems, at least for the know-how originated in infrared technologies, while spectroscopic detection is more developed by microwaves techniques knowledge.

Radiation at terahertz frequencies has unique properties that may be advantageous for Security applications, because it can penetrate many non-conducting materials, but unlike X-rays it is non-ionizing. The short pulses produced by laser techniques can allow radar-like imaging in three dimensions, as well as the simultaneous collection of spectroscopic information as in magnetic resonance imaging (MRI) or in optical spectroscopy. This is important because many substances have characteristic intermolecular vibrations at THz frequencies that can be used to characterize them completely. Many molecules absorb light of this spectral region in a very characteristic way and thanks to this kind of "spectroscopic fingerprint", terahertz radiation can be used for bio-chemical detectors playing an important role for medical imaging and, being a non-ionizing radiation with energy considerably lower than that of roentgen radiation and therefore it is not dangerous. On the other hand, its wavelength is shorter than that of microwave radiation, which means that it can be used to create higher resolution images.

Therefore Terahertz technology, which is a non-ionizing radiation, is really a powerful technique in security screening applications thanks to the following properties [39–43].

- Spectroscopy allows detecting and identifying different chemicals, thanks to their characteristic spectral signatures, even when hidden inside dress clothing.
- 2D Terahertz imaging capable of making visible metals and even plastics and ceramic, materials that are hard to be detected using backscatter X-ray.
- High resolution 3D imaging thanks to the extremely short pulses used in pulsed THz techniques like in the radar technology (e.g. layers of powder can be detected and resolved inside a mail envelope).

Multispectral (IR-THz) Systems for Security applications will have high growth especially coupled to new solid state tuneable Laser Sources with innovative, new technologies for Multispectral Multi-domain Smart Sensors:

- IR/THz Imaging/Pattern Security Applications,
- IR/THz Spectroscopy for Security Applications,
- Stand-Off Laser Detection IR/THz Spectroscopy.
- Gas chromatography mass spectroscopy (GC-MS).

1.5 Conclusions

Terroristic events, based on new types of threats and explosives, have pushed towards the developments of new techniques of detection and alarm employing different parts of the electromagnetic spectrum, particularly extending from infrared to terahertz radiation, that is the E.M. band between infrared and microwave. Also in these new developments there is a priority to the developing of imaging systems, at least for the know-how originated in infrared technologies, while spectroscopic detection is mostly developed by microwaves techniques. Recently the emerging Terahertz technologies [11], thanks to their non ionizing radiation and their detection capability of hidden objects in clothing and in packaging containers and luggage, coupled to the spectroscopic detection of plastic explosives and other chemical and biological agents, are the most promising technologies for integrated, efficient systems for security screening and counter-terrorism. Radiation at terahertz frequencies has unique properties that may be advantageous for security applications, in fact, can penetrate many non-conducting materials, but unlike X rays is non-ionizing, and can allow radar-like imaging in three dimensions thanks to the extremely short pulses used in pulsed terahertz techniques, as well as the simultaneous collection of spectroscopic information like Infrared. This is important because many substances have characteristic intermolecular vibrations at Far IR-THz frequencies that can be used to characterize them as molecules like IR spectroscopy can detect bond vibrations of single molecules. Many molecules absorb light of this spectral region in a very characteristic way – they can be considered to have an "optical fingerprint". Because of this, terahertz radiation can be used for chemical detectors. It also plays an important role for medical imaging; one the one hand, it is non-ionizing radiation, its energy is considerably lower than that of roentgen radiation, therefore it is not dangerous. On the other hand, its wavelength is shorter than that of microwave radiation, which means that it can be used to create higher resolution images.

Terahertz technologies, thanks to their non ionizing radiation and their detection capability of hidden objects in clothing and in packaging containers and luggage, coupled to the spectroscopic detection of plastic explosives and other chemical and biological agents, are very promising technologies for integrated, efficient systems for security screening and counter-terrorism-attacks. Far-infrared/micro-waves techniques and Imaging Hyper Spectral Sensors will be mostly developed for their selectivity and specificity to detect differences in the spectra of two or more species of chemical warfare agents (CWA) allowing detection of explosives with high sensitivity and very low false alarm rate. Smart Sensors Alarm Systems is expected. Multispectral (IR-THz) Systems for Security applications will have high growth especially coupled to new solid state tuneable Laser Sources with innovative, new technologies for Multi-Spectral Multi-Domain Smart Sensors.

Lastly, key technologies recently arising in new THz sources, e.g. THz quantum cascade lasers allow forecasting of an impressive growth of THz time-domain spectroscopy for the realization of strategically important medical applications in diagnostics and prognosis especially in cancer diseases; Moreover the understanding of THz interaction with polycrystalline powder materials will open new technological scenarios of potential security/screening applications besides new tools for pharmaceutical industry. Evidence of this growing interest is evidenced by important R&D EC Programmes (FP5 "TERAVISION", "WANTED", FP6 "TERANOVA" and FP7 "PEOPLE", Projects TERACOMP e TERACAM) and the rapid growth in industrial interest with the creation of active, high tech-Companies, e.g. NTT (Italy), Qinetiq, QMC, Picometrix, TeraView Thruvision (UK), Synview (Germany), TrayScience (Canada), Traycer, Toptica(US), etc. And overall a high number of R&D Centres, mainly linked to Universities and National Research Councils, which have achieved impressive results in THz new Sources.

Finally although no real mass-market application can be highlighted, because the technology does not meet yet the user requirements especially in easiness of use and costs, many THz applications have now emerging and showing a rapid growth, especially if costs fall and technology becomes more portable/compact, that is in presence of these future expected developments:

- Development of cheaper, more compact, imaging systems, e.g. based on THz QCLs, etc.;
- Development of high sensitivity, compact, cheap, multi-pixel detectors, working at room temperature (e.g. microbolometers);
- Development of on-chip technologies: micro and nanostructures;
- Development of THz spectra of biological materials (e.g. proteins), through design of appropriate experimental apparatus;
- Wider use of THz technology across the medical, biological and physical science.

Finally, the questions if THz is a solution looking for the right problems? Or are THz applications now looking for the right technology? Probably both answers are true, also if the new developments in THz QCLs might give positive answers to both questions.

Main References to Be Consulted

Martyn Chamberlain "Where optics meets electronics: recent progress in decreasing the terahertz gap". Philosophical Transactions of the Royal Society of London A (2004) Vol. 362, pp. 199–213

Peter H. Siegel "Terahertz technology". IEEE Transactions on Microwave Theory and Techniques" (March) Vol. 50, No. 3, p. 910

Edmund Linfield (School of Electronic and Electrical Engineering, University of Leeds) "Terahertz Background & Overview Terahertz: Solutions Looking for the Right Problems?". www.etn-uk.com

Richard Dudley, Mira Naftaly (National Physical Laboratory, Teddington, UK) "Introduction to Terahertz the Ins and Outs of Terahertz Technologies, What It Does and How It Works". www.npl.co.uk/17

Martyn Chamberlain (Physics Department, Durham University), **Mark Sherwin** (Institute for THz Science and Technology ITST/UCSB University of California) "Terahertz Spectroscopy Comes of Age". www.itst.ucsb.edu

Michael Shur (ECSE Physics and Centre for Integrated Electronics, Rensselaer Polytechnic Institute, Troy, NY) "Terahertz Electronics". http://www.csmantech.org/Digests/2008/2008%20Papers/2.2.pdf

D. T. Emerson (National Radio Astronomy Observatory Tucson, ArizonLab) "The work of Jagadis Chandra Bose: 100 years of mm research". IEEE Transactions on Microwave Theory and Techniques (December 1997) Vol. 45, No. 12, pp. 2267–2273

V. P. Wallace, E. MacPherson, J. A. Zeitler, C. Reid "Three-dimensional imaging of optically opaque materials using nonionizing terahertz radiation". Journal of the Optical Society of America A (2008) Vol. 25, No. 12

"Terahertz Sensing Technology" Vol. 1 (Selected Topics Electronics and Systems Vol. 30)

(eds) **D. Woolard, W. Leorop, M. S. Shur** (World Scientific, Singapore, 2003)

"Terahertz Sensing Technology" Vol. 2 (Selected Topics Electronics and Systems Vol. 32)

(eds) **D. Woolard, W. Leorop, M. S. Shur** (World Scientific, Singapore, 2004)

References

1. Barr ES (1960) Historical survey of early development of the IR region. Am J Phys 28:42–54
2. Corsi C (2007) IR technologies: history lessons new perspectives. In: Strojnic M, SPIE (eds) Proceedings of AITA IX conference Mexico, 4–19
3. Corsi C (2010) History highlights and future trends of infrared sensors. J Mod Opt 57(18):1663–1686, 09 April, Tutorial review
4. Herschel W (1800) Experiments on the refrangibility of the visible rays of the sun. Philos Trans R Soc Lond 90:284–292
5. Langley SP (1880 May–1881 June) The bolometer and radiant energy. Proc Am Acad Arts Sci 16:342–358
6. US 1158967 (1913) An application filed May. Patented 2 Nov 1915
7. Radar – Wikipedia, the free encyclopaedia en.wikipedia.org/wiki/Radar
8. Kostenko AA, Nosich AI, Tishchenko IA (2003) Radar prehistory, soviet side. In: Proceedings of IEEE APS international symposium 2001, vol 4, Boston, 8–13 July 2001, p 44
9. Belrose JS (1994) The sounds of a spark transmitter: telegraphy and telephony. Radio Scientist Online, 22 Dec
10. Marconi G (1901) British Patent 18 105, applied for Sept 10
11. www.minhas.net – Jagadish Chandra Bose
12. Bose JC (1901, Patented Mar 29) Detector for electrical disturbances, US Patent 755840. Application filed, 30 Sept 1904
13. Emerson DT (1997) The work of J. Chandra Bose: 100 years of millimetre-wave. IEEE Trans Microw Theory Tech 45(12):2267–2273
14. Pearson GL, Brattain WH (1998) History of semiconductor research. Proc IRE 43:1794–1806
15. Sengupta DL et al (1998) Centennial of the semiconductor diode detector. Proc IEEE 86(1):235–243
16. Asif Islam Khan (Pre-1900) Semiconductor research and semiconductor device applications. Bangladesh University of Engineering and Technology 28/3 BUET Quarters, Dhaka-1000, Bangladesh
17. Sen AK (1997) Sir J.C. Bose and radio science. Microwave symposium digest. IEEE MTT S Int 2:557–560
18. Bose JC (1904) Detector for electrical disturbances, US Patent 755840. Application filed 30 Sept 1901, Patented Mar 29, 1904
19. Shahidul I Khan, Asif I Khan (2004) An appreciation of Bose's seminal research on semiconductors. 3rd International Conference on Electrical & Computer Engineering ICECE 2004, 28–30 December, Dhaka, Bangladesh
20. Bose JC (1997) On the selective conductivity exhibited by certain polarizing substance. Repr Proc IEEE 86(1):244–247, Microwave Symposium Digest. IEEE MTT S Int 8–13 June 1997, 2:557–560
21. Rubens H, Nichols EF (1897) Heat rays of great wave-length. Phys Rev 4(314):1897
22. Nichols EF, Tear JD (1923) Joining the infrared and electric wave spectra. Phys Rev 21:3–78
23. Chamberlain M (2004) Where optics meets electronics: recent progress in decreasing the terahertz gap. Philos Trans R Soc Lond A 362:199–213
24. Fleming W (1974) High resolution submillimeter-wave Fourier-transform spectrometry of gases. IEEE Trans Microw Theory Tech MTT-22:1023–1025
25. Kerecman J (1973) The tungsten – P type silicon point contact diode. IEEE MTT-S Int Microw Symp Dig 30–34
26. Ashley R, Palka FM (1973) Transmission cavity and injection stabilization of an X-band transferred electron oscillator. IEEE G-MTT-Int Microw Symp Dig 4–6 June 1973, 181–182
27. Siegel PH (2002) Terahertz technology. IEEE Trans MTT 50(3):910–928
28. Gordon JP, Zeiger HJ, Townes CH (1955) The maser – new type of microwave amplifier, frequency standard, and spectrometer. Phys Rev 99:1264

29. Shur M (2008) Terahertz electronics. CS MANTECH conference, 14–17 April 2008, Chicago, IL, USA
30. Ulrich R (1967) Far-infrared-properties of metallic mesh and its complementary structure. Infrared Phys 7:37
31. Corsi C, Fiocco G, Magjar G (1970) Internal Technical Report. Int Rep ESRIN
32. Evenson K, Jennings D, Petersen F (1984) Tunable far infrared spectroscopy. Appl Phys Lett 44:576
33. Chamberlain M (2004) Where optics meets electronics: recent progress in decreasing the terahertz gap. Philos Trans R Soc Lond A 362:199–213
34. Schawlow L, Townes CH (1958) Infrared and optical masers. Phys Rev 112(6):1940–1849
35. Köhler R, Tredicucci A, Beltram F, Beere H, Linfield E, Davies A, Ritchie D, Iotti R, Rossi F (2002) Terahertz semiconductor-heterostructure laser. Nature 417:156–159, 9 May
36. Jiang Z, Zhang X-C (1995) E.O. measurement of THz field pulses with a chirped optical beam. Appl Phys Lett 67:3523
37. Wood RA, Han C and Kruse PW (1992) Integrated uncooled infrared detector imaging arrays. In: Proceedings of IEEE solid state sensors and actuators workshop, Hilton Head Island, SC, USA, June 1992
38. Corsi C, Liberatore N, Mengali S, Mercuri A, Viola R, Zintu D (2007) Advanced applications to security of IR smart microbolometers. In: Proceedings of SPIE, vol 6739, Florence, Italy, 17 September 2007
39. Sheen DM et al (2008) Active imaging at 350 GHz for security applications. In: Proceedings of SPIE, vol 6948. Passive millimetre-wave imaging technology XI, 69480M, 18 Apr 2008
40. Wallace VP, MacPherson E, Zeitler JA, Reid C (2008) Three-dimensional imaging of optically opaque materials using non-ionizing terahertz radiation. J Opt Soc Am A 25(12):3120–3133
41. Kemp MC, Taday PF, Cole BE, Cluff JA, Fitzgerald AJ, Tribe WR (2003) Security applications of terahertz technology. Proc SPIE 5070:44–52
42. Tribe WR, Newnham DA, Taday PF, Kemp MC (2004) Hidden object detection: security applications of terahertz technology. Proc SPIE 5354:168–176
43. Baker C et al (2007) Detection of concealed explosives at a distance using terahertz technology. Proc IEEE 95:1559–1565
44. Maiman TH (1960) Stimulated optical radiation in ruby. Nature 167:494
45. Williams BS (2007) Quantum Cascade Lasers, PhD Thesis M.I.T: 1973 nature photonics. 1:517–525
46. Woolard D, Leorop W, Shur MS (eds) (2003) Terahertz sensing technology, vol 1, Selected topics electronics and systems, vol 30. World Scientific, Singapore
47. Woolard D, Leorop W, Shur MS (eds) (2004) Terahertz sensing technology, vol 2, Selected topics electronics and systems, vol 32. World Scientific, Singapore
48. Faist J et al (1994) Quantum cascade laser. Science 264:553–556
49. Semtsiv MP et al (2007) Short-wavelength InP based strain-compensated quantum-cascade laser. Appl Phys Lett 90:051111
50. Vitiello M et al (2007) Terahertz quantum cascade lasers with large wall-plug efficiency. Appl Phys Lett 90:191115
51. Fathololoumi S et al (2012) Terahertz quantum cascade lasers operating up to 200 K with optimized oscillator strength and improved injection tunneling. OSA Opt Express 20(4): 3866–3876
52. Odaa N et al (2008) Detection of THz radiation from quantum cascade lasers using VO_x microbolometers FPAs. In: Proceedings of SPIE, vol 6940. Infrared technology and applications XXXIV, Orlando, USA

Chapter 2
Far-Infrared Semiconductor Detectors and Focal Plane Arrays

Antony Rogalski

Abstract The detection of far-infrared (far-IR) and sub-mm-wave radiation is resistant to the commonly employed techniques in the neighbouring microwave and IR frequency bands. In this wavelength detection range the use of solid state detectors has been hampered for the reasons of transit time of charge carriers being larger than the time of one oscillation period of radiation. Also the energy of radiation quanta is substantially smaller than the thermal energy at room temperature and even liquid nitrogen temperature.

Development of focal plane arrays started in 1970s last century and has revolutionized imaging systems in the next decades. This chapter presents progress in far-IR and sub-mm-wave semiconductor detector technology of focal plane arrays during the past 20 years. Special attention is given on recent progress in the detector technologies for real-time uncooled THz focal plane arrays such as Schottky barrier arrays, field-effect transistor detectors, and microbolometers. Also cryogenically cooled silicon and germanium extrinsic photoconductor arrays, and semiconductor bolometer arrays are considered.

2.1 Introduction

THz radiation is frequently treated as the spectral region within frequency range $\nu \approx 0.1$–10 THz ($\lambda \approx 3{,}000$–$30\ \mu$m) and it is partly overlapping with loosely treated submillimeter (sub-mm) wavelength band $\nu \approx 0.1$–3 THz ($\lambda \approx 3{,}000$–$100\ \mu$m). THz electronics spans the transition range from radio-electronics to photonics.

The THz region of the electromagnetic spectrum has proven to be one of the most elusive. Being situated between infrared light and microwave radiation, THz

A. Rogalski (✉)
Institute of Applied Physics, Military University of Technology, 2 Kaliskiego Street, 00-908 Warsaw, Poland
e-mail: rogan@wat.edu.pl

C. Corsi and F. Sizov (eds.), *THz and Security Applications*, NATO Science for Peace and Security Series B: Physics and Biophysics, DOI 10.1007/978-94-017-8828-1_2,

radiation is resistant to the techniques commonly employed in these well established neighbouring bands. Historically, the THz technology development was driven by astronomers and planetary scientists, but other potential uses, particularly in medical and security applications, have led to increased activity by the mainstream physics and engineering community in recent times. Because diffraction is important at these frequencies, THz systems cannot be successfully designed using traditional optical techniques alone. In the low frequency THz region (below 0.5 THz), the radiation can penetrate most nonmetallic and nonpolar substances and because of it the imaging applications are a substantial part of technological development in this direction.

The past 20 years have seen a revolution in THz systems, as advanced materials research provided new and higher-power sources, and the potential of THz for advanced physics research and commercial applications was demonstrated. Numerous recent breakthroughs in the field have pushed THz research into the centre stage. As examples of milestone achievements can be included the development of THz time-domain spectroscopy (TDS), THz imaging, and high-power THz generation by means of nonlinear effects. Researches evolved with THz technologies are now received increasing attention, and devices exploiting this wavelength band are set to become increasingly important in diverse range of human activity applications (e.g., security, biological, drugs and explosion detection, gases fingerprints, imaging, etc.). The interest in the THz range is attracted by the fact that this range is the place where different physical phenomena are revealed which frequently calls for multidisciplinary special knowledge in this research area. Nowadays, the THz technology is also of much use in fundamentals science, such as nanomaterials science and biochemistry. This is based on the fact that THz frequencies correspond to single and collective excitations in nanoelectronic devices and collective dynamics in biomolecules. In 2004 *Technology Review*'s editors selected THz technology as one of "10 emerging technologies that will change your world" [3].

2.2 Trends in Developments of Far-Infrared Detectors

The detection of THz radiation is resistant to the commonly employed techniques in the neighbouring microwave and infrared (IR) frequency bands. In THz detection the use of solid state detectors has been hampered for the reasons of transit time of charge carriers being larger than the time of one oscillation period of THz radiation. Also the energy of radiation quanta is substantially smaller than the thermal energy at room temperature and even liquid nitrogen temperature.

Detector development is at the heart of all current plants. There are exist a large variety of traditional deeply cooled mm and sub-mm wavelength detectors (mainly bolometers) as well as new propositions based on optoelectronic quantum devices, carbon nanotube bolometers, plasma wave detection by field effect transistors, and hot electron room temperature bipolar semiconductor bolometers [38, 39].

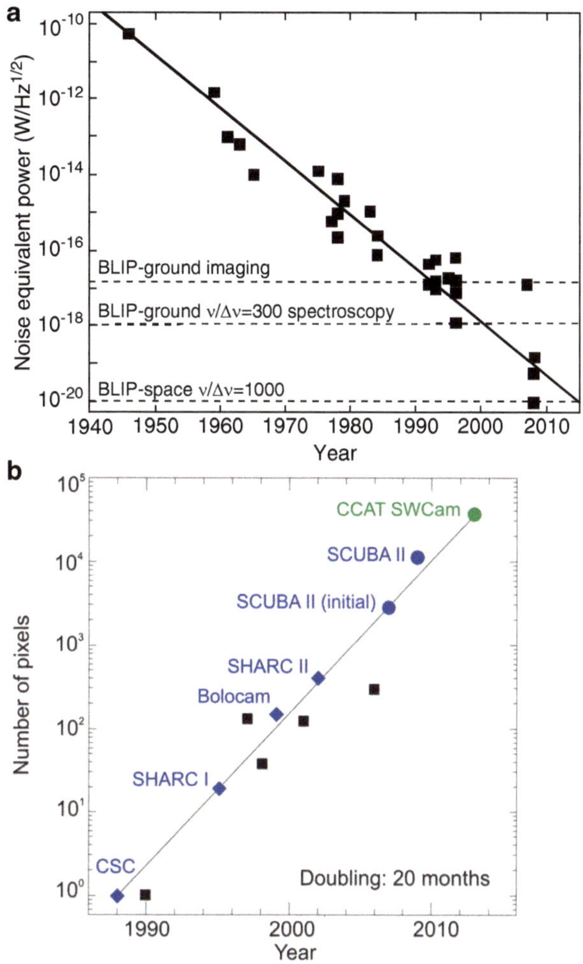

Fig. 2.1 Trends in development of THz detectors: (**a**) improvement of the bolometer NEP-value for more than half a century (sensitivity doubled every 2 years over the past 70 years), (**b**) detector arrays have doubled in format every 20 months over the past 10 years; *green symbol* indicates current expectation (Reproduced from http://www.ipac.caltech.edu/DecadalSurvey/farir.html)

Progress in THz detector sensitivity has been impressive in a period more than half century what is shown in Fig. 2.1a in the case of bolometers used in far-IR and sub-mm-wave astrophysics (http://www.ipac.caltech.edu/DecadalSurvey/fa-rir.html). The *NEP* value has decreased by a factor of 10^{11} in 70 years, corresponding improvements by factor of 2 every 2 years. Individual detectors achieved photon noise limited performance for ground-based imaging in the 1990s. The photon noise from astrophysical sources, achievable in space with a cold telescope, $\sim 10^{-18}$ W/$\sqrt{\text{Hz}}$, is now within demonstrated sensitivities. In present decade, the studies of inflation via cosmic microwave background (CMB) polarization will be driven not by detector sensitivity but by array formats. Far-infrared spectroscopy from a cold telescope however requires sensitivity $\sim 10^{-20}$ W/$\sqrt{\text{Hz}}$ to reach the astrophysical photon noise limit. Achieving this sensitivity in working detector

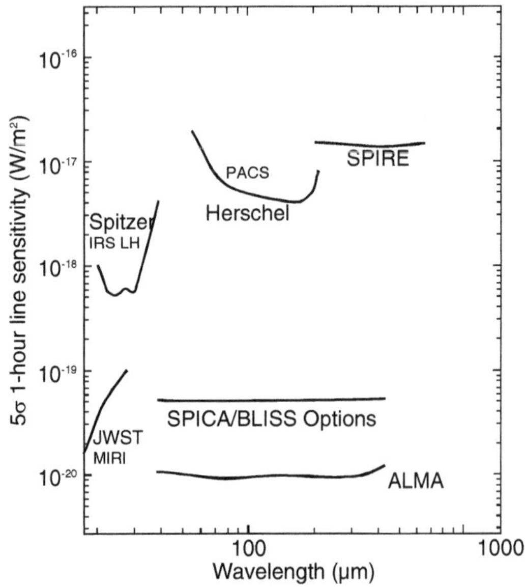

Fig. 2.2 Sensitivity of far-IR spectroscopy platforms (Reproduced from http://www.ipac.caltech.edu/DecadalSurvey/farir.html [16])

arrays remains a challenge for the coming decade, as the number of the detectors in the array is a key parameter that determines the information capabilities of the system and the speed improvement in obtaining a complete imaging or spectrum when observing galactic objects.

The development of pixel arrays has been comparably revolutionary. Figure 2.1b shows increasing in number of pixels in the period over last two decades. Detector arrays have doubled in format every 20 months over the past 10 years producing arrays with pixels now numbering in the thousands. Steady increase in overall observing efficiency is expected in near future, which by now has reached factors in the range of 10^{12} in comparison to capabilities in the early 1960s.

NASA has historically been the leading US agency for promoting the development of long wavelength detector technologies. Figure 2.2 shows the sensitivities of currently-planned or active far-IR/sub-mm spectroscopic facilities in near future.

As last figure shows, the James Webb Space Telescope (JWST) operates at wavelengths below about 27 μm. The Atacama Large Millimeter/submillimeter Array (ALMA) operating through a number of submillimeter atmospheric windows as well as ~650 μm, will have sensitivities at least 100 times higher than Herschel spanning the intervening 60–650 μm wavelength range. ALMA and JWST are currently scheduled to start operations within the next few years. The Space Infrared Telescope for Cosmology and Astrophysics (SPICA), with launch envisioned in 2017, will provide two-to-three orders of magnitude increase in sensitivity in comparison with Herschel that will bring far-IR/sub-mm sensitivity into line with those of JWST and ALMA. The ambitious requirements of future space missions are summarized in Ref. [3].

2.3 General Classification of Detectors

2.3.1 Photon Detectors

In photon detectors the radiation is absorbed within the material by interaction with electrons either bound to lattice atoms or to impurity atoms or with free electrons. The observed electrical output signal results from the changed electronic energy distribution. The fundamental optical excitation processes in semiconductors are illustrated in Fig. 2.3. The photon detectors show a selective wavelength dependence of response per unit incident radiation power. They exhibit both good signal-to-noise performance and a very fast response. But to achieve this, the photon detectors require cryogenic cooling. This is necessary to prevent the thermal generation of charge carriers. The thermal transitions compete with the optical ones, making non-cooled devices very noisy.

Depending on the nature of the interaction, the class of photon detectors is further sub-divided into different types. The most important are: intrinsic detectors, extrinsic detectors, photoemissive (Schottky barriers). Different types of detectors are described in details in monograph *Infrared Detectors* [37]. Figure 2.5 shows spectral detectivity characteristics of different types of detectors [37, 41].

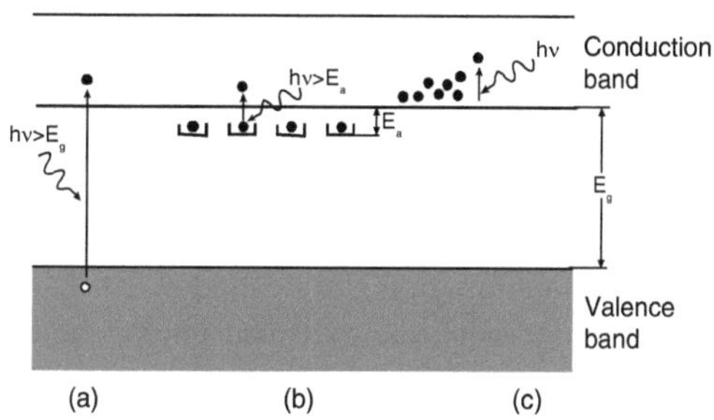

Fig. 2.3 Fundamental optical excitation processes in semiconductors: (*a*) intrinsic absorption, (*b*) extrinsic absorption, (*c*) free carrier absorption

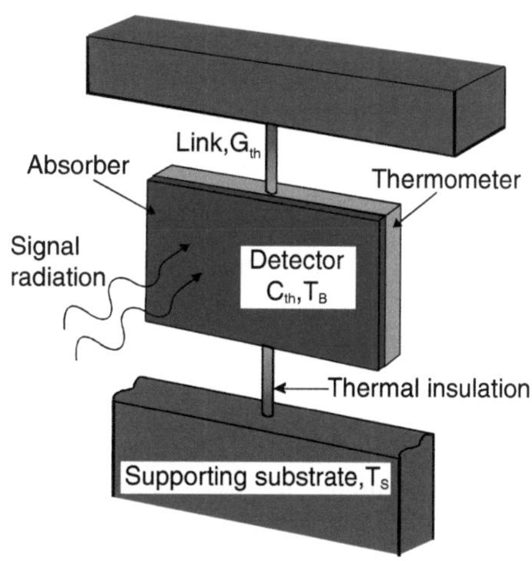

Fig. 2.4 Schematic diagram of thermal detector

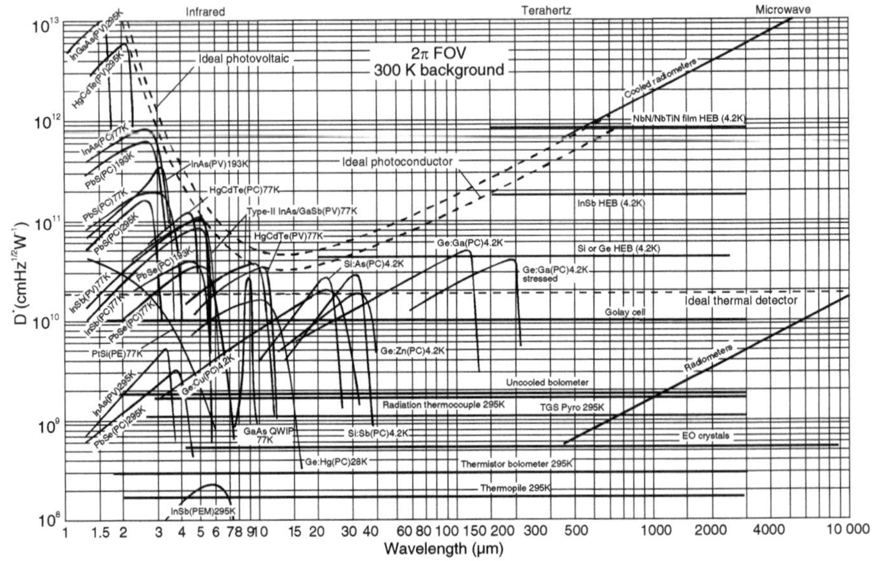

Fig. 2.5 Comparison of the D^* of various available detectors when operated at the indicated temperature. *PC* photoconductive detector, *PV* photovoltaic detector, *PEM* photoelectromagnetic detector, and *HEB* hot electron bolometer

Photoconductors that utilize excitation of an electron from the valence to conduction band are called intrinsic detectors. Instead those which operate by exciting electrons into the conduction band or holes into the valence band from impurity states within the band (impurity-bound states in energy gap, quantum wells or quantum dots), are called extrinsic detectors. In comparison with intrinsic photoconductivity, the extrinsic photoconductivity is far less efficient because of limits in the amount of impurity that can be introduced into semiconductor without altering the nature of the impurity states (see Fig. 2.6). Intrinsic detectors are most common at the short wavelengths, below 20 μm. It is interesting to note the presence of a marked gap in the wavelength coverage around 40 μm.

A key difference between intrinsic and extrinsic detectors is that extrinsic detectors require much cooling to achieve high sensitivity at a given spectral response cutoff in comparison with intrinsic detectors. Low-temperature operation is associated with longer-wavelength sensitivity in order to suppress noise due to thermally induced transitions between close-lying energy levels. The long wavelength cutoff can be approximated as

$$T_{max} = \frac{300\ \mathrm{K}}{\lambda_c\,[\mu m]} \tag{2.1}$$

The general trend is illustrated in Fig. 2.7 for five high performance detector materials suitable for low-background applications: Si, InGaAs, InSb, HgCdTe

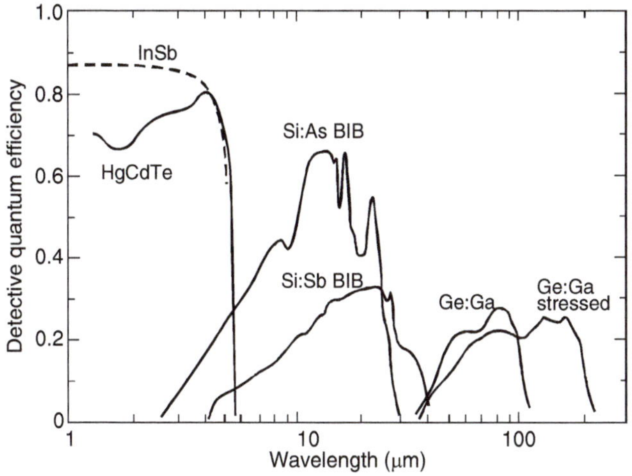

Fig. 2.6 Typical quantum efficiencies of semiconductors used in the wavelength range between 1 and 300 μm

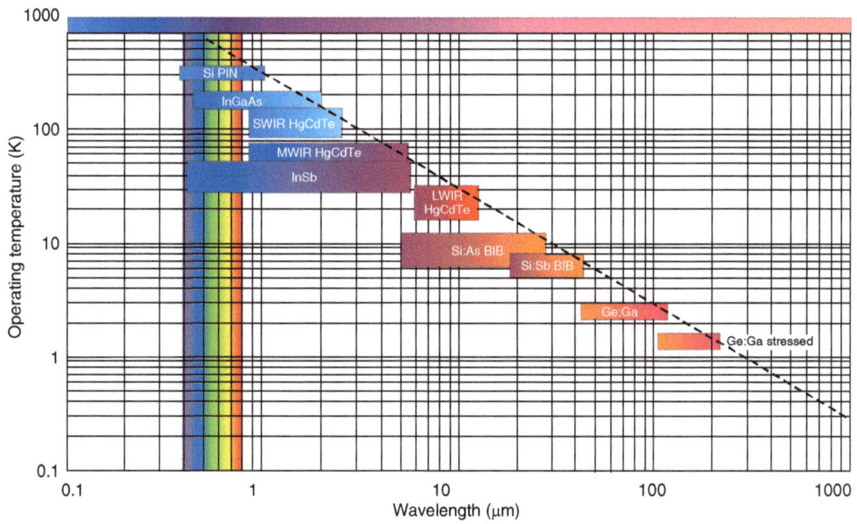

Fig. 2.7 Operating temperatures for low-background material systems with their spectral band of greatest sensitivity. The *dashed line* indicates the trend toward lower operating temperature for longer wavelength detection

photodiodes, and Si:As blocked impurity band (BIB) detectors; and extrinsic Ge:Ga unstressed and stressed detectors. Terahertz photoconductors are operated in extrinsic mode. One advantage of photoconductors is their current gain which is equal to the recombination time divided by the majority-carrier transit time. This current gain leads to higher responsivity than is possible with nonavalanching photovoltaic detectors. However, series problem of photoconductors operated at low temperature is nonuniformity of detector element due to recombination mechanisms at the electrical contacts and its dependence on electrical bias.

2.3.2 *Thermal Detectors*

The second class of detectors is composed of thermal detectors. In a thermal detector shown schematically in Fig. 2.4, the incident radiation is absorbed to change the material temperature, and the resultant change in some physical property is used to generate an electrical output. The detector is suspended on lags, which are connected to the heat sink. The signal does not depend upon the photonic nature of the incident radiation. Thus, thermal effects are generally wavelength independent; the signal depends upon the radiant power (or its rate of change) but not upon its spectral content. Since the radiation can be absorbed in a black surface coating, the spectral response can be very broad. Attention is directed toward three approaches which have found the greatest utility in infrared technology, namely, bolometers, pyroelectric and thermoelectric effects. In pyroelectric detectors a change in the internal electrical polarization is measured, whereas in the case of thermistor bolometers a change in the electrical resistance is measured.

Usually bolometer is a thin, blackened flake or slab, whose impedance is highly temperature dependent. Bolometers may be divided into several types. The most commonly used are the metal, the thermistor, and the semiconductor bolometers. A fourth type is the superconducting bolometer. This bolometer operates on a conductivity transition in which the resistance changes dramatically over the transition temperature range.

The key trade-off with respect to conventional uncooled thermal detectors is between sensitivity and response time. The detector sensitivity is often expressed by noise equivalent temperature (*NEDT*) represented by the temperature change, for incident radiation, that gives an output signal equal to the rms noise level. The thermal conductance is an extremely important parameter, since the noise equivalent temperature difference (*NEDT*) is proportional to $(G_{th})^{1/2}$, but the thermal response time of the detector, τ_{th}, is inversely proportional to G_{th}. Therefore, a change in thermal conductance due to improvements in material processing technique improves sensitivity at the expense of time response.

2.4 Detectors for Room Temperature THz Imaging

Particular attention in development of THz imaging systems is devoted to the realization of sensors with a large potential for real-time imaging while maintaining a high dynamic range and room-temperature operation. CMOS process technology is especially attractive due to their low price tag for industrial, surveillance, scientific, and medical applications. However, CMOS THz imagers developed thus far have mainly operated single detectors based on lock-in technique to acquire raster-scanned imagers with frame rates on the order of minutes. With this mind, much of recent developments are directed towards three type of focal plane sensors:

- Schottky barrier diodes (SBDs) compatible with CMOS process,
- field effect transistors (FETs) relay on plasmonic rectification phenomena, and
- adaptation of infrared bolometers to the THz frequency range.

An important issue for a FPA is pixel uniformity. It appears however, that the production of monolithically-integrated detector arrays encounters so many technological problems that the device-to-device performance variations and even the percentage of non-functional detectors per chip tend to be unacceptably high. To solve the problems a number of research groups are working on.

The performance of monolithically integrated detector arrays with other room temperature THz detectors is summarized in Table 2.1. In general, the *NEP* of SBD and FET detectors is better than that of Golay cells and pyroelectric detectors around 300 GHz. Both the pyroelectric and the bolometer FPAs with detector response times in the millisecond time range are not suited for heterodyne operation. FET detectors are clearly capable in heterodyne detection with improving sensitivity. Diffraction aspects predicts FPAs for higher frequencies (0.5 THz and above) and in conjunction with large f/# optics.

Below, a short description of three kinds of uncooled THz detectors is presented.

2.4.1 Schottky Barrier Diodes

In spite of achievements of other kind of detectors for THz waveband, the Schottky barrier diodes (SBDs) are among the basic elements in THz technologies. They are used either in direct detection and as nonlinear elements in heterodyne receiver mixers operating in temperature range of 4–300 K. The cryogenically cooled SBDs were used in mixers preferably in 1980s and early 1990s and then they have been replaced widely by superconductor-insulator-superconductor (SIS) or hot electron bolometer (HEB) mixers, in which mixing processes are similar to that observed in SBDs, but, e.g., in SIS structures the rectification process is based on quantum-mechanical photon-assisted tunnelling of quasiparticles (electrons). The nonlinearity of SBD *I-V* characteristic (the current increases exponentially with the applied voltage) is the prerequisite for mixing to occur.

Historically first Schottky-barrier structures were pointed contacts of tapered metal wires (e.g., a tungsten needle) with a semiconductor surface (the so-called crystal detectors). Widely used were, e.g., contacts p-Si/W. At operation temperature $T = 300$ K they have $NEP \approx 4 \times 10^{-10}$ W/Hz$^{1/2}$. Also pointed tungsten or beryllium bronze contacts to n-Ge, n-GaAs, n-InSb were used. In the mid 1960s Young and Irvin [42] developed the first lithographically defined GaAs Schottky diodes for high frequency applications. Their basic diode structure was next replicated by a variety of groups.

Table 2.1 Comparison of uncooled THz detector performance

Technology		NEP ($pW/Hz^{1/2}$)	Respon. (kV/W)	Frequency (THz)	Monolithic integration	Optics	Reference
SBD	2 × 2 array, 130-nm CMOS, 450 × 450 μm pitch	66[a] 73[b]	0.063[a] 20[b]	0.28	Yes	–	[14]
	4 × 4 array, 130-nm CMOS, 500 × 500 μm pitch	29[a]	355	0.28	Yes	Without Si lens	[15]
	GaAs	20	0.4[a]	0.8	No	–	[17]
		1.5	3.8[a]	0.15			
	ErAs/InAlGaAs	1.4	6.8[a]	0.1	No	–	[6]
Hetero. backward diode	InAs/AlSb/AlGaSb	0.18	49.7	0.094	No	–	[43]
MOSFET	Silicon	100[a]	0.033[a]	0.7	Yes	–	[40]
	3 × 5 array, 250-nm CMOS, 150 × 200 μm pitch	300[b]	80[b]	0.65	Yes	Si lens	[32]
	32 × 32 array, 65-nmSOI CMOS, 80 × 80 μm pitch	100[b]	140[b]	0.856	Yes	Si lens	[2]
Bolometer	320 × 240 array, 23.5 × 23.5 μm pitch, NbN	<100		3	Yes	Lens	[31]
	320 × 240 array, 49 × 49 μm pitch, α-Si	<30		2.5	Yes	Lens	[28]
	160 × 120 array, 52 × 52 μm pitch, VO_x	70		3	Yes	Mirrors and lens	[5]
Golay cell	–	200–400	0.1–45	0.2–30	No	–	[30]
Pyroelectric	–	400	150[b]	0.1–30	No	–	[36]

[a] Without amplifier
[b] With amplifier

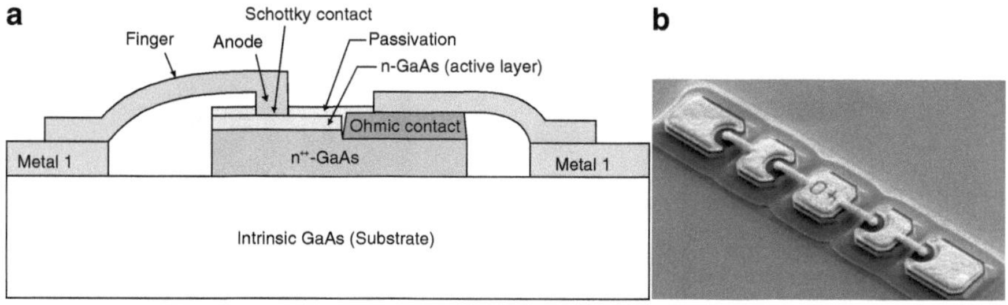

Fig. 2.8 GaAs Schottky barrier diode: (**a**) schematic of a planar diode and (**b**) a four-diode chip array (Reproduced from Crowe et al. [8])

Due to limitation of whisker technology, such as constraints on design and repeatability, starting in the 1980s, the efforts were made to produce planar Schottky diodes (see Fig. 2.8a). The air-bridged fingers replace the now obsolete whisker contact. This design has been the most important steps toward a practical Schottky diode mixer for THz frequency applications, with several thousand diodes on a single chip and where parasitic losses such as the series resistance and the shunt capacitance are minimized. To achieve good performance at high frequencies the diode area should be small. Reducing junction area one reduces junction capacities to increase operating frequency. But at the same time one increases the series resistance.

The planar Schottky diodes are fabricated on highly doped GaAs substrate ($\approx 5 \times 10^{18}$ cm^{-3}) with an ohmic contact on the back side and a thin GaAs epitaxial layer, with thickness of 300 nm–1 μm, is grown on the top of the substrate. Holes filled with a metal (Pt) in the SiO_2 insulating layer on top of the epitaxial layer define the anode area (0.25–1 μm). The state-of-the-art devices have anode diameters about 0.25 μm and capacitances 0.25 fF.

In the case of the discrete diode chip, the diodes are flip-chip mounted into the circuit with either solder or conductive epoxy. Using advanced technology, the diodes are integrated with many passive circuit elements (impedance matching, filters and waveguide probes) onto the same substrate. By improving the mechanical arrangement and reducing loss, the planar technology is pushed well beyond 300 GHz up to several THz. For example, Fig. 2.8b shows photographs of a bridged four-Schottky diodes' chip arrayed in a balanced configuration to increase power handling.

The typical Schottky diodes usually have high low-frequency noise levels due to introduction of oxides, contaminants, and damage to the junction in the fabrication process. More recently, an alternative method of Schottky barrier formation has been elaborated by molecular beam epitaxy (MBE) in-situ deposition of a semimetal (ErAs) on semiconductor (InGaAs/InAlAs on InP substrates) to reduce the imperfections that give rise to excess low-frequency noise, particularly l/f noise [6]. Excellent *NEP* performance for this III–V semiconductor SBD has been reported (1.4 pW/Hz$^{1/2}$ at 100 GHz). By using interband tunnelling, a heterojunction

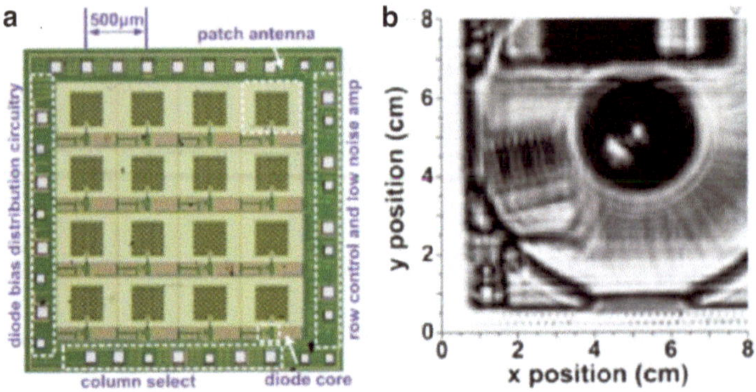

Fig. 2.9 CMOS SBD 280-GHz imager: die photos of the array (**a**) and an image of music greeting card obtained (Reproduced from Han et al. [15])

backward diode demonstrated 49.7 kV/W responsivity and 0.18 pW/Hz *NEP* at 94 GHz [43]. Recently Han et al. [15] have demonstrated fully functional CMOS imager operating near or in the sub-millimeter-wave frequency range (see Fig. 2.9). The 4 × 4 array increases the imaging speed by 4–8 times, due to fewer mechanical scan steps.

2.4.2 Field Effect Transistor Detectors

The channel of a field effect transistor (FET) can act as a resonator for plasma waves with a typical wave velocity 10^8 cm/s. The fundamental frequency of this resonator depends on its dimensions and, for gate length of a micron or less, can reach the THz range. The use FETs as detectors of THz radiation was first proposed by Dyakonov and Shur in 1993 [10] on the basis of formal analogy between the equations of the electron transport in a gated two-dimensional transistor channel and those of shallow water, or acoustic waves in music instruments. As a consequence, hydrodynamic-like phenomena should exist also in the carrier dynamics in the channel. Instability of this flow in the form of plasma waves was predicted under certain boundary conditions. Figure 2.10 schematically shows the resonant oscillation of plasma waves in gated region of FET.

Nonlinear properties of plasma wave excitations (the electron density waves) in nanoscale FET channels enable their response at frequencies appreciably higher than the device cutoff frequency, what is due to electron ballistic transport. In the ballistic regime of operation, the momentum relaxation time is longer that the electron transit time. The FETs can be used both for resonant (resonant case of high electron mobility, when plasma oscillation modes are excited in the channel) and non-resonant (broadband) THz detection [23] in the case of low mobility, where plasma oscillations are overdamped. The physical mechanism supporting

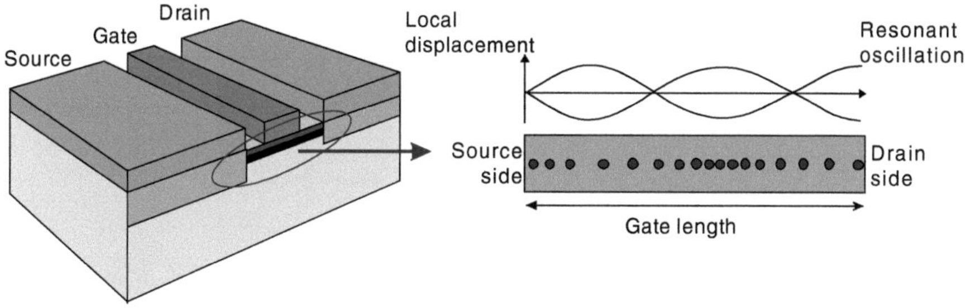

Fig. 2.10 Plasma oscillations in a field effect transistor

the development of stable oscillations lies in the refection of plasma waves at the borders of transistor with subsequent amplification of the wave's amplitude. Plasma excitations in FETs with sufficiently high electron mobility can be used for emission as well as detection of THz radiation.

The detection by FETs is due to nonlinear properties of the transistor, which lead to the rectification of the ac current induced by the coming radiation. As a result, a photoresponse appears in the form of a dc voltage between source and drain. This voltage is proportional to the radiation intensity (photovoltaic effect). Even in the absence of an antenna, the THz radiation is coupled to the FET by contact pads and bonding wires. A big progress in sensitivity can be obtained by adding a proper antenna or a cavity coupling.

The transistor receivers operate in wide temperature range up to room temperatures. Different material systems are used in fabrication FET, HEMT, and MOSFET devices including: Si, GaAs/AlGaAs, InGaP/InGaAs/GaAs, and GaN/AlGaN. Currently, the most promising application appears to be the broadband THz detection and imaging in the non-resonant regime.

The large-scale interest in using FETs as THz detectors started around 2004 after the first experimental demonstration of sub-THz and THz detection in silicon-CMOS FETs [24]. Two years later it was shown that Si-CMOS FETs can reach a *NEP* value competitive with the best conventional room temperature THz detectors [40]. At present the advantages of Si-CMOS FET technology (room temperature operation, very fast response time, easy on-chip integration with read-out electronics and high reproducibility) lead to the straight-forward array fabrication.

Recently, the first CMOS FPA used to capture transmission-mode THz video streams in real-time without the need for raster scanning and source modulation has been fabricated. A camera with 32×32 pixel array fully integrated in a 65-nm CMOS process technology has been demonstrated [2] (see right side of Fig. 2.11). Each 80-μm array pixel consists of a differential on-chip ring antenna coupled to NMOS direct detector operated well-beyond its cut-off frequency. The camera chip has been packed together with a 41.7-dBi silicon lens in a $5 \times 5 \times 3$ cm^3 camera module. In continuous-wave illumination the camera achieves an responsivity of 100–200 kV/W and a total *NEP* of 10–20 nW/Hz$^{1/2}$ up to 500 fps at 856 GHz. In non-video mode a maximum responsivity is 140 kV/W and a minimum *NEP* was reached at 856 GHz for a 5-kHz chopping frequency.

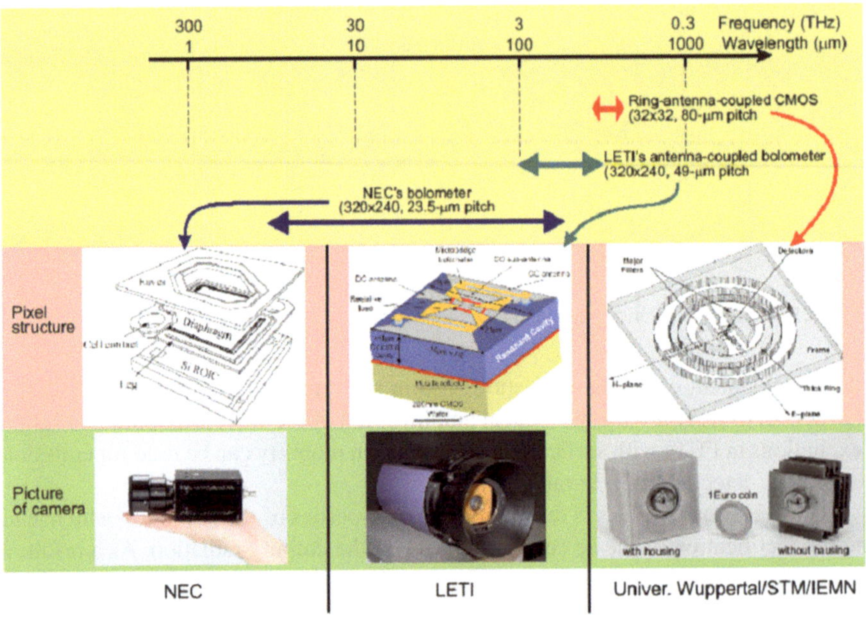

Fig. 2.11 Development status of uncooled THz focal plane arrays

2.4.3 Microbolometers

An impressive promising technology is also coming from commercially available microbolometer arrays. Adaptation of infrared microbolometers to the THz frequency range after the successful demonstration of active THz imaging in 2006 [25] entailed that in the period 2010–2011 three different companies/organizations announced cameras optimized for the >1-THz frequency range: NEC (Japan) [31], INO (Canada) [5] and Leti (France) [28]. The number of vendors is expected to increase soon.

The experimental active THz imaging arrangement is shown in Fig. 2.12. The most of papers presented characterizations of FPAs using monochromatic THz sources such as quantum cascade lasers (QCLs) or far infrared optically pumped lasers delivering mW-range powers. As shows Fig. 2.12, the reflected beam backlights an object with a maximum area and the transmitted light is collected by a camera lens. The focal plane is positioned behind the camera lens, making the object plane in front of the lens. Also shown is the modified reflection mode setup, where a specular reflection is collected by the repositioned lens and camera.

Different designs of THz bolometer pixels have been proposed. NEC's pixel is divided into two parts (see left side of Fig. 2.11), silicon Si readout integrated circuit (ROIC) in the lower part and suspended microbridge structure in the upper part. The microbridge has a two-storied structure. The 1st floor is composed of a diaphragm and two legs, while the eaves structure is formed on the diaphragm to increase the

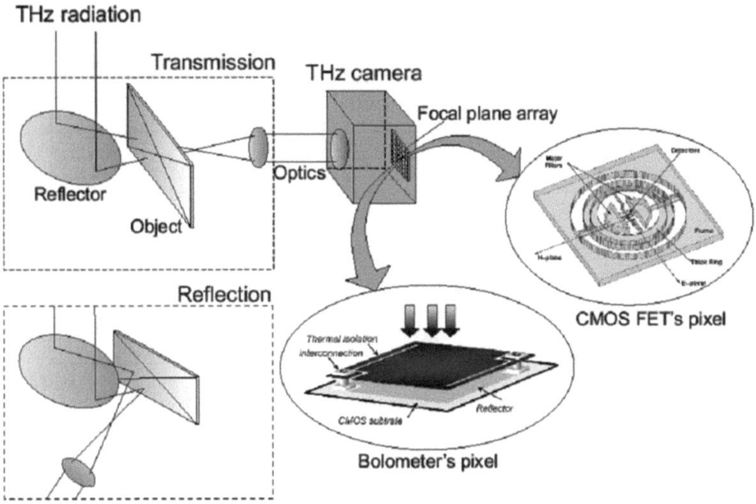

Fig. 2.12 Experimental setup of THz imaging system. Cutaway depicts alternative reflection mode setup

sensitive area and fill factor. The diaphragm and the eaves absorb THz radiation. The diaphragm is composed of VO_x bolometer thin film, SiN_x passivation layers and TiAlV electrodes, while the eaves structure is composed of SiN_x layer and TiAlV thin film THz absorption layer.

A schematic of one Leti's pixel of amorphous silicon microbolometer array is shown in top centre of Fig. 2.11. The 50-μm pitch is associated with quasi-double-bowtie antennas to a thermometer microbridge structure derived from the standard IR bolometer. The membrane is suspended over the substrate by arms and pillars. In order to enhance the antenna gain, an equivalent quarter-wavelength resonant cavity is realized under antennas with an 11-μm thick SiO_2 layer deposited over the metallic reflector. To ensure electric contact between the bolometer pillars and CMOS metal upper contacts, the vias are etched through 11-μm cavity and then metalized.

Figure 2.13 summarized the *NEP* values for bolometer FPAs fabricated by three vendors. The FPAs optimized for 2–5 THz exhibit impressive *NEP* values below 100 pW/Hz$^{1/2}$ (also the order of tens of pW/Hz$^{1/2}$ at video rate 20–30 fps). It can be seen that wavelength dependence of *NEP* is quite flat below 200 μm. Further improvement of performance is possible by increasing number of pixels, modification of antenna design while preserving pixel pitch, ROIC and technological stack.

Recently new microbolometric approach for design of array pixel has been proposed – using of Si FET and exploiting the dependence of the channel's conductance on temperature [7]. It should be mentioned about Tracer Systems Inc. FPA design and fabrication based on antenna-coupled tunnelling diodes for the frequency regime 0.6–1.2 THz. FPAs with a size up to 120×120 pixels, are

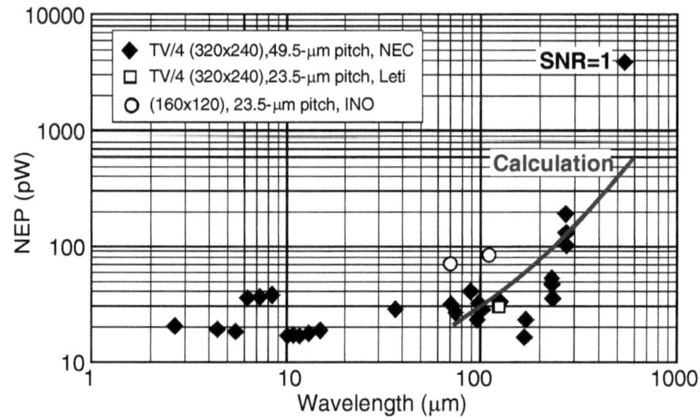

Fig. 2.13 Spectral dependence of *NEP* for bolometer THz FPAs

specified with an *NEP* of 5 nW/Hz$^{1/2}$. Heating by radiation in HgCdTe layers can be also used for designing of uncooled THz/sub-THz detectors with appropriate for active imaging characteristics and possibility to be assembled into arrays [9].

2.5 Extrinsic Detectors

Historically, an extrinsic photoconductor detector based on germanium was the first extrinsic photodetector. Extrinsic photodetectors are the principal detectors operating in the range $\lambda > 20$ μm. The spectral range of particular photodetectors is determined by the doping impurity and by the material into which it is introduced. For the shallowest impurities in GaAs, the long wavelength cutoff of photoresponse is around 300 μm.

Research and development of extrinsic IR photodetectors have been ongoing for more than 50 years. In the 1950s and 1960s, germanium could be made purer than silicon; doped Si then needed more compensation than doped Ge and was characterized by shorter carrier lifetimes than extrinsic germanium. Today, the problems with producing pure Si have been largely solved, with the exception of boron contamination. Si has several advantages over Ge; for example, three orders of magnitude higher impurity solubilities are attainable, hence thinner detectors with better spatial resolution can be fabricated from silicon. Si has lower dielectric constant than Ge, and the related device technology of Si has now been more thoroughly developed, including contacting methods, surface passivation and mature MOS and CCD technologies. Moreover, Si detectors are characterized by superior hardness in nuclear radiation environments.

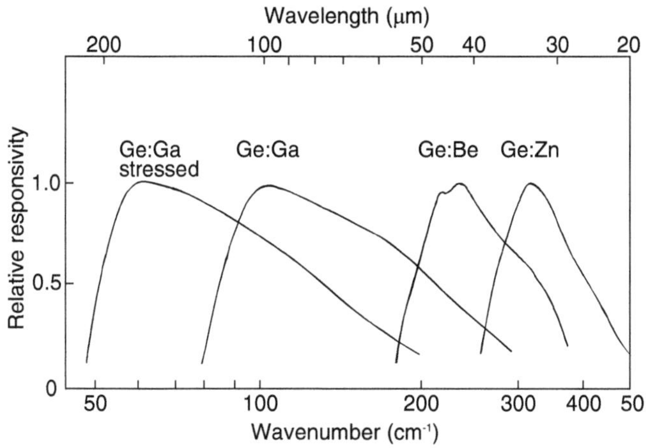

Fig. 2.14 Relative spectral response of some germanium extrinsic photoconductors (Reproduced from Leotin [26])

2.5.1 Extrinsic Germanium Photoconductors

Silicon detectors have largely supplanted germanium extrinsic detectors for both high and low background applications where comparable spectral response can be obtained. However, for wavelengths longer than 40 μm there are no appropriate shallow dopants for silicon; therefore germanium devices are still of interest for very long wavelengths. Germanium photoconductors have been used in a variety of infrared astronomical experiments, both airborne and space-based at wavelength ranging from 3 to more than 200 μm. Very shallow donors, such as Sb, and acceptors, such as B, In or Ga, provide cut-off wavelengths in the region of 100 μm. Figure 2.14 shows the spectral response of the extrinsic germanium photoconductors doped with Zn, Be, Ga and of stressed gallium doped germanium [26]. Despite a large amount of effort recently in the development of very sensitive thermal detectors, germanium photoconductors remain the most sensitive detectors for wavelength shorter than 240 μm.

Ge:Ga photoconductors are the best low background photon detectors for the wavelength range from 40 to 120 μm. Since the absorption coefficient for a material is given by the product of the photoionization cross-section and the doping concentration, it is generally desirable to maximize this concentration. The practical limit occurs when the concentration is so high that impurity band conduction results in excessive dark current. For Ge:Ga the onset of impurity banding occurs at approximately 2×10^{14} cm^{-3}, resulting in an absorption coefficient of only 2 cm^{-1} and typical values of quantum efficiency range from 10 to 20 %.

There are a number of problems with the use of germanium. To control dark current the material must be lightly doped and therefore absorption lengths become long (typically 3–5 mm). Because the diffusion lengths are also large (typically 250–300 μm), pixel dimensions of 500–700 μm are required to minimize crosstalk. In space applications, large pixels imply higher hit rates for cosmic radiation. This in turn implies very low readout noise for arrays operated in low background limit, what is difficult to achieve for large pixels with large capacitance and large noise. Due to small energy band gap, the germanium detectors must operated well below the silicon "freeze-out" range (4.2 K).

Application of uniaxial stress along the [100] axis of Ge:Ga crystals reduces the Ga acceptor binding energy, extending the cutoff wavelength to $\approx$240 μm [22]. At the same time, the operating temperature must be reduced to less than 2 K (see Fig. 2.7). In making practical use of this effect, it is essential to apply and maintain very uniform and controlled pressure to the detector so that the entire detector volume is placed under stress without exceeding its breaking strength at any point. A number of mechanical stress modules have been developed. The stressed Ge:Ga photoconductor systems have found a wide range of astronomical and astrophysical applications [13, 35].

The standard planar hybrid architecture, commonly used to construct near and mid-infrared focal-plane arrays [37], is not suitable for far IR detectors where readout glow, lack of efficient heat dissipation, and thermal mismatch between the detector and the readout could potentially limit their performance. Usually, the far-infrared arrays have a modular design with many modules stacked together to form a 2-dimensional array.

The Infrared Astronomical Satellite (IRAS), the Infrared Space Observatory (ISO), and for the far-infrared channels the Spitzer-Space Telescope (Spitzer) have all used bulk germanium photoconductors. In Spitzer mission a 32×32-pixel Ge:Ga unstressed array was used for the 70-μm band, while the 160 μm band had a 2×20 array of stressed detectors. The detectors are configured in the so-called Z-plane to indicate that the array has substantial size in the third dimension.

An innovative integral field spectrometer, called the Field Imaging Far-Infrared Line Spectrometer (FIFI-LS) that produces a 5×5 pixel image with 16 spectral resolution elements per pixel in each two bands was constructed at the Max Planck Institut für Extraterrestrische Physik under the direction of Albrecht Poglitsch. This array was developed for the Herschel Space Observatory and SOFIA – see Fig. 2.15. To accomplish this, the instrument has two 16×25 Ge:Ga arrays, unstressed for the 45–110 μm range and stressed for the 110–210 μm range. The low-stressed blue detectors has a mechanical stress on the pixels which is reduced to about 10 % of the level needed for the long-wavelength response of the red detectors. Each detector pixel is stressed in its own subassembly, and a signal wire is routed to preamplifiers housed nearby what obviously limits this type of array to much smaller formats than are available without this constraints.

The Photodetector Array Camera and Spectrometer (PACS) is one of the three science instruments on ESA's far infrared and sub-millimetre observatory – Herschel Space Laboratory. Apart from two Ge:Ga photoconductor arrays, it employs two filled silicon bolometer arrays with 16×32 and 32×64 pixels, respectively, to perform integral-field spectroscopy and imaging photometry in the

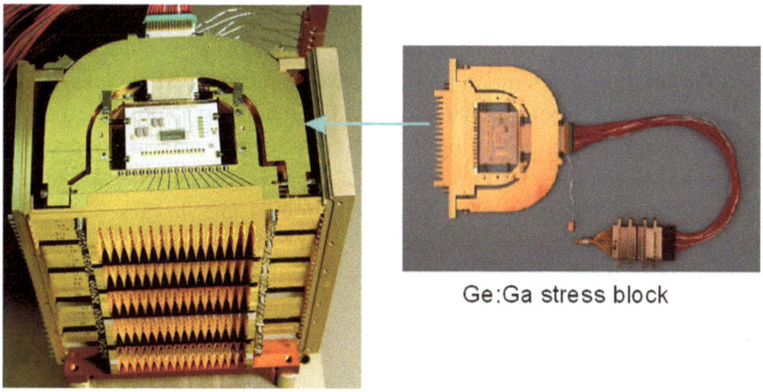

Fig. 2.15 PACS photoconductor focal plane array. The 25 stressed and low-stress modules of PACS instrument (corresponding to 25 spatial pixels) in the *red* and *blue arrays* are integrated into their housing (Reproduced from http://fifi-ls.mpg-garching.mpg.dr/detector.html [19])

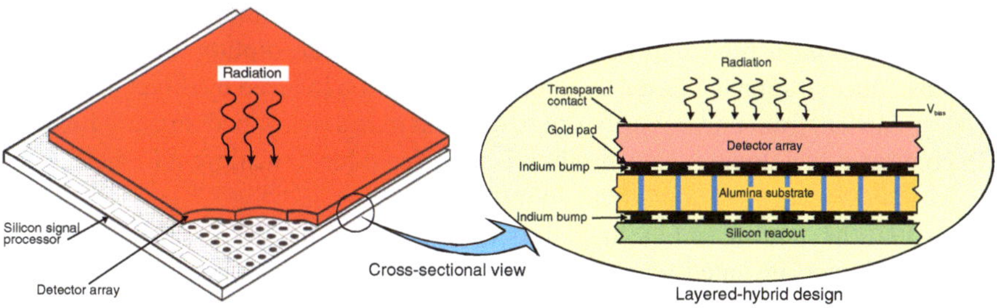

Fig. 2.16 The layered-hybrid design suitable for large format far IR and sub-mm arrays

60–210 μm wavelength regime. Median *NEP* values are 8.9×10^{-18} W/Hz$^{1/2}$ for the stressed and 2.1×10^{-17} W/Hz$^{1/2}$ for the unstressed detectors, respectively. The detectors are operated at ~1.65 K. The readout electronics is integrated into the detector modules – each linear module of 16 detectors is read out by a cryogenic amplifier/multiplexer circuit in CMOS technology but operates at temperature 3–5 K.

As is mentioned above, the standard hybrid focal plane array (FPA) architecture is not generally suitable for far-IR arrays (although also this architecture is used) primarily because glow from the readout is sensed by the detector, degrading its performance. In response, a new layered-hybrid structure was introduced to alleviate these problems and make possible the construction of large format far-IR FPAs (see Fig. 2.16) [11, 12]. In this design, an intermediate substrate is placed between the detector and the readout, which is pixelized on both sides in a format identical to that of the array and the electrical contact between corresponding pixel pads are made through embedded vias. The substrate material must be chosen to have sufficient IR-blocking property, high thermal conductivity, and an expansion coefficient that is between that of germanium and silicon. Alumina (Al_2O_3) and aluminium nitrite (AlN) have these properties and are possible choices as substrate materials. Blocking of the readout glow from reaching the detector, provide more efficient heat dissipation, improve temperature uniformity across the array, and mitigate the thermal mismatch between the detector and the readout. In addition, the substrate serves as a fanout board providing a simple and robust way to connect the FPA to the external electronics with no additional packaging requirement. This layered-hybrid architecture has been used for fabrication of Ge:Sb FPA ($\lambda_c \approx 130$ μm). It is predicted, that using this structure very large format FPAs with sensitivities better than 10^{-18} W/Hz$^{1/2}$ could be realized, fulfilling the technology goals of the upcoming astronomical instruments.

2.6 Blocked Impurity Band Detectors

One of the major problems in the design of extrinsic photoconductors is that the doping concentration is driven by conflicting requirements: the doping concentration needs to be as high as possible to get high photon-absorption coefficients (the doping concentration is limited by hopping conduction induced by direct transfer of charge carriers from one impurity site to the next in heavily-doped semiconductors). In contrast a low doping concentration is also desirable to achieve a low electrical conductivity which in turn reduces Johnson noise.

In 1979 M. Petroff and D. Stapelbroeck, working at the Rockwell International Science Center, invented what they referred to as Blocked-Impurity Band (BIB) detectors [33]. These detectors were developed to provide significantly reduced nuclear radiation sensitivity and improved performance compared to extrinsically doped silicon photoconductors. BIBs have some excellent properties which make them extremely useful for astronomical applications.

BIB detectors overcome the limitation of the doping density present in a standard extrinsic photoconductor by placing a thin intrinsic (undoped) silicon blocking layer between a heavily doped IR active layer and a planar contact (see Fig. 2.17). The active region of detector structure, usually based on epitaxially grown n-type material, is sandwiched between a higher doped degenerate substrate electrode and an undoped blocking layer. Doping of active layer is high enough for the onset of an impurity band in order to display a high quantum efficiency for impurity ionization (in the case of Si:As BIB, the active layer is doped to $\approx 5 \times 10^{17}$ cm^{-3}). The device exhibits a diode-like characteristic, except that photoexcitation of electrons takes place between the donor impurity and the conduction band. The heavily doped n-type IR-active layer has a small concentration of negatively charged compensating acceptor impurities. In the absence of an applied bias, charge neutrality requires an equal concentration of ionized donors. Whereas the negative charges are fixed at acceptor sites, the positive charges associated with ionized donor sites (D^+ charges) are mobile and can propagate through the IR-active layer via the mechanism of hopping between occupied (D^0) and vacant (D^+) neighbouring sites. A positive bias to the transparent contact creates a field that drives the pre-existing D^+ charges

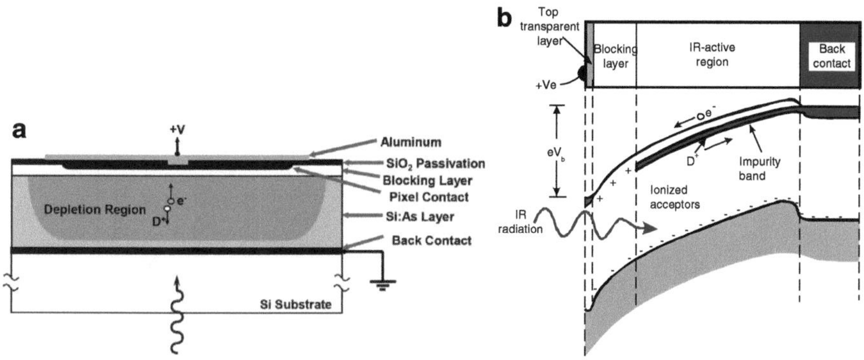

Fig. 2.17 Si:As BIB detector: (**a**) layer structure and (**b**) energy band structure of a positively biased detector

towards the substrate, while the undoped blocking layer prevents the injection of new D^+ charges. A region depleted of D^+ charges is therefore created, with a width depending on the applied bias and on the compensating acceptor concentration.

BIB detectors effectively use the hopping conductivity associated with "impurity banding" in relatively heavily doped semiconductors. Because of the presence of the blocking layer, BIB detectors do not follow the usual photoconductor model. The behaviour of BIB detectors is closer to that of a reverse-biased photodiode except that photoexcitation of electrons occurs from the donor impurity band to the conduction band. The gap between the impurity band and the conduction band is narrow; therefore, the response of a BIB detector extends to the VLWIR region of the spectrum. Silicon-based BIB detectors doped with arsenic and antimony have the materials of choice at wavelengths from 5 to 40 μm (see Fig. 2.18a). Conventionally designed and processed Si:As BIB detectors have a cut-off wavelength of about 28 μm. Thermal excitation of electrons across the narrow bandgap leads to dark current and the detectors must be operated at temperatures sufficiently low ($T < 13$ K) to limit the dark current.

The design of BIB detectors offers a number of advantages over conventional extrinsic photoconductors: the high absorption coefficient of the absorbing layer means that detectors with comparatively small active volumes can be made, providing low susceptibility to cosmic rays without compromising quantum efficiency. Also, due to the heavy doping of the active layer, the impurity band increases in width, therefore effectively decreasing the energy gap between the impurity band and the conduction band. As a result, BIB detectors typically offer spectral responses extended towards longer wavelengths compared to bulk-type photoconductors with the same dopant. Ultimately BIB devices also provide better noise performance compared to conventional photoconductors.

DRS Technologies has demonstrated a longer cut-off wavelength of the BIB detectors (see Fig. 2.18b). The far infrared extended BIB (FIREBIB) detectors are fabricated by further increasing of donor doping. Average overlap of donor wave

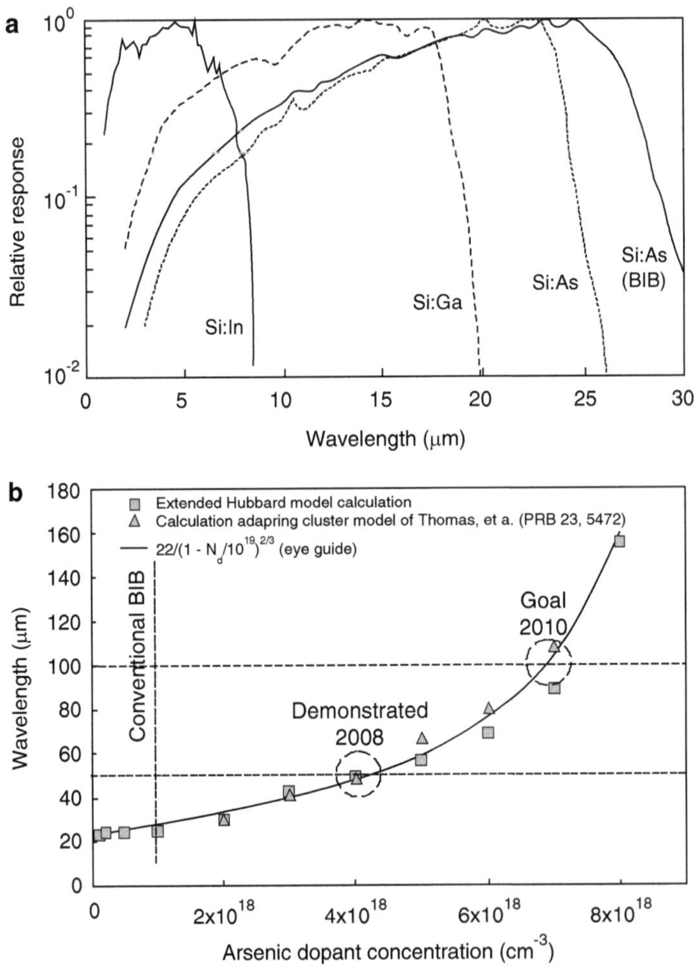

Fig. 2.18 Si:As BIB detectors: (**a**) examples of extrinsic silicon detector spectral response. Shown are Si:In, Si:Ga, and Si:As bulk detectors and a Si:As IBC (Reproduced from Norton [29]); (**b**) cut-off wavelength vs. donor doping (Reproduced from Hogue et al. [18])

functions becomes large, resulting in the impurity conduction band becoming wider and the gap between the impurity band and the conduction band becoming narrower. The narrowed bandgap results in longer wavelength response. The points according two models are represented in Fig. 2.18b. The proposed broadband detector in the 10- to 50-μm (goal 3- to 100-μm) is based on the As doped Si BIB detectors to operate at 10–12 K.

The main application of BIB arrays today is for ground- and space-based far-infrared astronomy – Si-As BIB performance are gathered in Table 2.2. The arrays

Table 2.2 Performance of Si:As BIB FPAs fabricated in several formats for both ground and space based applications

Parameter	Si:As BIB	Phoenix	MIRI	Aquarius-1k
Applications/ users	Ground-based telescopes ESO, Univer. of Tokyo	Space telescopes JAXA	Space telescopes JWST, NASA	Ground-based telescopes ESO, Univer. of Arizona
Format	320 × 240	1,024 × 1,024 1,024 × 1,024	1,024 × 1,024	1,024 × 1,024
Pixel size	50 μm	25 μm	25 μm	30 μm
ROIC type	DI	SFD	SFD	SFD
Fill factor	≥95 %	≥95 %	≥98 %	≥98 %
ROIC input referred noise	$<1{,}000e_{\text{-RMS}}$	$6\text{–}20e_{\text{-RMS}}$	<10–30e	Low gain $<1{,}000e_{\text{-RMS}}$ High gain $<100e_{\text{-RMS}}$
Integration capacity	7 or 20 × 10^6 e-	3 × 10^6 e-	2 × 10^5 e-	1 or 11 × 10^6 e-
Max. frame rates	100–500 Hz	0.1 Hz	0.1 Hz	120 Hz
Number of outputs	16 or 32	4	4	16 or 64
Packaging	LCC	LCC	Module	Module – 2 side buttable

Reproduced from Mills et al. [27]

should be operated under the most uniform possible conditions, in the most benign and constant environment possible. Array performance is strongly affected by background levels. Extrinsic silicon arrays for high background applications are less developed than that for low background applications.

The largest extrinsic infrared detector arrays are manufactured for astronomy owing to investments from NASA and the National Science Foundation. At present Raytheon Vision Systems (RVS), DRS Technologies, and Teledyne Imaging Sensors (formerly Rockwell Scientific Company) supply the majority of IR arrays used in astronomy between them the most important are BIB detector arrays. Impressive progress has been achieved especially in Si:As BIB array technology with formats as large as 2,048 × 2,048 and pixels as small as 18 μm. Figure 2.19 shows the evolution of BIB detector arrays at RVS.

Nowadays, attempts have been also made to provide a detector technology for operation in the far-IR by switching from silicon to a semiconductor material that would provide a shallower impurity band. Both Ge-based and GaAs-based BIB systems have been attempted, with greater success achieved in germanium. The smaller binding energy of shallow donors in GaAs compared to Ge results in response at wavelengths exceeding 300 μm without uniaxial stress.

Fig. 2.19 Evolution of BIB FPAs at RVS. From *left* to *right*: SIRTF 256 × 256, CRC774 320 × 240, Aquarius-1k 1,024 × 1,024, and Phoenix 2,048 × 2,048 devices (Reproduced from Mills et al. [27])

2.7 Semiconductor Bolometers

Cooled silicon bolometers demonstrate broadband and nearly flat spectral response in the 1–3,000 μm wavelength range. They are easier to fabricate with high operability, good uniformity, and lower cost, but has low operating temperature (4.2–0.3 K). During fabrication of bolometers, their area, operating temperature, thermal time constant, and thermal conductance are adjusted to meet the specific design requirements. The present day technology exist to produce arrays of hundreds of pixels that are operated in many experiments including NASA pathfinder ground based instruments, and balloon experiments such as BOOMERANG, MAXIMA and BAM.

The thermistors are typically fabricated by lithography on membranes of Si or SiN. The impedance is selected to a few MΩ to minimize the noise in JFET amplifiers operated at about 100 K. Limitation of this technology is assertion of thermal mechanical and electrical interface between the bolometers at 100–300 mK and the amplifiers at ≈100 K. Usually, JFET amplifiers are sited on membranes which isolate them so effectively that the environment remains at much low temperature (about 10 K) – see Fig. 2.20. In addition, the equipment at 10 K is itself thermally isolated from nearby components at 0.1–0.3 K. There are not practical approaches to multiplexing many such bolometers to one JFET amplifier. Current arrays require one amplifier per pixel and are limited to a few hundred pixels.

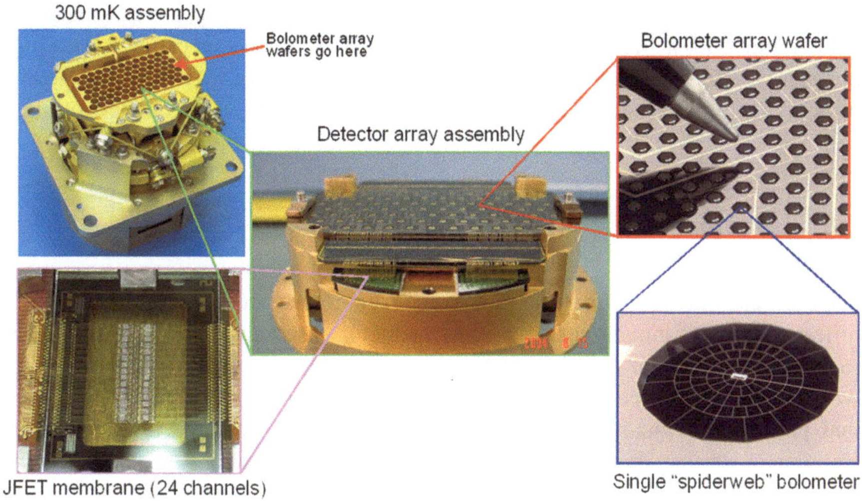

Fig. 2.20 Bolometer array of the Spectral and Photometric Imaging Receiver (*SPIRE*) (Reproduced from http://herschel.jpl.nasa.gov/spireInstrument.shtml [20])

In bolometer metal films that can be continuous or patterned in a mesh absorb the photons. The patterning is designed to select the spectral band, to provide polarization sensitivity, or to control the throughput. Different bolometer architectures are used. In close-packed arrays and spider web, the pop-up structures or two-layer bump bonded structures are fabricated. Agnese et al. have described a different array architecture, which is assembled from two wafers by indium bump bonds [1]. Other types of bolometers are integrated in horn-coupled arrays. To minimize low frequency noise an AC bias is used.

With the development of low-noise readouts that can operate near the bolometer temperature, the first true high-performance bolometer arrays for the far IR and sub-mm spectral ranges are just becoming available. For example, the Herschel/PACS instrument uses a 2,048-pixel array of bolometers [4] and is an alternative to JFET amplifiers. The architecture of this array is vaguely similar to the direct hybrid mid-infrared arrays, where one silicon wafer is patterned with bolometers, each in the form of a silicon mesh. The development of silicon micromachining has enabled substantial advances in bolometer construction generally and is central to making large-scale arrays. This technology allows the construction of very large arrays suitable for higher background applications. Further details are in Billot et al. [4].

In the normal bolometer, the crystal lattice absorbs energy and transfers it to the free carriers via collisions. However, in hot electron bolometer the incident radiation power is absorbed directly by free carriers, the crystal lattice temperature remaining essentially constant. Note that this mechanism differs from photoconductivity in that free-electron mobility rather than electron number is created by incident light. This mechanism offers submicrosecond response and broad far-infrared coverage out to millimeter wavelengths but requires liquid-helium cooling.

First bolometer with "hot electrons" (hot electron bolometer – HEB) was low temperature bulk n-InSb [34]. Currently, this detector uses a specially shaped high purity n-type InSb crystal which may be coupled directly to a very low noise preamplifier [21] (http://www.infraredlaboratories.com/InSb_Hot_e_Bolometers.html).

2.8 Conclusions

Far-infrared and sub-mm-wave radiation technologies are actively spreading in different areas of human activity. Semiconductor-based detectors have been used to detect IR radiation since the early days of space astronomy and are now the workhorse of IR space missions. Progress in their technology achieves the level that performance many discrete and low pixel arrays operated at low or sub-Kelvin temperature are close to ultimate performance at low background. Detector development is at the heart of all current plants. There are existed a large variety of traditional deeply cooled mm and sub-mm wavelength detectors as well as new propositions based on novel optoelectronic quantum devices.

The leading position of far-infrared semiconductor-based is challenged by the promise of large-format, high-performance arrays and will be undoubtedly continue to play an important role in the future. Particular attention in development of far-infrared/THz imaging systems will be devoted to the realization of sensors with a large potential for real-time imaging while maintaining a high dynamic range and room-temperature operation. CMOS process technology is especially attractive due to their low price tag for industrial, surveillance, scientific, and medical applications.

References

1. Agnese P, Buzzi C, Rey P et al (1999) New technological development for far-infrared bolometer arrays. Proc SPIE 3698:284–290
2. Al Hadi R, Sherry H, Grzyb J et al (2012) A 1 k-pixel video camera for 0.7–1.1 terahertz imaging applications in 65-nm CMOS. IEEE J Solid State Circ 47:2999–3012
3. Arone D (2004) 10 emerging technologies that will change your world. Technol Rev (February): 32–50
4. Billot N, Agnese P, Augueres JL et al (2006) The Herschel/PACS 2560 bolometers imaging camera. Proc SPIE 6265:62650D
5. Bolduc M, Terroux M, Tremblay B et al (2011) Noise-equivalent power characterization of an uncooled microbolometer-based THz imaging camera. Proc SPIE 8023:80230C-1–10
6. Brown ER, Young AC, Zimmerman J et al (2006) High-sensitivity, quasi-optically-coupled semimetal-semiconductor detectors at 104 GHz. Proc SPIE 6212:621205
7. Corcos D, Brouk I, Malits M et al (2011) The TeraMOS sensor for monolithic passive THz imagers. In: 2011 IEEE international Conference on Microwaves, Communications, Antennas and Electronic Systems (COMCAS), Tel Aviv, 8–9 November
8. Crowe TW, Porterfield DP, Hesler JL (2005) Terahertz sources and detectors. Proc SPIE 5790:271–280

9. Dobrovolsky V, Sizov FF (2010) THz/sub-THz bolometer based on the electron heating in a semiconductor waveguide. Opto Electron Rev 18:250–258
10. Dyakonov M, Shur MS (1993) Shallow water analogy for a ballistic field effect transistor: new mechanism of plasma wave generation by the dc current. Phys Rev Lett 71:2465–2468
11. Farhoomand J, Sisson DL, Beeman JW (2008) Viability of layered-hybrid architecture for far IR focal-plane arrays. Infrared Phys Technol 51:152–159
12. Farhoomand J, Sisson DL, Beeman JW (2010) Latest progress in developing large format Ge arrays for far-IR astronomy. Proc SPIE 7741:77410A-1–8
13. Haller EE, Breeman JW (2002) Far infrared photoconductors: recent advances and future prospects. Far-IR, Sub-mm & MM Detector Technology Workshop, Monterey
14. Han R, Zhang Y, Coquillat D et al (2011) A 280-GHz diode detector in 130-nm digital CMOS. IEEE J Solid State Circ 46:2602–2612
15. Han R, Zhang Y, Kim Y et al (2012) 280GHz and 860GHz image sensors using Schottky-barrier diodes in 0.13μm digital CMOS. In: IEEE international solid-state circuits conference, San Francisco, 19–23 February, pp 254–255
16. Harwit M, Helou G, Armus L et al (2012) Far-infrared/submillimeter astronomy from space tracking an evolving universe and the emergence of life. http://www.ipac.caltech.edu/DecadalSurvey/farir.html
17. Hesler JL, Crowe TW (2007) Responsivity and noise measurements of zero-bias Schottky diode detectors. In: Proceedings of the 18th international symposium on Space Terahertz Technology, Pasadena
18. Hogue H, Atkins E, Reynolds D et al (2011) Update on blocked impurity band detector technology from DRS. Proc SPIE 7780(778004):1–10
19. http://fifi-ls.mpg-garching.mpg.dr/detector.html
20. http://herschel.jpl.nasa.gov/spireInstrument.shtml
21. http://www.infraredlaboratories.com/InSb_Hot_e_Bolometers.html
22. Kazanskii AG, Richards PL, Haller EE (1977) Far-infrared photoconductivity of uniaxially stressed germanium. Appl Phys Lett 31:496–497
23. Knap W, Dyakonov MI (2013) Field effect transistors for terahertz applications. In: Saeedkia (ed) Handbook of terahertz technology. Woodhead Publishing, Cambridge, pp 121–155
24. Knap W, Teppe F, Meziani Y et al (2004) Plasma wave detection of sub-terahertz and terahertz radiation by silicon field-effect transistors. Appl Phys Lett 85:675
25. Lee AWN, Williams BS, Kumar S et al (2006) Real-time imaging using a 4.3-THz quantum cascade laser and a 320 × 240 microbolometer focal-plane array. IEEE Photon Technol Lett 18:1415–1417
26. Leotin J (1986) Far infrared photoconductive detectors. Proc SPIE 666:81–100
27. Mills R, Beuville E, Corrales E et al (2011) Evolution of large format impurity band conductor focal plane arrays for astronomy applications. Proc SPIE 8154:81540R-1–10
28. Nguyen D-T, Simoens F, Ouvrier-Buffet J-L et al (2012) Broadband THz uncooled antenna-coupled microbolometer array—electromagnetic design, simulations and measurements. IEEE Trans Terahertz Sci Technol 2:299–305
29. Norton PR (1991) Infrared image sensors. Opt Eng 30:1649–1663
30. OAD-7 Golay detector operating manual, QMC Instruments Ltd., Cardiff, UK, 4 Jan 2005
31. Oda N (2010) Uncooled bolometer-type terahertz focal-plane array and camera for real-time imaging. Comptes Rendus Phys 11:496–509
32. Öjefors E, Pfeiffer UR, Lisauskas A et al (2009) A 0.65 THz focal-plane array in a quarter-micron CMOS process technology. IEEE J Solid State Circ 44:1968–1280
33. Petroff MD, Stapelbroek MG (1986) Blocked-impurity-band detectors. US Patent No.4,566,960. Filled 23 Oct 1980, granted 4 Feb 1986
34. Phillips TG, Jefferts KB (1973) A low temperature bolometer heterodyne receiver for millimeter wave astronomy. Rev Sci Instrum 44:1009–1014
35. Poglitsch A, Waelkens C, Bauer OH et al (2008) The Photodetector Array Camera and Spectrometer (PACS) for the Herschel Space Laboratory. Proc SPIE 7010:701005

36. Pyroelectric Detector, product sheet for model SPH-62. Spectrum Detector Inc., Lake Oswego, OR. www.spectrumdetector.com
37. Rogalski (2011) Infrared detectors, 2nd edn. CRC Press, Boca Raton
38. Sizov F (2010) THz radiation sensors. Opto Electron Rev 18:10–36
39. Sizov F, Rogalski A (2010) THz detectors. Prog Quantum Electron 34:278–347
40. Tauk R, Teppe F, Boubanga S et al (2006) Plasma wave detection of terahertz radiation by silicon field effects transistors: responsivity and noise equivalent power. Appl Phys Lett 89:253511
41. Ueda T, An Z, Komiyama S (2011) Temperature dependence of novel single-photon detectors in the long-wavelength infrared range. J Infrared Millim Terahertz Waves 32:673–680
42. Young DT, Irvin JC (1965) Millimeter frequency conversion using Au-n-type GaAs Schottky barrier epitaxial diodes with a novel contacting technique. Proc IEEE 53:2130–2132
43. Zhang Z, Rajavel R, Deelman P et al (2011) Sub-micro area heterojunction backward diode millimeter-wave detectors with 0.18 pW/Hz$^{1/2}$ noise equivalent power. IEEE Microw Wirel Compon Lett 21:267–269

Chapter 3
Uncooled Rectification and Bolometer Type THz/Sub-THz Detectors

F. Sizov, M. Sakhno, A. Golenkov, V. Petryakov, Z. Tsybrii, V. Reva, and V. Zabudsky

Abstract Responsivity (*R*) and noise equivalent power (*NEP*) of long channel unbiased silicon field effect transistors (FETs) as THz/sub-THz detectors with account of some parasitic effects were considered. These parameters and their radiation frequency ν dependences are compared with those of contemporary zero biased Schottky barrier diodes (SBDs) THz/sub-THz detectors and hot-electron bolometers (HEBs) on the base of narrow-gap mercury cadmium telluride (MCT). To describe and compare the known experimental data for *Si*-FET and SBD detectors the models unification was developed.

3.1 Introduction

Detectors are among the critical components of THz/sub-THz imaging systems. To make such systems cost-effective it is desirable to use un-cooled detectors and arrays. For their applications, it is important to estimate an ultimate possible *NEP* of the detector or array that defines their use in active or passive imaging systems.

SBD detectors have long been used since 1940s [1] for microwave detection and mixing because of their high sensitivity and ability to operate at ambient or cryogenic temperatures. The study of FET as THz detectors was initiated by Dyakonov-Shur [2] publication. In Dobrovolsky and Sizov [3] it was shown that MCT-HEB can be used as uncooled THz/sub-THz detector. Review of other THz detector types can be found in Sizov [4]. Main advantages of FET detectors lies in CMOS mature technology.

FET and SBD detectors are fast and their operation rate in THz/sub-THz systems is limited only by the read-out electronics. At low radiation powers FET, SBD and

F. Sizov (✉) • M. Sakhno • A. Golenkov • V. Petryakov • Z. Tsybrii • V. Reva • V. Zabudsky
Department of Physics and Technology of Low-Dimensional Systems, Institute of Semiconductor Physics, Nauki Av., 41, 03028 Kiev, Ukraine
e-mail: sizov@isp.kiev.ua

C. Corsi and F. Sizov (eds.), *THz and Security Applications*, NATO Science for Peace and Security Series B: Physics and Biophysics, DOI 10.1007/978-94-017-8828-1_3,

MCT-HEB detectors are square-law detectors in which the response is proportional to the incoming power, and they can be assembled into arrays.

One of the principal factors that limit the performance of these three types of THz/sub-THz detectors is the presence of parasitic effects. Important also is accounting of the crosstalk between the adjacent sensitive elements in arrays of detectors placed with antennas deposited on thick high dielectric permittivity ε substrates. The latter one can be suppressed by thinning the substrates with high ε or using the substrates with lower ε.

To describe and compare the known experimental data for Si FET and SBD detectors the model for FET detectors similar to model for SBD was developed. The purpose of this investigation is to estimate the ultimate performance of responsivity R, electrical NEP^{el} and optical NEP^{opt} of FET and SBD detectors within the similar models. Some results on HEB-MCT detectors and arrays are also presented. Three solid-state detector systems on different substrates were considered: MCT microbolometers (CdZnTe, GaAs substrates), Si FETs, GaAs SBDs or III-V zero-bias SBDs.

Estimations have shown that these three types of detectors in $\nu \sim 150$–300 GHz spectral region can be used mostly for active imaging. In this region the resolution, as shown below, is relatively good, and e.g. dry clothes, cardboard boxes and envelopes are transparent for radiation. Moreover, in this region the radiation power of existing sources is relatively high and can be used successfully with multielement detector arrays to get real-time imaging.

3.2 FET Detectors

Interest to FET as THz detectors has begun after the theoretical considerations of Dyakonov-Shur [2] based on hydrodynamic approximation. Analyzing FETs operation it is necessary to clarify the phrase like "FET works at frequency much larger than cut-off frequency". Within the small signal approximation of FET operation there are the cut-off frequency ν_{co} at which the input current exceeds the output one, and the unit gain frequency at which the input power exceeds the output one. These frequencies characterize the linear response of the transistor but not the non-linear behavior. The latter can be observed at larger than ν_{co} frequencies. This phenomenon was theoretically foreseen in Dyakonov-Shur [2].

In fact THz detection is connected with generation of static component and distortion analysis is concerned with generation of high order harmonics. So they investigate the same phenomena from different points of view. Distortion analysis of FET at very high frequency with detailed explanation of common assumptions is given e.g. in Pu and Tsividis [5].

Usually FET is divided into intrinsic and extrinsic parts [6]. The intrinsic part is the channel. The extrinsic one is the rest of FET. It includes parasitic capacitances, gate resistance, source and drain resistance, etc. Parasitic effects become more

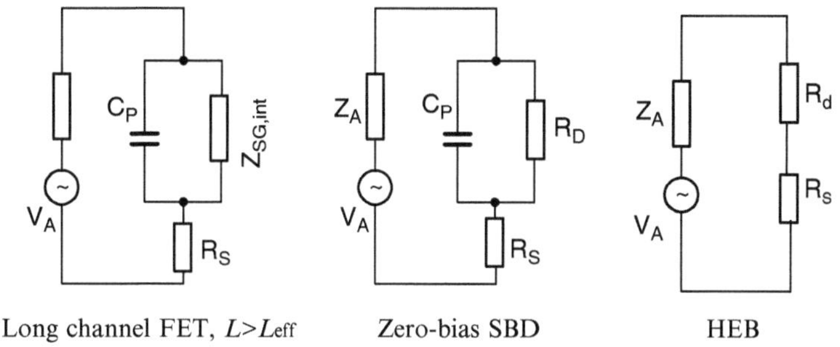

Fig. 3.1 Schematic representations with basic parasitic components. Z_A is the antenna impedance; V_A is the antenna voltage amplitude; $R_S = R_G + R_{source}$ in FET is the active series (parasitic) resistance of FET, where R_G is the gate active resistance; R_d is MCT-HEB active resistance, R_S in SBD and MCT-HEB is the series parasitic active resistance; R_D is SBD differential active resistance; C_P is the parasitic reactance (usually capacitive), $Z_{GS,int}$ is internal source-gate impedance

important at higher frequencies. In Fig. 3.1 the equivalent electrical circuits with antennas and only active and capacitance parasitics are shown for three types of THz/sub-THz detectors considered here.

To describe FET THz sensitivity there exist two models. One is based on Dyakonov-Shur theory [2]. The second one is so called distributed resistive self-mixing model (see e.g. [7]). Both these models concerning MOSFETs (MOS – metal-oxide semiconductor) are based on drift current equation and assumption of strong inversion. The Dyakonov-Shur model is based on hydrodynamic approximation and should not be applicable for inverse layer charge dynamic for MOSFET for weak and moderate inversion regions (were the maximum sensitivity is observed) due to the dominant diffusion current in these regions (see Fig. 3.2). In Boppel et al. [8] the drift equation with some form of inverse layer charge dependence approximation was used to describe rectification in all inversion regions and extending the model [2] to describe the FET impedance, responsivity and *NEP*.

From starting expressions of Dyakonov-Shur theory only *I*-*V*-dependence for the strong inversion region (where the drift current dominates) can be obtained. However, it appears that resulting expressions can describe experimental data in all inversion regions [8–10]. Also it was found that resistive mixing and distributed resistive self-mixing are similar [8]. Resulting expression for detected voltage is relatively simple [9] and is based on static *I*–*V*-dependence. All this results suggests that detection in THz range is similar to detection in e.g. MHz range to some extent. Nevertheless, there is unsolved question of proportionality constant. The shape of detected voltage dependence on gate voltage can be predicted accurately [8, 9], but proportionality constant (e.g. the absolute value of detected voltage in maximum) cannot be forecasted.

FET or SBD responses as mm-wave/THz detectors are conditioned by signal rectification due to *I*–*V*-dependence nonlinearities. For FET description the approach

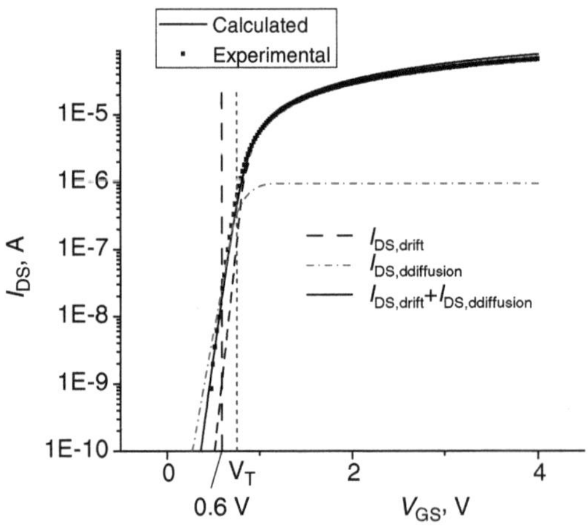

Fig. 3.2 Comparison of drift $I_{DS,drift}$ and diffusion $I_{DS,diffusion}$ currents in different operation regions of *Si* FET that were experimentally investigated ($W/L = 20/2$ (μm)) at $V_{GS} \approx 0.6$ V (at signal maximum), $V_{DS} = 0.05$ V. $I_{DS,diffusion}/I_{DS,drift} \approx 20$ according to the universal model [6]. $V_T = 0.758$ V, $n = 1.625$ is the sub-threshold turn-on swing factor

similar to describe SBD properties [1] was used. The intrinsic current responsivity (FET channel itself) is the ratio of detected current $<I_{DS}>$ to power absorbed in the intrinsic part $P_{in,int}$ of device. Then the FET current internal responsivity

$$R_{I,\text{int}} = \langle I_{DS} \rangle / P_{in,\text{int}}. \tag{3.1}$$

The intrinsic current and voltage responsivities are connected by channel resistance

$$R_{V,\text{int}} = R_{CH} R_{I,\text{int}}. \tag{3.2}$$

To find $R_{I,int}$ the detected current $\langle I_{DS} \rangle$ and input power $P_{in,int}$ should be known. The expression for $<I_{DS}>$ based on static *I*–*V*-dependence was obtained in Sakowicz et al. [9]. The resulting expression was experimentally verified (with precision to some constant factor). However, that derivation uses assumption of strong inversion that is not applicable for weak and moderate inversion regions of *Si*-MOSFET.

Considering FET as THz/sub-THz detector as a rule one places an antenna between the gate and source and thus the THz signal is introduced into FET channel via the source. Detection occurs at short distance L_{eff} near the source [11]. Here is considered the "long-channel" FETs in which the channel length L is larger compared to the "detection effective channel length" L_{eff}, which in Si FETs at room temperature is ~<100 nm [12, 13].

What is better for uncooled detectors and arrays: good antenna impedance matching with worse detector *NEP* or lower (better) detector *NEP* but worse antenna impedance matching? In Fig. 3.1 the electrical circuits with antennas and only active and capacitance parasitics are shown. The typical values are: $|Z_A| \sim 50\ldots200$ Ω; *Si*–FET: $R_S \sim 200\ldots500$ Ω, $|Z_{GS,int}| \sim 500\ldots2{,}000$ Ω; SBDs: $R_S \sim 20\ldots100$ Ω, $|Z_{int}| \sim 1\ldots2$ kΩ; MCT–HEB: $R_S \sim 20\ldots100$ Ω, $|Z_{int}| \sim 100\ldots1{,}000$ Ω. $|Z_{int}|$ is the detectors internal impedance.

To estimate L_{eff} and intrinsic input impedance $Z_{GS,int}$ transmission line model is used. In it the channel and gate form a transmission line for a.c. voltage signal (with impedance per unit length $Z = R_{CH}/L$, and admittance per unit length $Y = j\omega C/L$ (see e.g. [14, 15])) with constant of propagation

$$k = \sqrt{\frac{R_{CH}}{L} \cdot \frac{j\omega C}{L}} \tag{3.3}$$

(typical $k > 10^5 \text{cm}^{-1}$) and characteristic impedance

$$Z_0 = \sqrt{\frac{R_{CH}}{j\omega C}}. \tag{3.4}$$

Here C is the capacitance between the gate and channel (it is dependent on the gate-source voltage V_{GS} and in the region of strong inversion $C = C_{ox}$). For estimations it is accepted $C \approx C_{ox}$ for any V_{GS} biases. Here $\omega = 2\pi\nu$ is a radiation circular frequency.

In the model developed for *Si*-MOSFET sensitivity it is assumed that a short part of FET channel near the source, where the rectification of THz signal occurs, operates similar to the whole FET. To describe *I*–*V*-dependences (drain-source current) of this part the dependence for the whole FET channel can be used with the proper coefficient (L_{eff}/L). As this part is rather short ($L_{eff} < 100$ nm) the quasi-static approximation can be applied.

To calculate $<I_{DS}>$ detected current the next Exp. can be written

$$\langle I_{DS} \rangle = \frac{1}{T} \int_0^T I_{DS} \left(V_{DS,\text{int}}, V_{GS,\text{int}} \right) dt. \tag{3.5}$$

This integral (3.5) is difficult to find analytically due to complex form of $I_{DS}(V_{DS}, V_{GS})$. For the case of low input power, Taylor expansion can be used. Then at series expansion of FET for current in Exp. (3.5) near the $V_{DS} = 0$, $V_{GS} = 0$ points it follows

$$\begin{aligned} I_{DS}\left(V_{GS}+\Delta V_{GS}, V_{DS}+\Delta V_{DS}\right) &= I_{DS}\left(V_{GS}, V_{DS}\right) + \frac{\partial I_{DS}}{\partial V_{GS}} \Delta V_{GS} + \frac{\partial I_{DS}}{\partial V_{DS}} \Delta V_{DS} \\ &+ \frac{1}{2}\frac{\partial^2 I_{DS}}{\partial V_{GS}^2} \Delta V_{DS}^2 + \frac{\partial^2 I_{DS}}{\partial V_{GS} \partial V_{DS}} \Delta V_{DS} \Delta V_{GS} + \frac{1}{2}\frac{\partial^2 I_{DS}}{\partial V_{DS}^2} \Delta V_{DS}^2 + \ldots \end{aligned} \tag{3.6}$$

Let's assume that (*i*) no V_{DS} bias (zero bias approximation) so, $I_{DS}(V_{DS} = 0, V_{GS}) = 0$ and thus $\partial^2 I_{DS}/\partial V^2{}_{GS} = 0$; (*ii*) transistor is symmetrical $I_{DS}(V_{DS}, V_{GS}) = -I_{DS}(V_{DS}, V_{GS})$ and thus

$$\frac{\partial^2 I_{DS}}{\partial V_{DS}^2} = 0. \tag{3.7}$$

Let's take for intrinsic (channel) part of FET $V_{GS} = V_{G0} + \Delta V_{GS}$, $V_{DS} = \Delta V_{DS}$, $\Delta V_{GS} = \Delta V_{GS,\mathrm{int}} \cos(\omega t)$, $\Delta V_{DS} = \Delta V_{DS,\mathrm{int}} \cos(\omega t + \Delta\phi)$, where $\Delta\phi$ is a phase shift, $\omega = 2\pi\nu$, ν is a radiation frequency. As the gate resistance is much smaller than the channel one, the potential across the gate can be taken as a constant, but the potential difference between the channel and gate exponentially decreases along the gate. Therefore, the a.c. voltage will exist at short part of the channel, and the rectified voltage will appear near the source. Then $\Delta V_{G0} = \Delta V_{D0} = \Delta V_{GS,\mathrm{int}}$, $\Delta\phi = 0$. The rest of transistor channel serves as a resistance for a detected rectified current. The equivalent circuit for detected d.c. current is the series connected current source with impedance $R_{\mathrm{CH}}(L_{\mathrm{eff}}/L)$, current $I_{\mathrm{DS}}(L/L_{\mathrm{eff}})$ and resistor $R_{\mathrm{CH}}(L\text{–}L_{\mathrm{eff}})/L$. This circuit is equivalent to current source with internal channel resistance R_{CH}. So, the detected voltage is equal to that one when the whole channel operates.

After time-averaging in Exp. (3.5) it follows

$$\langle I_{DS} \rangle \approx \frac{1}{2} \frac{\partial^2 I_{DS}}{\partial V_{DS} \partial V_{GS}} \bigg|_{V_{DS}=0} \Delta V_{GS,\mathrm{int}}^2. \tag{3.8}$$

The approach considered is very similar to resistive self-mixing approach [16]. To find $\partial^2 I_{\mathrm{DS}}/\partial V_{\mathrm{DS}} \partial V_{\mathrm{GS}}$ the dependence of $I_{DS}(V_{DS},V_{GS})$ should be known. The first approach is to use model provided by the manufacturer (usually BSIM3) to find $I_{DS}(V_{DS},V_{GS})$. However there is some random parameters variation between transistors. So, experimental data for $I_{DS}(V_{DS},V_{GS})$ should be measured and model be modified to fit the data. The second approach lies in using the experimental data for $I_{DS}(V_{DS},V_{GS})$ and then applying numerical differentiating (as was proposed in Sakowicz et al. [9]). Nevertheless, current in low inversion region is low leading to sufficient noise in measured data. Thus, the differentiating data are very noisy. The third approach applied here uses some relatively simple formula (see Exp. (3.9)) to fit experimental $I_{DS}(V_{DS},V_{GS})$ dependence.

The semi-empirical expression that is applicable in all current-voltage characteristics regions [6] for drain-source current I_{DS} on drain-source V_{DS} and gate-source V_{GS} voltages was used for experimental data description, and also for *R* and *NEP* evaluations

$$I_{DS}\left(V_{DS}, V_{GS}\right) = \frac{W}{L} \frac{\mu_n}{1 + u_a\left(\frac{V_{GS}+V_T}{T_{ox}}\right) + u_b\left(\frac{V_{GS}+V_T}{T_{ox}}\right)^2} \cdot C'_{ox}(2n)\varphi_t^2 \times \left[\ln\left(1 + e^{\frac{V_{GS}-V_T}{2n\varphi_t}}\right)^2 - \ln\left(1 + e^{\frac{V_{GS}-V_T-nV_{DS}}{2n\varphi_t}}\right)^2\right]. \tag{3.9}$$

Here μ_n is the electron mobility in the channel (it is the parameter u_0 in BSIM file), and u_a, u_b are the parameters of BSIM model, $V_T \sim 0.7$ V is the threshold voltage, $\varphi_t = (k_B \cdot T)/q$ is the thermal potential, k_B is the Boltzmann constant, W is the channel width, $n \sim 1.5$ is the sub-threshold turn-on swing factor, that is a function of V_{GS} and other FET parameters, T_{ox} is the oxide thickness and C'_{ox} is the oxide capacitance. In Fig. 3.2 the experimental and calculated (according to Exp. (3.9)) *I-V*-characteristics for one of the investigated *Si* MOSFETs are presented which demonstrate the importance of different regions consideration. As the signal responses have maximum at $V_{GS} \sim 0.6$ V for transistors investigated (see also e.g. Ojefors et al. [7], Tauk et al. [12], Schuster et al. [17]) it can be seen that the static current is mainly the diffusion current at these V_{GS} values, and thus, the signal response is mainly governed by the diffusion current component.

The absorbed power is connected to intrinsic input impedance $Z_{GS,int}$ and signal amplitude $\Delta V_{GS,int}$

$$P_{in,\text{int}} = \frac{1}{T}\int_0^T I_{GS,\text{int}}(t)V_{GS,\text{int}}(t)dt = \frac{1}{2}\frac{\text{Re}Z_{GS,\text{int}}}{|Z_{GS,\text{int}}|^2}\left|\Delta V_{GS,\text{int}}\right|^2. \tag{3.10}$$

To find $Z_{GS,int}$ the transmission line model is used. The impedance of transmission line with length L is

$$Z(L) = Z_0 \cdot \frac{Z_L + Z_0 \tanh(kL)}{Z_0 + Z_L \tanh(kL)}. \tag{3.11}$$

Here Z_L is impedance at the other side of transmission line, Z_0 is transmission line characteristic impedance (see Eq. (3.4)), k is the propagation constant (see Eq. (3.3)).

The internal source-gate impedance is equal to the transmission line characteristic impedance for the case of a long transistor. As $\tanh(kd) \approx 1$ for $\text{Re}(kd) \gg 1$ then

$$Z_{GS,\text{int}} = \frac{\alpha\,(1-j)}{\sqrt{\omega}}, \tag{3.12}$$

where $\alpha = (R_{CH}/2C_{ox})^{1/2}$ as $C \approx C_{ox}$.

The internal current responsivity is

$$R_{I,\text{int}} = 2\frac{\alpha}{\sqrt{\omega}}\frac{\partial^2 I_{DS}}{\partial V_{DS}\partial V_{GS}}\bigg|_{V_{DS}=0}. \tag{3.13}$$

The FET response for intrinsic part of transistor is

$$V_{\text{det,int}} = \left(\frac{\partial I_{DS}}{\partial V_{DS}}\right)^{-1}_{V_{GS}} \cdot \langle I_{DS} \rangle, \tag{3.14}$$

and if one takes into account that at $V_{DS}\to 0$ the differential resistance is close to the ordinary one

$$\left.\frac{\partial I_{DS}}{\partial V_{DS}}\right|_{V_{DS}=0} \approx \left.\frac{I_{DS}}{V_{DS}}\right|_{V_{DS}\to 0} = \sigma_{CH}, \tag{3.15}$$

then for detector internal voltage response can be written the expression

$$V_{\text{det}} = \frac{1}{2}\cdot\frac{1}{\sigma_{CH}}\cdot\frac{d\sigma_{CH}}{dV_{GS}}\cdot\Delta V_{GS,\text{int}}^2. \tag{3.16}$$

3.2.1 Power Transfer Coefficient

Parasitic effects in extrinsic part of transistor lead to power dissipation. So it is important to find how much power is dissipated in different parasitic effects. The impact of parasitic capacitance on FET operation was considered in Boppel et al. [8, 16], and the impact of parasitic resistance influence was taken into account [14, 18]. However, it is important to consider both parasitic resistance and capacitance. The equivalent complex circuit for FET, that includes a lot of parasitics, was considered in Tsividis and McAndrew [6] and Enz and Cheng [19]. As we are interested in the part between the source and gate it can be simplified to form shown in Fig. 3.1a.

The ratio of power that is absorbed in the intrinsic part of the transistor $P_{\text{in,int}}$ to power P_{in} that is absorbed in transistor as a whole (in its intrinsic and extrinsic parts) is

$$\eta = \frac{P_{in,\text{int}}}{P_{in}} = \left|\frac{Z_1}{Z_{GS}}\right|^2 \frac{\text{Re}Z_{GS,\text{int}}/|Z_{GS,\text{int}}|^2}{\text{Re}Z_{GS}/|Z_{GS}|^2}, \tag{3.17}$$

where $Z_1 = Z_{GS,\text{int}}\|X_P$, $Z_{GS} = Z_1 + R_S$, $X_P = (j\omega C_P)^{-1}$ (see Fig. 3.1). Here η is the power transmission coefficient, because it is the ratio of two powers. Thus, the electrical voltage responsivity can be written as

$$R_V^{el} = \eta\cdot R_{V,\text{int}}. \tag{3.18}$$

To see the detected signal dependence on radiation frequency ν let's present η in the next form. From Exps. (3.12) and (3.17) it can be found

$$\eta = \frac{1}{1 + (\omega/\omega_{c1})^{1/2} + \omega/\omega_{c2} + (\omega/\omega_{c3})^{3/2}}, \tag{3.19}$$

where the coefficients $\omega_{c1} = (R_S/\alpha)^{-2}$, $\omega_{c2} = (2\cdot C_P\cdot R_S)^{-1}$, $\omega_{c3} = (2C_P^2\cdot R_S\cdot\alpha)^{-2/3}$, $\alpha = (R_{CH}/2C_{ox})^{1/2}$ as $C\approx C_{ox}$ and coefficients $\omega_{ci} = 2\pi\nu_{ci}$.

Let's see why both R_S and C_P should be considered simultaneously. The impedance $Z_{GS} = R_S + X_P||Z_{GS,int}$. Considering estimations for 1-μm design rule *Si*-FET technology with $W = 20$ μm, $L = 2$ μm, $R_S \approx 200\ \Omega$, $C_P = 4$ fF, $C_{ox} = 35$ fF, $R_{CH} = 1.1$ MΩ, radiation frequency $\nu = 300$ GHz, internal input impedance $Z_{GS,\text{int}} = \sqrt{\frac{R_{CH}}{j\cdot 2\pi f\cdot C_{ox}}} \approx (3-3j)\,\text{k}\Omega$, impedance of parasitic capacitance $X_P = (j\cdot 2\pi\nu\cdot C_P)^{-1} = -j130\ \Omega$.

At first it seems R_S can be neglected due to $R_S << Z_{GS,int}$ and X_P due to its pure reactive nature. However, considering $X_P \parallel Z_{GS,\text{int}} = \frac{X_P Z_{GS,\text{int}}}{X_P + Z_{GS,\text{int}}} = (2.7 - j128)\,\Omega$ it is seen that $R_S \gg \text{Re}(X_P \| Z_{GS,\text{int}})$.

Thus, it's clear that parasitics have strong impact on FET performance because the main part of power equal to $(1-\eta = \frac{R_S}{R_S + \text{Re}(X_p \| Z_{GS,\text{int}})} \approx 98.6\%)$ is lost on them. Therefore, it's important to consider R_S and C_P at the same time. If hypothetically, there is no R_S than C_P is not an issue, if there is no C_P than R_S is not an issue.

Usually $\nu_{c1} > \nu_{c2} > \nu_{c3}$. Thus, let's consider FET operation at sufficiently large frequencies $\nu > \nu_{c3}$. Then from Eqs. (3.2), (3.13), (3.18), and (3.19)

$$R_V^{el} = 2\alpha R_{CH} \left.\frac{\partial^2 I_{DS}}{\partial V_{DS}\partial V_{GS}}\right|_{V_{DS}=0} \omega_{c3}^{3/2}\omega^{-2} = \frac{1}{C_p^2 R_S} R_{CH} \left.\frac{\partial^2 I_{DS}}{\partial V_{DS}\partial V_{GS}}\right|_{V_{DS}=0} \omega^{-2} \tag{3.20}$$

doesn't depend on parameter $\alpha = (R_{CH}/2C_{ox})^{1/2}$ (from the physical point of view the parasitic capacitance C_P shunts the internal part $Z_{GS,int}$, and thus the power is absorbed mainly in R_S).

This formula can be rewritten in the form similar to formula for SBD

$$R_V^{el} \approx R_{V0}^{el}\left(\frac{\omega_c}{\omega}\right)^2, \tag{3.21}$$

where $R_{V0}^{el} = R_{CH}^2 \left.\frac{\partial^2 I_{DS}}{\partial V_{DS}\partial V_{GS}}\right|_{V_{DS}=0}$, $\omega_c = \frac{1}{C_p\sqrt{R_S R_{CH}}}$.

For example, for FET investigated (1-μm technology, $W = 20$ μm, $L = 2$ μm), $\nu_c = 2.7\cdot 10^9$ Hz, $\left.\frac{\partial^2 I_{DS}}{\partial V_{DS}\partial V_{GS}}\right|_{V_{DS}=0} = 2\cdot 10^{-5}\frac{A}{V^2}$, $R_{CH} = 1.1\cdot 10^6\ \Omega$, $R_{V0}^{el} = 2\cdot 10^7 \frac{V}{W}$ the "zero" current responsivity $R_{I0}^{el} = R_{CH} \left.\frac{\partial^2 I_{DS}}{\partial V_{DS}\partial V_{GS}}\right|_{V_{DS}=0} \approx 21$ A/W, which is larger than the theoretical ultimate performance R_I for SBD detector (at $T = 300$ K, $R_I = 19.3$ A/W).

Next expressions can be useful for some estimations

$$NEP^{el} = \frac{\sqrt{4kTR_{CH}}}{R_V^{el}} \approx \sqrt{4kT}\frac{\sqrt{R_{CH}}C_p^2 R_S}{R_{I0}^{el}}\omega^2, \tag{3.22}$$

where the thermal noise, which is the most important [12] at $V_{GS} = 0$ for FET detectors is included.

Common model for dependence of parasitic effects on FET dimension used e.g. in BSIM3 model [20] is $C_p = C_x W$, $R_S = \frac{\rho_1}{W} + \rho_2 \frac{W}{L}$, $R_{CH} = \frac{L}{W} R_{CH0}$. Using this model with Eq. (3.22) one has

$$NEP^{el} = \frac{\sqrt{4kTR_{CH0}}}{R_{I0}^{el}} \omega^2 \left(C_x^2 \rho_1 \sqrt{LW} + C_x^2 \rho_2 W^{5/2} L^{-1/2} \right). \tag{3.23}$$

Here R_{CH0} (Ω) is the threshold channel resistance, C_{x} (F/m), ρ_1 ($\Omega \cdot$ m), ρ_2 ($\Omega/\square$) are the technological parameters for given design rules. For example for 1-μm and 0.35-μm design rules the parameter $R_{\text{CH0}} = 1.1 \cdot 10^7\ \Omega$ ($V_{\text{GS}} = 0.6$ V) and $R_{\text{CH0}} = 1.3 \cdot 10^6\ \Omega$ ($V_{\text{GS}} = 0.5$ V), respectively. The parameters C_{x} and ρ_2 are the same ($C_{\text{x}} = 2 \cdot 10^{-10}$ F/m, $\rho_2 = 16\ \Omega/\square$ for 1-μm technology, and $\rho_2 = 12\ \Omega/\square$ for 0.35-μm technology), and $\rho_1 = 9 \cdot 10^{-4}$ and $4.1 \cdot 10^{-4}\ \Omega \cdot$ m, respectively. From Exp. (3.23) the electrical noise equivalent power decreases with W decrease. So, for a given L value the transistor with less W value will have lower (better) NEP^{el}.

Optimizing this expression with technological constraints (1 μm $\leq L \leq$ 5 μm, 1 μm $\leq W \leq$ 20 μm), the optimal FET dimensions can be evaluated (and they are $L = 1$ μm and $W = 1$ μm). For example for 1.0-μm CMOS technology one can evaluate "zero" current responsivity $R_{I0}^{el} \approx 21\ A/W$ and the optimal electrical $NEP^{\text{el}} \sim 2.8 \cdot 10^{-13}$ W/Hz$^{1/2}$ at $\nu \approx 77$ GHz. For 0.35-μm CMOS technology these values are $R_{I0}^{el} \approx 21 A/W$ and $NEP^{\text{el}} \approx 2.5 \cdot 10^{-13}$ W/Hz$^{1/2}$. Better NEP for 0.35-μm is due to lower oxide thickness that leads to lower channel resistance. So, electrical NEP NEP^{el} is smaller for FET with lower W.

3.2.2 Antenna Transfer Coefficient

Using FET as a detector with antenna it is important to find the transfer coefficient (antenna matching coefficient) η_{a}, which is the ratio of power that is absorbed in transistor P, to max possible part of power $P_{\max}$ that can be taken off from the antenna. Let's consider the case when transistor is connected to antenna (with the input impedance Z_{ant}) directly (assuming antenna connection to gate-source).

The max power $P_{ant,\max}$ that can be supplied to FET channel from antenna at plane wave illumination is [21, 22]

$$P_{ant,\max} = G \frac{\lambda^2}{4\pi} I_0. \tag{3.24}$$

Here G is the antenna gain, λ is the wavelength in vacuum (air), and I_0 is the radiation intensity. This power is taken when load is perfectly matched with antenna.

The voltage on antenna terminals U_{ant} is

$$|U_{ant}|^2 = 2 \frac{\lambda^2 G}{\pi} \text{Re} Z_{ant} I_0 \tag{3.25}$$

In the case $Z_{GS} \neq Z^*_{ant}$ the power on antenna load is

$$P = \frac{1}{2}|U_{ant}|^2 \left| \frac{Z_{GS}}{Z_{GS} + Z_{ant}} \right|^2 \frac{\mathrm{Re}Z_{GS}}{|Z_{GS}|^2} = \frac{\lambda^2 G I_0}{\pi} \frac{\mathrm{Re}Z_{GS}\mathrm{Re}Z_{ant}}{|Z_{GS} + Z_{ant}|^2}. \tag{3.26}$$

Then for transfer coefficient η_a one can write

$$\eta_a = \frac{P}{P_{ant,\max}} = \frac{4\mathrm{Re}Z_{ant}\mathrm{Re}Z_{GS}}{|Z_{ant} + Z_{GS}|^2}. \tag{3.27}$$

Thus, the optical responsivity (as the ratio of output signal to power maximally collected from antenna) can be defined as

$$R_V^{opt} = R_V^{el}\eta_a \tag{3.28}$$

3.2.3 *Voltage Divider Transfer Coefficient*

When measuring the voltage at the transistor drain-source by voltmeter, the FET channel and voltmeter input circuit form the voltage divider. The voltage part at voltmeter (the voltage divider transfer coefficient η_L from transistor to voltmeter) is [9, 23]: $\eta_L = |(1 + R_{CH}/Z_{load})|^{-1}$, where $Z_{load} = R_{load} \| \frac{1}{2\pi f_{\mathrm{mod}} C_{load}}$ is the impedance of the measuring device.

The measured resulting voltage by voltmeter at FET transistor

$$V_{\det} = P_{ant,\max} R_{V,\mathrm{int}} \eta \eta_a \eta_L \tag{3.29}$$

To obtain the "measured" responsivity R_V^{meas} some area S_{meas} is needed. It's assumed that power is collected from this area which usually is uncertain: it can be the diffraction limited area, pixel area, antenna area, etc. That's why it is difficult to compare the responsivities cited in various papers. Really, it is a "hazy" procedure to obtain the performance of detector itself R_V^{el} from the performance of a system R_V^{meas} without knowing the antenna parameters. Frequently S_{meas} is taken equal to pitch size. The measured responsivity R_V^{meas}

$$R_V^{meas} = \frac{V_{\det}}{S_{meas} I_0} = \frac{G\lambda^2}{4\pi S_{meas}} \eta_a \eta \eta_L R_{V,\mathrm{int}} \tag{3.30}$$

The resulting responsivity that can be measured (R_V^{meas}) for one of the FETs investigated as a detector is shown in Fig. 3.3. One can see the exponential signal decay at $V_{GS} < 0.4$ V that is connected with increase ($R_{CH} \sim \exp(-V_{GS}/n\phi_t)$) of the channel resistance and thus, reduce the matching coefficient η_L with constant voltmeter input resistance.

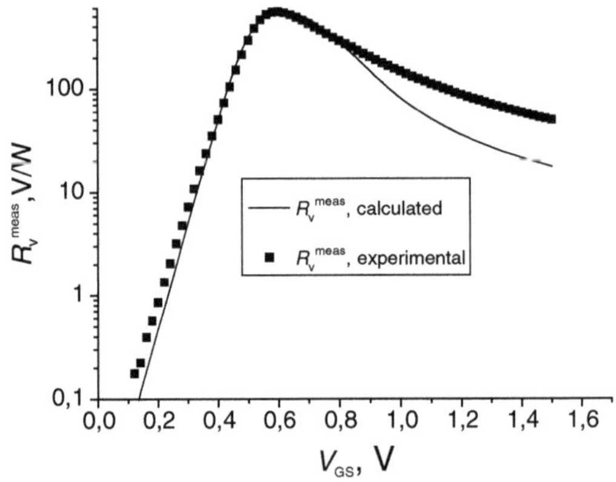

Fig. 3.3 Comparison of experimental (*points*) and calculated (*solid line*) voltage responsivity vs the gate voltage V_{GS} for *Si* MOSFET with $W = 20\ \mu m$, $L = 20\ \mu m$. To fit calculations and experimental data the latter ones were multiplied by certain coefficient as the power introduced to FET channel is uncertain because of unknown antenna gain G and antenna input impedance

One can see that the influence of coefficients $\eta\eta_a\eta_L$ product, which determines the power absorbed by FET detector, is crucial for determination of registered detector signals and thus, for responsivity R and *NEP* evaluations. When taking the values of these coefficients estimations $\eta \sim 0.1$ from Exp. (3.19) for radiation frequency $\nu \approx 77$ GHz, $\eta_a \sim 0.2$ at $Z_{ant} \sim (100–j \cdot 100)\ \Omega$ from Exp. (3.27), and FET impedance $Z_{FET} \sim (200–j \cdot 500)\ \Omega$, $\eta_L \sim 1$ at voltmeter $R_{input} \sim 10\ M\Omega$, and antenna gain $G \sim 1$ one has with $R_{V,int}$ at $V_{GS} \approx 0.6$ V the value of $R_V{}^{meas} \sim 4 \cdot 10^3$ V/W. In Schuster et al. [17] the voltage sensitivities observed in CMOS detectors obtained $R_V^{meas} \approx 6 \cdot 10^3$ V/W at $\nu \approx 300$ GHz and it seems that this larger value at higher frequency range is connected with larger gain $G > 1$ due to the resonances in thick substrates used and good antenna matching.

3.3 SBD Detectors

SBD THz/sub-THz detectors are mainly manufactured on the base of III-V semiconductors. For quasi-optical SBD detectors over a frequency range from ~150 to ~400 GHz the typical responsivity values are from ~300 to 1,000 V/W. In Fig. 3.1b is shown the SBD equivalent schematic representation which is frequently used in estimations of SBD parameters [1, 24].

Contrary to FET detector the SBD responsivity can be found in analytical form in the case of no bias (without Taylor expansion usage). The SBD current signal I_d is an averaged quantity over the period T

$$I_d = \langle I \rangle = \frac{1}{T}\int_0^T I(t)dt = \frac{1}{T}\int_0^T I_s\left(e^{\frac{\Delta V \cos(\omega t)}{n\phi_t}} - 1\right)dt = I_s\left(I_0\left(\frac{\Delta V}{n\phi_t}\right) - 1\right). \tag{3.31}$$

The power that is absorbed in SBD intrinsic part is

$$P_{in,\text{int}} = \frac{1}{T}\int_0^T I(t)V(t)dt = \frac{1}{T}\int_0^T I_s\left(e^{\frac{\Delta V\cos(\omega t)}{n\phi_t}} - 1\right)\Delta V\cos(\omega t)\,dt$$
$$= I_s\Delta V I_1\left(\frac{\Delta V}{n\phi_t}\right). \tag{3.32}$$

Then for internal current responsivity $R_{\text{I,int}}$ it follows

$$R_{I,\text{int}} = \frac{I_d}{P_{\text{in,int}}} = \frac{I_s\left(I_0\left(\frac{\Delta V}{n\phi_t}\right) - 1\right)}{I_s\Delta V I_1\left(\frac{\Delta V}{n\phi_t}\right)}, \tag{3.33}$$

and in small signal voltage approximation

$$R_{I,\text{int}}\big|_{\Delta V\to 0} = \frac{1}{2n\phi_t} \tag{3.34}$$

Here I_0, I_1 are the modified Bessel functions. In small signal voltage approximation $\Delta V < n\phi_t$ is realized the square-law SBD response. Sometimes an approach for square-law SBD response that is based on Taylor expansion is used (see e.g. [1])

$$R_{I,\text{int}} = \frac{\left.\frac{\partial^2 I}{\partial V^2}\right|_{V=0}}{2\left.\frac{\partial I}{\partial V}\right|_{V=0}} = \frac{1}{2n\cdot\phi_t}. \tag{3.35}$$

At $T = 300$ K and $n = 1$, $R_{\text{I,int}} = 19.3$ A/W is the max possible internal current responsivity of zero-bias SBD. And for measured at low frequencies the voltage responsivity R_V (when SBD is connected to a high impedance load such as e.g. oscilloscope) $R_{\text{V,int}} \approx R_{\text{I,int}} \cdot R_{\text{D}}$, where R_{D} is the zero-bias differential resistance.

In the square-law region the SBD current responsivity R_I (defined by ratio of current through the SBD to power absorbed in SBD) under the power with radiation frequency ν for circuit shown in Fig. 3.1b can be written [1] as

$$R_I^{el} = R_{I,\text{int}} \cdot \eta \tag{3.36}$$

where η is a power transfer coefficient (the ratio of power dissipated in intrinsic part to total power)

$$\eta = \frac{1}{(1 + R_S/R_D)\cdot\left(1 + (\nu/\nu_c)^2\right)}, \tag{3.37}$$

where the cut-off frequency

$$\nu_c = \frac{(1 + R_S/R_D)^{1/2}}{2\pi \cdot C_P \cdot (R_S \cdot R_D)^{1/2}} \tag{3.38}$$

With "good" values of different parameters $R_S \approx 10\ \Omega$, $R_D \approx 3 \cdot 10^3\ \Omega$ and typical zero-bias capacitance for small area SBDs $C_P \approx 5$ fF, $\nu_c \approx 184$ GHz, at frequency $\nu \approx 100$ GHz the responsivity $R_I \approx 15$ A/W ($n = 1$).

At $R_D \approx 6.5$ kΩ and $n = 1.6$ [25] the zero bias SBD internal responsivity can be estimated as $R_{I,int} \approx 11.7$ A/W giving for internal voltage responsivity $R_{V,int} \approx R_{I,int} \cdot R_D \approx 75 \cdot 10^3$ V/W. For comparison in one of the investigated FETs (Si MOSFET, 1.0-μm design rules) the estimated internal responsivity $R_{V,int} \approx 243 \cdot 10^3$ V/W and $Z_{GS,int} \approx 5.7 \cdot (1\text{–}j) \cdot 10^3\ \Omega$ at $V_{GS} = 0.6$ V.

3.4 NEP of FET and SBD Detectors

The detector noise equivalent power *NEP* can be determined as $NEP = N/R_V$, where N is the noise level. One should discriminate the electrical, measured and optical responsivities which, in turn, define the electrical (NEP^{el}), measured (NEP^{meas}) and optical (NEP^{opt}) noise equivalent powers, respectively, that are calculated from the noise N and responsivity R_V levels.

The electrical responsivity R_V^{el} characterizes the detector itself, the optical responsivity R_V^{opt} characterizes the detector+antenna circuit, and the measured responsivity R_V^{meas} characterizes the detector+antenna+(method of measurement and e.g. lenses used) arrangement. From these parameters different *NEP* values can be estimated. Thus, for different devices similar quantities should be compared (R_V^{el} with R_V^{el}, but not R_V^{el} with R_V^{opt}). The same refers to NEP^{el} and NEP^{opt}. The quantity R_V^{el} is the maximum responsivity that can be gained at optimal detector matching with antenna and optical system.

Accounting only for the case of no voltage applied to FET source-drain contacts and zero bias SBD detectors, the noise is only the thermal one. Then for FET electrical $NEP^{FET,el}$ it follows

$$NEP^{FET,el} = \frac{\sqrt{4kTR_{CH}}}{R_V^{FET,el}}. \tag{3.39}$$

The similar expression can be written for SBD detector with changing the channel resistance R_{CH} for FET to SBD resistance R_D and SBD electrical responsivity.

The results of calculation NEP^{el} are depicted in Fig. 3.4. The performance increase with an advance of technology is observed (it is due to the thinner oxide at better technology and thus, the lower channel resistance at the same level of parasitics).

The results for NEP^{el}, that can be compared with experimental data, assuming constant $Z_{ant} = (100\text{–}j100)\ \Omega$ for both types of detectors, are shown in Fig. 3.5.

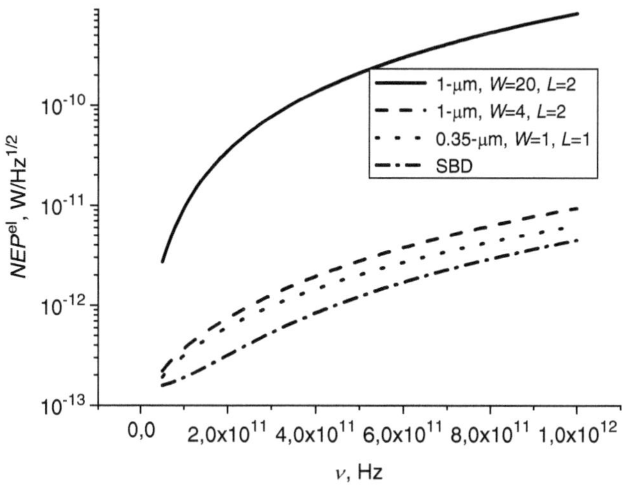

Fig. 3.4 Comparison of calculated electrical NEP^{el} for different devices. FET *NEP* improvement performance when going from 1-μm technology, $W = 20$ μm, $L = 2$ μm to 0.35-μm technology, $W = 1$ μm, $L = 1$ μm, is seen. SBD has better performance

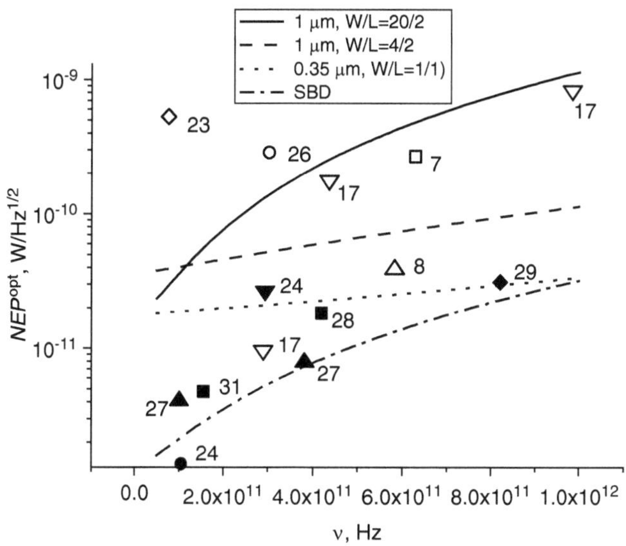

Fig. 3.5 Comparison of calculated optical NEP^{op} for a few devices with the antenna impedance $Z_{ant} = (100\text{–}j \cdot 100)\ \Omega$. *Open marks* are data for FET detectors and *filled marks* are for SBD detectors. Numbers at experimental data signs mean the numbers of Refs (Data of Sizov et al. [23] were obtained without antennas use)

As one can see the model used for SBD detectors describes well the known experimental data for NEP^{opt} but for FET ones it is seen a sufficient scatter in experimental data. The reason of it seems is mainly due to non-optimized technologies for FETs as detectors for THz/sub-THz radiation.

Comparing Figs. 3.4 and 3.5 one can point out the appreciable difference between NEP^{el} and NEP^{opt} that is due to antenna mismatch in both types of detectors that suppress the *NEP* values about an order. Several times difference between $NEP^{opt} = 43 \cdot 10^{-12}$ W/Hz$^{1/2}$ at $\nu \sim 570$ GHz and estimated $NEP^{el} = 9 \cdot 10^{-12}$ W/Hz$^{1/2}$ was pointed out e.g. in Boppel et al. [30]. It can be also pointed out from Fig. 3.5 that for FET advanced technologies NEP^{opt} will be lower (better) compared to that one for SBD detectors in high frequency region of $\nu \sim >1$ THz.

3.5 MCT-HEB Detectors

Theory of MCT uncooled microbolometers with intrinsic conductivity was developed in Dobrovolsky and Sizov [3, 31]. It was shown that encouraging NEP values ($NEP \sim 10^{-11}$ W/Hz$^{1/2}$) can be obtained. The character of response temperature dependences for samples with different types and values of conductivity, energy gap can be changed sufficiently but the use of theoretical model developed qualitatively explains of response behavior taking into account various components of the output signal: Dember effect, thermoelectromotive contribution and free carrier concentration changes.

Narrow-gap MCT technologies are well developed now, and today this material is one of the basic semiconductors for photon detectors from near IR (wavelength $\lambda \sim 1.5$ μm) to long IR ($\lambda \sim 20$ μm) and is used in large scale arrays with silicon CMOS readouts. However, in spite of achievable band gap $E_g \sim 0$ these semiconductors can't be used as photon detectors in THz/sub-THz region because of high thermal generation rate even at temperatures $T \sim 1$ K [32]. When considering MCT–HEB as THz/sub-THz detectors here it was shown than they are sensitive at least in the range of $\nu \approx 37$ GHz... 1.6 THz though at high frequencies their *NEP* is only $NEP \sim 10^{-8}$ W/Hz$^{1/2}$.

Since low dielectric constant and thickness are required for substrates with deposited antennas [21, 22] here the fused quartz substrate and other dielectric substrates were chosen as materials with sufficient hardness and relatively small permittivity ($\varepsilon < 4.8$). The thickness of quarts substrates were ~200 μm that provide the estimated ~60 % radiation efficiency for target frequency.

To decrease the influence of hot electron bolometer on antenna parameters in hybrid structures the HEB dimensions should be much less than the antenna ones. Area of antennas optimized for frequency in $\nu \sim 128$–144 GHz range was about 1.5 mm^2. Total bolometer area (thinned to $d \sim 20$ μm) was ~30×30 μm with contact pads 0.6×0.2 mm^2 to be used for hybridization with antennas on quartz substrates. The active resistance of MCT–HEB detectors was within $R \sim 50 \ldots 300\ \Omega$.

Here to increase the antenna effectiveness and broaden an operating frequency range MCT-HEBs together with lens-coupled antennas were applied too. To reduce the crosstalk between the sensitive elements, the investigations were carried on 16 element bi-linear array on relatively low permittivity substrate (fused quartz)

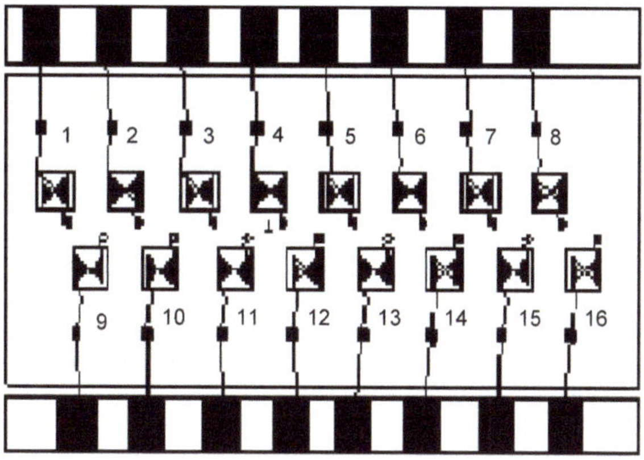

Fig. 3.6 Hybrid bi-linear 16 element MCT detector antenna array for 130–145 GHz frequency range

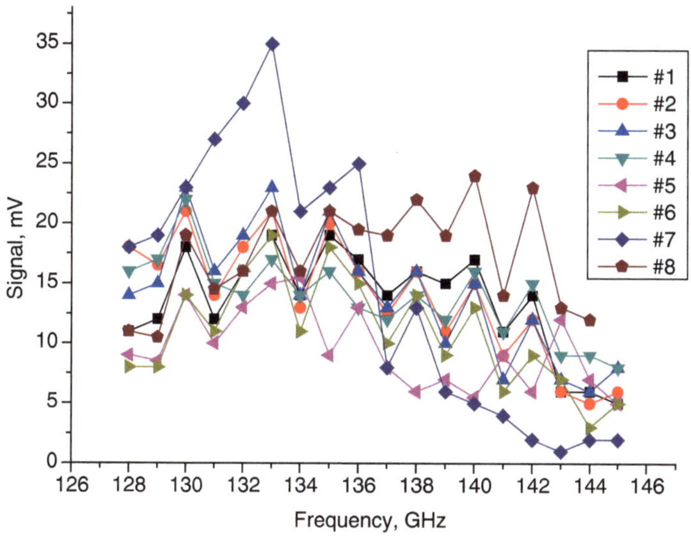

Fig. 3.7 Frequency dependence of output signals for first eight elements of bi-linear array

intended for applications in 130–145 GHz frequency range. Schematic of such an array realized is shown in Fig. 3.6. By squares are indicated separate sensitive elements with GaAs SBDs or MCT-HEB in their center and bow-tie antennas on the fused quartz substrates with dimensions $2 \times 2 \times 0.2$ mm (with sufficiently lower permittivity compared to GaAs, Si or MCT substrates).

Frequency dependences of the registered signals for first eight elements of bi-linear array are shown in Fig. 3.7.

To verify modeling of sensitive elements with bow-tie antennas and silicon microlenses, two small number MCT microbolometers in arrays with different "immersion" depth were manufactured (Fig. 3.8). Detectors were backside attached (see Fig. 3.8b) to microlenses with diameter $\varnothing = 4.4$ mm (for $\nu \sim 120$–150 GHz

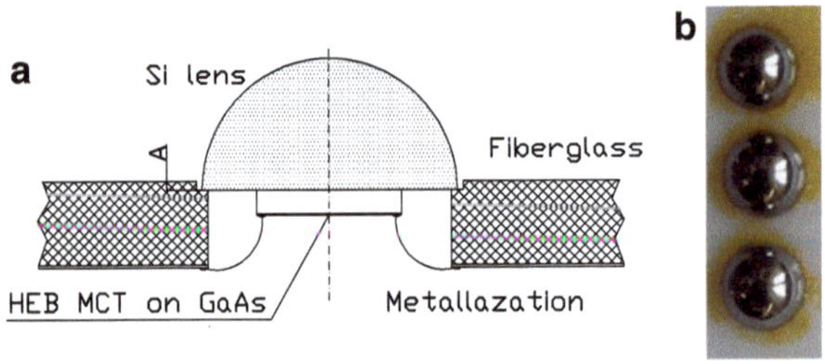

Fig. 3.8 Schematic of fiber glass wafer with microlens and sensitive element (**a**), microlenses and sensitive elements (MCT microbolometers on the back side of microlenses) with antennas on GaAs substrate and fiber glass wafer (**b**)

spectral region). Bow-tie antennas on 2×2 mm GaAs substrates have dimensions within 1.5×1.5 mm and sensitive elements were $\sim 30 \times 40$ μm.

The lenses were immersed into the fiber glass wafer with much lower dielectric permittivity compared to MCT permittivity. Without silicon lenses the output signals were about from 5–7 times lower.

The experiments performed have shown that electron heating by electromagnetic waves in MCT layers can be used for designing of uncooled THz/sub-THz detectors with appropriate for active imaging characteristics and these detectors can be manufactured within the well established MCT technologies. And from this point of view this narrow-gap semiconductor can be considered as a material for THz/sub-THz detectors which can be designed as arrays. It seems that here exist some points that can simplify FPAs fabrication on the base of MCT. First, it is small and easily adjusted (by choose of configuration, composition, carriers concentration) active resistance of detecting element with neglecting reactive component that simplifies matching with antenna. Second, MCT-HEBs are easy to fabricate and are quite homogeneous to provide good uniformity of detecting elements.

3.6 Imaging

To check the possibility of active imaging with this type of detector it was used common experimental layout based on irradiation transmission through object and two-coordinate mechanical scanning.

The designed with diameters $\varnothing \approx 120$ mm hyperbolic lenses allowed to get diffraction limited spots for wavelength $\lambda \sim 2$ mm at objects and detectors displacements with Airy disk diameter $D \approx 2.44 \cdot \lambda \cdot \mathrm{F/\#} \sim 5$ mm (for optical system f-number $F/\# \approx 1$) giving the possible system resolution ~2.5 mm. In Airy disk there is concentrated ~84 % of radiation power from the source. Tunable backward-wave oscillator (BWO with output power ~15 mW) operating in frequency range $\nu \approx 128$–152 GHz was used as a radiation source. Output electrical signal of detectors was measured by lock-in amplifier. Dynamical range of the system was

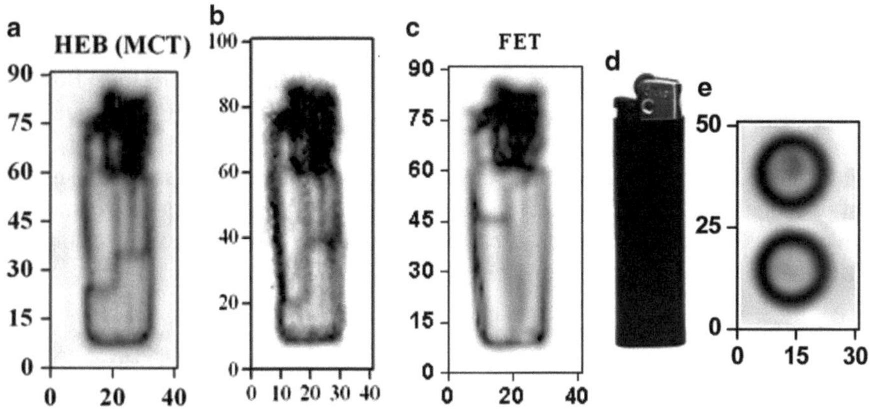

Fig. 3.9 Lighters and medical pills imaging in transmission mode at $\nu \approx 150$ GHz: (**a** and **b**) uncooled MCT-HEB, (**c**) uncooled Si FET detector, and (**d**) visible region. (**a** and **c**) lighters in envelope, (**b**) lighter in envelope seeing through the sheet of gypsum plasterboard of thickness $d = 12$ mm, (**e**) two medical pills in envelope (in the *upper* pill small ($\sim 2 \times 2$ mm) dielectric item ($d \sim 0.3$ mm) is imbedded). *Dark rings* around pills seem arise due to the phase differences between the beams in air and in pills

about 60 dB. Room temperature noise equivalent power of MCT-HEB detectors at $\nu \sim 150$ GHz was estimated as $NEP \sim 2.5 \cdot 10^{-10}$ W/Hz$^{1/2}$.

The pictures of objects monitored at $\nu \sim 150$ GHz in an envelope that is non-transparent in visible region are shown in Fig. 3.9. Here the image obtained with FET-based detector is presented too. From Fig. 3.9 one distinctly can observe the different fluid (liquid gas) levels even behind the gypsum plasterboard (Fig. 3.9b).

From Fig. 3.9 it can be concluded that MCT-HEB and FET detectors can be used in sub-THz region (where relatively high output power of $\sim$10–30 mW sources are available) for active imaging with relatively good resolution.

3.7 Conclusions

The analysis of silicon FETs, SBDs as THz/sub-THz detectors at zero drain-source ($V_{DS} = 0$) or SBD zero bias ($V_D = 0$) was considered. Both the drift and diffusion current components for MOSFETs were taken into account. It is shown that parasitic effects mainly limit the FET and SBD detectors performance. The considerations have shown that with advance of FET technology the development of FET detector performance can be improved with advancing of the design rules due to lowering the parasitic effects and increase of oxide capacitance per unit area.

In spite of the fact that at the moment the *NEP* of *Si*-MOSFET THz/sub-THz detectors in the range of $\nu \sim 100$–300 GHz are an order worse compared to SBD ones, due to more advanced technologies applied in *Si* CMOS area, these detectors can have some benefits because of well-developed silicon technologies and the possibility of proper integrated detector design and fabrication compared to technologies of III-V ternary compounds applied to SBD detectors. Both of these detectors seem can be applied mostly for active imaging at frequencies $\nu > 150$ GHz.

As concerning MCT-HEBs it is shown that MCT layers can be used for designing of uncooled THz/sub-THz detectors and arrays with appropriate for active imaging characteristics and these detectors can be manufactured within the well established MCT technologies.

References

1. Cowley AM, Sorensen HO (1966) Quantitative comparison of solid-state microwave detectors. IEEE Trans Microw Theory Tech 14:588–602
2. Dyakonov M, Shur M (1996) Detection, mixing, and frequency multiplication of terahertz radiation by two-dimensional electronic fluid. IEEE Trans Electron Device 43:380–387
3. Dobrovolsky V, Sizov F (2007) A room temperature, or moderately cooled, fast THz semiconductor hot electron bolometer. Semicond Sci Tech 22:103–106
4. Sizov F (2010) THz radiation sensors. Opto-Electron Rev 18:10–36
5. Pu L-J, Tsividis YP (1990) Harmonic distortion of the four-terminal MOSFET in non-quasistatic operation. Circuits, devices and systems. IEE Proceed G 137:325–332
6. Tsividis Y, McAndrew C (2011) Operation and modeling of the MOS transistor. Oxford University Press, New York
7. Ojefors E, Pfeiffer UR, Lisauskas A, Roskos HG (2009) A 0.65 THz focal-plane array in a quarter-micron CMOS process technology. IEEE J Solid State Circ 44:1968–1976
8. Boppel S, Lisauskas A, Mundt M et al (2012) CMOS integrated antenna-coupled field-effect transistors for the detection of radiation from 0.2 to 4.3 THz. IEEE Trans Microw Theory Tech 60:3834–3843
9. Sakowicz M, Lifshits MB, Klimenko OA et al (2011) Terahertz responsivity of field effect transistors versus their static channel conductivity and loading effects. J Appl Phys 110:054512
10. Lisauskas A, Boppel S, Mundt M et al (2013) Subharmonic mixing with field-effect transistors: theory and experiment at 639 GHz high above fT. IEEE Sensors J 13:124–132
11. Knap W, Videlier H, Nadar S et al (2010) Field effect transistors for terahertz detection – silicon versus III–V material issue. Opto-Electron Rev 18:225–230
12. Tauk R, Teppe F, Boubanga S et al (2006) Plasma wave detection of terahertz radiation by silicon field effects transistors: responsivity and noise equivalent power. Appl Phys Lett 89:253511
13. Gutin A, Kachorovskii V, Muraviev A, Shur M (2012) Plasmonic terahertz detector response at high intensities. J Appl Phys 112:014508
14. Preu S, Kim S, Verma R et al (2012) An improved model for non-resonant terahertz detection in field-effect transistors. J Appl Phys 111:024502
15. Pozar DM (2011) Microwave engineering. Wiley, New York
16. Boppel S, Lisauskas A, Roskos HG (2013) Terahertz array imagers: towards the implementation of terahertz cameras with plasma-wave-based silicon MOSFET detectors. In: Saeedkia D (ed) Handbook of terahertz technology for imaging, sensing and communications. Woodhead Publishing Limited, Oxford, pp 231–271
17. Schuster F, Coquillat D, Videlier H et al (2011) Broadband terahertz imaging with highly sensitive silicon CMOS detectors. Opt Express 19:7827–32

18. Knap W, Kachorovskii V, Deng Y et al (2002) Nonresonant detection of terahertz radiation in field effect transistors. J Appl Phys 91:9346
19. Enz C, Cheng Y (2000) MOS transistor modeling for RF IC design. IEEE J Solid State Circ 35:186–201
20. Liu W (2001) MOSFET Models for SPICE simulation, including BSIM3v3 and BSIM4. Wiley, New York
21. Balanis CA (2005) Antenna theory analysis and design. Wiley, Hoboken
22. Volakis HL (2007) Antenna engineering handbook. McGraw-Hill, New York
23. Sizov F, Golenkov A, But D et al (2012) Sub-THz radiation room temperature sensitivity of long-channel silicon field effect transistors. Opto Electron Rev 20:194–199
24. Brown ER, Young AC, Zimmerman J et al (2007) Advances in Schottky rectifier performance. IEEE Microw Magaz 8:54–59
25. Kazemi H, Nagy G, Tran L et al (2007) Ultra sensitive ErAs/InAlGaAs direct detectors for millimeter wave and THz imaging applications. In: 2007 IEEE/MTT-S international microwave symposium. IEEE, Honolulu, pp 1367–1370
26. Pleteršek A, Trontelj J (2012) A self-mixing n-MOS channel-detector optimized for mm-wave and THz signals. J Infrared Millim THz Waves 33:615–626
27. Chahal P, Morris F, Frazier G (2005) Zero bias resonant tunnel Schottky contact diode for wide-band direct detection. IEEE Electron Device Lett 26:894–896
28. Liu L, Hessler JL, Hu H et al (2010) A broadband quasi-optical terahertz detector utilizing a zero-bias Schottky diode. IEEE Microw Wirel Compon Lett 20:504–506
29. Han R, Zhang Ya, Kim Y et al (2012) 280GHz and 860GHz image sensors using Schottky-barrier diodes in 0.13 μm digital CMOS. In: IEEE international solid-state circuits conference, pp 254–256
30. Boppel S, Lisauskas A, Krozer V, Roskos HG (2011) Performance and performance variations of sub-1 THz detectors fabricated with 0.15 μm CMOS foundry process. Electron Lett 47: 661–662
31. Dobrovolsky V, Sizov F (2010) THz/sub-THz bolometer based on electron heating in a semiconductor waveguide. Opto Electron Rev 18:250–258
32. Sizov FF, Reva VP, Golenkov AG, Zabudsky VV (2011) Uncooled detectors challenges for THz/sub-THz arrays imaging. J Infrared Millim THz Waves 32:1192–1206

Chapter 4
High Power, Narrow-Linewidth, and Compact Terahertz Sources for Room Temperature Applications

Manijeh Razeghi, Quanyong Lu, Neelanjan Bandyopadhyay, Steven Slivken, and Yanbo Bai

Abstract We present the recent progress of room temperature terahertz (THz) sources based on intracavity difference-frequency generation from mid-infrared quantum cascade lasers. The active cores based on the single-phonon resonance scheme are designed with a giant THz nonlinearity. Integrated dual-period distributed-feedback grating is used to purify the THz spectra. Room temperature single-mode operation is obtained from the modal-phase matched THz sources at 3.3–4.6 THz, with a THz power of 65 μW at 4 THz. A wider THz spectral range of 1–4.6 THz is obtained from the Čerenkov phase-matched THz sources, where a maximum power of 32 μW at 2.6 THz is demonstrated. The THz power of Čerenkov device is further improved by epi-down mounting with a high THz power of 215 μW at 3.5 THz. The rapid development renders this type of THz sources promising local oscillators for many astronomical and medical applications.

4.1 Introduction

The terahertz (THz) region of the electromagnetic spectrum ($\nu \sim 0.1$–10 THz) has a number of applications in biological engineering and security [1]. Being a non-ionizing radiation, it is environmentally-friendly, enabling a safer exposure than conventional X-ray based techniques. Many molecules and solids have strong and distinct spectral signatures at terahertz frequencies, which makes terahertz technology important for both science and commercial applications related to spectroscopy and imaging [2]. However, this spectral range is the so called "THz gap" between the radio and optical frequency ranges. It still suffers from the lack of high performance compact sources at room temperature. The electronic devices on

M. Razeghi (✉) • Q.Y. Lu • N. Bandyopadhyay • S. Slivken • Y. Bai
Center for Quantum Devices, Department of Electrical Engineering and Computer Science, Northwestern University, Evanston, IL 60208, USA
e-mail: razeghi@eecs.northwestern.edu

C. Corsi and F. Sizov (eds.), *THz and Security Applications*, NATO Science for Peace and Security Series B: Physics and Biophysics, DOI 10.1007/978-94-017-8828-1_4,

the lower THz frequency side ($\nu < 1$ THz), such as Gunn diode, resonance tunneling diode, and impact ionization avalanche transit-time diodes are able to deliver room temperature THz signal with a compact device size [3–5]. However, this category of sources has limited bandwidth, poor power efficiency and low output power levels at the super THz frequencies ($\nu > 2$ THz). For the optical devices, the THz quantum cascade lasers (QCLs) have emerged as the most powerful semiconductor THz sources in the 2–5 THz range, however, the operational temperature is still below 200 K [6–8]. On the other hand, the InP-based mid-infrared (mid-IR) QCLs have gained significant advancements in the recent years [9–12]. As a result, the THz source based on intracavity difference-frequency generation (DFG) [13] from the mid-IR QCLs is free from the temperature issue that is suffered by the THz QCLs based on direct optical transition, and it did not take too long to achieve room temperature operation [14].

Continuous progress in the higher THz power and conversion efficiency is driven by the advances in multiple areas, including dual active regions based on single-phonon resonance design engineered with giant nonlinearities [15], Čerenkov phase-matching scheme [16], integrated dual-period distributed feedback (DFB) gratings [17, 18], and epi-down mounting and packaging [19]. In this paper, we break down and discuss the aspects responsible for the final device performance. Optimization of dual-core active region design, dual-period DFB grating design, and Čerenkov phase-matching design with experimental results are discussed in Sects. 4.2, 4.3, and 4.4, respectively, and finally a conclusion is given with Sect. 4.5.

4.2 Dual-Core Active Region Designs

Giant nonlinearities in quantum well structures originate from the interaction of the strongly coupled quantum states. Among these, the nonlinear DFG [20, 21] is of special interest due to its capability in producing THz emission, and is free from the temperature dependent carrier transport and gain issues that hamper traditional THz QCLs. For a mid-infrared (mid-IR) QCL featuring a giant second-order nonlinear susceptibility ($|\chi^{(2)}|$) in its active region and dual wavelength operation, THz emission can be generated from an extremely compact device at room temperature. The THz power strongly relies on high powers from the mid-IR sources (W_1, W_2), and a giant $|\chi^{(2)}|$. From the first demonstration of THz DFG in QCLs [13] to the subsequent works, dual-cores featuring double-phonon-resonance (DPR) and bound-to-continuum (BTC) depopulation schemes, has been used. Among the two core designs, only the BTC core is designed with a giant nonlinear susceptibility and the DPR core merely acts as a mid-IR source with little THz generation. Meanwhile, the mid-IR QCL technology has made tremendous progress in power and efficiency, thanks to the improved material quality, waveguide, and especially the elaborate quantum designs [22–24]. Out of these, QCLs based on the single-phonon-resonance (SPR) design has demonstrated the highest power and efficiency in the shorter mid-IR wavelength range (4–6 μm) [11]. The strong coupling of

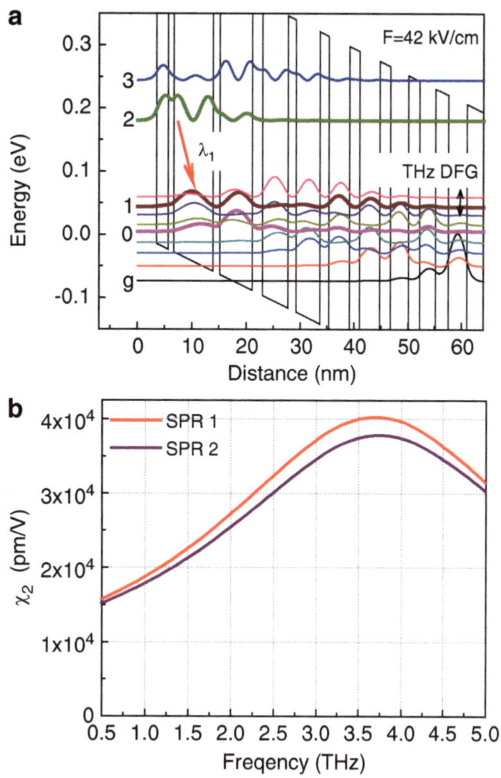

Fig. 4.1 (**a**) Conduction band and wave function diagram for the SPR design featuring a giant nonlinear susceptibility $|\chi^{(2)}|$. (**b**) The calculated $|\chi^{(2)}|$ for SPR1 and 2 as functions of the THz frequency

the lower lasing level and the upper injector level indicates that the SPR design in the long mid-IR wavelength range can be designed with giant nonlinearity without much compromise on its high power feature. This is because the long-wave QCL structure based on lattice matched material system generally have large dipole strengths and smaller intersubband transition broadening than their long-wave counterpart based on strain-balanced material system.

Figure 4.1a shows the SPR design in the long mid-IR wavelength $\lambda_1 \sim 9.0$ μm. The design is shown at the high field when the maximum mid-IR power is obtained. State 0 is separated by about one optical-phonon energy from state 1, which helps to depopulate state 1 with ultrafast optical-phonon scattering. The THz DFG nonlinearity (Fig. 4.1b) is calculated through a summation over all possible transitions. The major contributions stem from the adjacent states that are within 10–20 meV of the lower lasing state 1. Compared with the BTC structure in Belkin et al. [14] near roll-over voltage, using the same simulation parameters, the lifetime of the upper lasing state 2 caused by phonon scattering is increased from 0.39 ps for the BTC design to 0.46 ps for the SPR design, while the lifetime of the lower lasing state 1 is maintained at 0.1 ps. The nonlinear susceptibility at 4 THz is calculated to be around $|\chi^{(2)}| \sim 4 \times 10^4$ pm/V. The other laser core at $\lambda_2 \sim 10.2$ μm designed with the same SPR scheme (layer thickness was increased about 3 %) shows a similar

$|\chi^{(2)}| \sim 3.8 \times 10^4$ pm/V. A diagonal design is used to make the gain bandwidth of each laser core (~250 cm^{-1}) significantly larger than the energy separation of the two wavelengths (~130 cm^{-1}), so that the overall gain spectrum of the dual core structure is expected to exhibit a single peak. This allows for wide-range gap-free THz wavelength tailoring by using dual-period DFB gratings. Figure 4.1b shows the calculated nonlinear susceptibilities with $|\chi^{(2)}| \sim 1.5 - 4 \times 10^4$ pm/V in a range of 0.5–5 THz for the two SPR structures, which provides the basis for tunable THz emission over a wide frequency range.

4.3 THz Sources at 3.3–4.6 THz with Integrated Dual-Period Grating Designs

In a typical multimode Fabry-Pérot (FP) cavity, the light intensity spreads out among different mid-IR frequencies, and the total power is the sum over many small W_i components. As such, the product W_iW_j will be small and the THz spectrum will rather broad ($\Delta\nu \sim 0.5$–1 THz). In order to purify and tune the THz spectrum, all the mid-IR power needs to be concentrated on the two mid-IR frequencies with single mode operation, and their frequency positions need to be controllable and tunable. The most straightforward way to this end is to use the integrated dual-period DFB grating to purify and tune the mid-IR spectra. While the design strategy of dual-period DFB grating shares some similarity to the single-period DFB like sufficient coupling strength for single mode operation, balancing the coupling strength, waveguide loss, and modal gain of the two grating components is of special importance.

Double exposure holographic lithography (HL) or electron-beam (e-beam) lithography has used to define the dual-period DFB grating for both mid-IR and THz spectral purifications. Figure 4.2a shows the simulated superpositioned HL interferograms with equal-dose double exposures. As it is difficult to accurately reproduce the photo resist profile with anisotropic dry etching processing, the pattern was simply truncated into binary-equivalent square-shaped structures when transferred into SiO_2 mask, as shown in Fig. 4.2a. Figure 4.2c shows the calculated coupling coefficients (κ_1 for λ_1 and κ_2 for λ_2) for the case of equal exposure intensity. Clearly, κ_1 is much smaller than κ_2. This is because the longer wavelength λ_2 has a more expanded mode profile (see the inset of Fig. 4.2c) and has a stronger interaction with the surface gratings than the shorter wavelength λ_2. To obtain similar coupling coefficients, the grating shape is reconstructed by changing the dose ratio of the two exposures with a similar strategy as in Lu et al. [25]. The HL interferograms with a dose ratio of 1.5:1, the truncated grating profile are shown in Fig. 4.2c. Around a grating depth of 200 nm (shaded region in Fig. 4.2c), the coupling coefficients of these two wavelengths are similar (6–10 cm^{-1}). This design offers sufficient coupling strength for 2–3 mm cavities with high-reflection (HR) coatings. For the case of e-beam lithography, the grating profile will be similar to the superpositioned

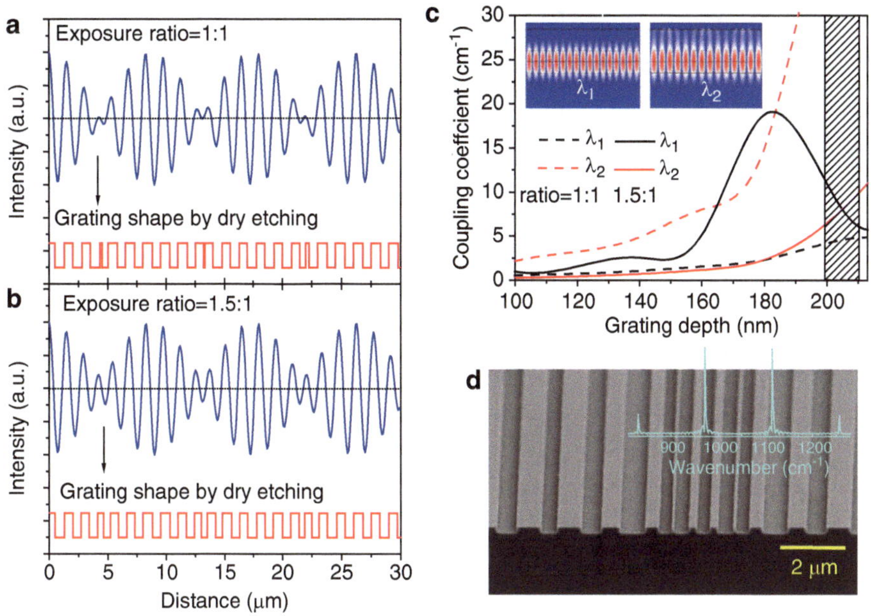

Fig. 4.2 Superpositioned HL interferograms with exposure dose ratio of 1:1 (**a**) and 1.5:1 (**b**) and the truncated square gratings after dry etching. (**c**) Coupling coefficients of the truncated gratings in (**a**), (**b**) with different grating depths. The *shaded* region is the targeted grating depth. The *insets* in (**c**) are two eigenmodes for the dual-period grating. (**d**) SEM image of a dual-period grating after dry etching. *Inset*: Fourier analysis of the corresponding gratings

interferogram of the two rectangular waves, as shown in Fig. 4.2d. Double e-beam exposure with the same exposure doses is used for two gratings with different period components, which naturally balances the coupling strengths for the two wavelength via their different grating duty cycles. The Fourier analysis of the extrapolated grating shape from the scanning electron microscope SEM picture shown in the inset of Fig. 4.2d gives two distinct peaks with a THz energy spacing around 4 THz. The other two satellite peaks with the same energy spacing away from the two main peaks correspond to the high-order Fourier series of the grating shape.

The QCL structure was grown by gas-source molecular beam epitaxy on an n-InP substrate ($\sim 1.5 \times 10^{17}$ cm^{-3}). The growth started with a 5-μm InP buffer layer ($\sim 2 \times 10^{16}$ cm^{-3}). The laser cores consist of 30 stages of single-phonon resonance (SPR) structures designed for the wavelength of $\lambda_1 \sim 9.0$ μm, and another 30 stages of SPR structure designed for the wavelength of $\lambda_2 \sim 10.2$ μm. The growth ended with a 3.5-μm InP cladding layer ($\sim 2 \times 10^{16}$ cm^{-3}) and a 200-nm InP contact layer ($\sim 5 \times 10^{18}$ cm^{-3}). The wafer was processed into double-channel geometries with a 16-μm ridge width tapered into 60 μm with a taper angle of 1° toward one end. A row of ten devices with dual-period DFB gratings spanning a DFG frequency range from 3.3 to 4.6 THz are defined by electron beam lithography. The gratings

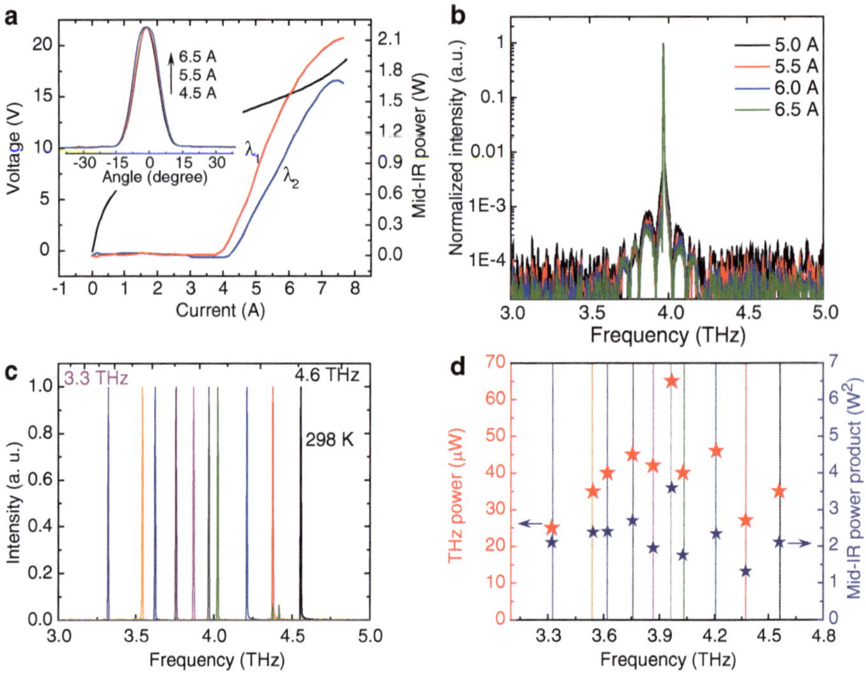

Fig. 4.3 (**a**) *P-I-V* characterization for the two wavelengths of a device operating at 4 THz. *Inset*: far fields at different currents. (**b**) Normalized THz spectra at different currents. (**c**) The THz spectra of the devices emitting from 3.3 to 4.6 THz. (**d**) THz powers and mid-IR power products of the devices emitting at different THz frequencies. The *dashed lines* are used for guidance

were transferred into the cap layer with a grating depth of 200 nm, following the process in Lu et al. [26]. The surface grating is defined in the non-tapered region. The substrate was lapped and polished to around 150 μm for backside metallization. The device with 3-mm cavity length is HR and anti-reflection (AR) coated with Si_3N_4/Au (400/1,000 nm) and Y_2O_3 (1,000 nm), respectively.

Figure 4.3a shows the power-current-voltage (*P-I-V*) characterizations for the two wavelengths with the frequency spacing of 4 THz. Compared with results in Belkin et al. [14], the total output power is enhanced by a factor of 2 and the input electrical power is reduced by about 40 %. The far fields (inset of Fig. 4.3a) of the total mid-IR power are single lobed and nearly diffraction-limited, which indicates that all the wavelengths are working in their fundamental transverse modes. This device exhibits stable single mode operation around 4 THz ($\lambda \sim 75.4$ μm), as shown in Fig. 4.3b. The spectrum linewidth is about 5 GHz, which is mainly limited by the resolution of the FTIR (0.125 cm^{-1}). The side-mode suppression ratio (SMSR) is as high as 40 dB at high currents (>6.0 A), and the THz spectral position is very stable as the current changes. The THz current tuning rate is 0.60 GHz/A, which is about one order of magnitude smaller than the mid-IR tuning rates (5.4 and 6.0 GHz/A for λ_1 and λ_2).

A row of ten devices with dual-period DFB gratings designed with varied THz frequencies are tested and characterized. Single mode operations spanning a spectral range of 3.3–4.6 THz from these devices are demonstrated, as shown in Fig. 4.3c, with a tuning range of 1.3 THz and a mean SMSR above 30 dB. Figure 4.3d shows the maximum THz powers and mid-IR power products for each frequency. The THz power and conversion efficiency both peak at 4 THz with 65 μW and 23 $\mu W/W^2$ respectively, and decrease to 25 μW and 11 $\mu W/W^2$ at 3.3 THz. The varied power conversion efficiencies are mainly attributed to the wavelength-dependent phase mismatching and waveguide loss. While the mid-IR and THz modes are nearly modal-phase matched around 4 THz, their modal phases become increasingly mismatched as the THz frequency shifts away from the matching condition. Besides, the THz loss increases rapidly as THz frequency decreases from 5 to 3 THz. The overall contribution from these effects makes the THz power conversion efficiency varying significantly in this range. THz generation outside of this range becomes less efficient due to the narrow band modal phase matching scheme.

4.4 THz Sources at 1–4.6 THz with Čerenkov Phase-Matching Scheme

The above THz DFG in QCLs takes place within a highly lossy waveguide with a modal phase matching scheme. Due to a faster frequency dependent effective index in the THz range with respect to the mid-IR index ($n_{mid\text{-}IR}$), the modal phase matching is only satisfied within a relatively narrow frequency range for a certain waveguide. In contrast to this narrow-band modal phase matching design, the Čerenkov scheme [27] has been used as an effective broad-band phase matching method for THz generation in the externally pumped optical rectification or DFG setups [28]. Recently, the Čerenkov phase-matching scheme has been demonstrated for the mid-IR second harmonic generation (SHG) [29] and THz DFG [16] in QCL structures. In the Čerenkov configuration, the THz index (n_{THz}) in the QCL active region is higher than the mid-IR index, the fundamental mid-IR wave propagates faster than the second-order DFG THz wave. The phase matching condition is fulfilled when the THz wave travels at an angle $\theta_C = \cos^{-1}(n_{mid\text{-}IR}/n_{THz})$ with respect to the mid-IR wave, θ_C is the Čerenkov angle.

This broad-band Čerenkov phase matching scheme is used to take full advantage of the wide bandwidth of the THz nonlinear medium based on SPR design. The QCL epilayer structure together with a 1-μm InP bottom contact layer and 100-nm InGaAs etch-stop layer are grown by gas-source molecular beam epitaxy on a semi-insulating InP substrate. Both active sections feature giant nonlinear susceptibilities of $|\chi^{(2)}| = 3.8\text{–}4.0 \times 10^4$ pm/V for DFG at high operating fields where maximum mid-IR powers are expected.

The sample was processed into ridge waveguide by inductive coupled plasma (ICP) etching. The device geometry is similar to that of the modal phase-matched

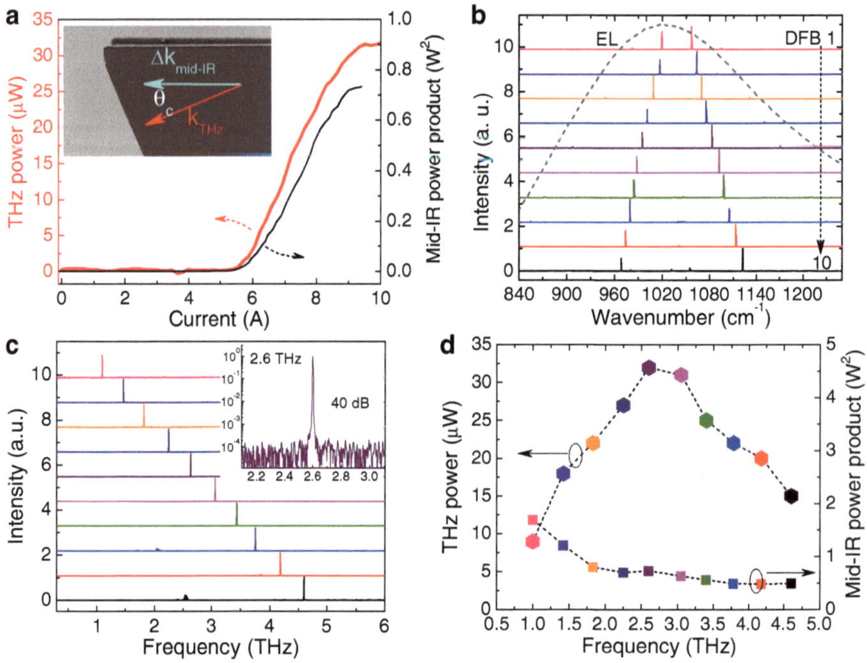

Fig. 4.4 (**a**) THz power and mid-IR-power products characteristics of a device operating at 2.6 THz. *Inset*: SEM image of the device close to the laser front facet with the Čerenkov scheme depicted. (**b**) The mid-IR spectra of different DFB designs with varied frequency differences between λ_1 and λ_2 and the EL spectrum. (**c**) The THz spectra of the devices emitting from 1.0 to 4.6 THz. (**d**) THz powers and mid-IR power products of the devices emitting at different THz frequencies

device. Due to the semi-insulating substrate, the top and bottom contacts are fabricated on the same side of the wafer, with a distance of 25–30 μm, which introduces a resistance of 0.2–0.4 Ω. The backside was polished to keep the substrate thickness around 340 μm. The inset of Fig. 4.4a is the SEM image of the device near the front facet with the Čerenkov scheme depicted. Considering the Čerenkov angle $\theta_C = 22°$ in the substrate calculated with $n_{THz} = 3.67$, $n_{mid\text{-}IR} = 3.25$ and $n_{sub} = 3.5$, where n_{sub} is the substrate index, the front facet is polished into 30° to facilitate the THz output. A row of ten devices with dual-period DFB gratings spanning a THz frequency range from 1.0 to 4.6 THz with a step of 0.4 THz is defined by electron beam lithography.

The device with the highest THz output power is at a frequency of 2.6 THz, as shown in Fig. 4.4a. A maximum THz peak power of 32 μW with THz power conversion efficiency ($\eta = W/W_1W_2$) about 45 μW/W^2 at high currents are obtained. Compared to the above results with modal phase matching, the improved conversion efficiency is attributed to the improved active design and an effective THz extraction with the Čerenkov phase-matching scheme, while the reduced power is related to the lower mid-IR pumping power (~50 %) and smaller THz photon energy.

A row of ten devices with different dual-period DFB gratings designed are tested and characterized. Figure 4.4b is the mid-IR spectra from different DFB grating designs and the EL spectrum from a reference mesa structure. The mid-IR spectra with various frequency differences are intentionally blue shifted by 25–35 cm^{-1} for a better balance the gain/loss between the two wavelengths. Single mode operations from 1.0 to 4.6 THz are demonstrated as shown in Fig. 4.4c, with a frequency step-tuning range ($\Delta\nu$) of 3.6 THz and a mean SMSR above 30 dB. This range represents 1.28 times that of the central frequency. The inset of Fig. 4.4c shows the spectrum at 2.6 THz with SMSR ~40 dB at high currents. For some of the devices with mid-IR frequency spacing close or lager than 3.78 THz, there are FP modes appearing in between the two DFB modes as seen in Fig. 4.4b and the corresponding THz spectra in Fig. 4.4c. These additional FP modes with intensities much weaker than DFB modes can be removed by increasing the DFB coupling strength.

Figure 4.4d shows the THz powers and mid-IR power products for each frequency. The THz power peaks at 2.6 THz with 32 μW and the conversion efficiency peaks at 3.1 THz with 50 $\mu W/W^2$. At 1.0 THz, the power and conversion efficiency are 8 μW and 5 $\mu W/W^2$, respectively. As mentioned before, the single-lobed EL spectrum of the design allows for gap-free THz generation down to 1 THz with a potential into sub-THz range. Although the conversion efficiency decreases at lower frequencies due to the decreasing of $|\chi^{(2)}|$ (Fig. 4.1b), the THz power output can still benefit from the higher mid-IR powers as the two mid-IR wavelengths move closer towards the gain peak. An alternative to the nonlinear design so that the $|\chi^{(2)}|$ maxima is shifted to a lower frequency (e.g., ~1 THz) is expected to enable sub-THz emission with good power and efficiency. On the high frequency side, the multilayer transmittance in the waveguide from the active region to the substrate decreases above 4.5 THz, which makes the power conversion in this range less efficient. Further reducing the doping and thickness of the bottom contact layer and buffer layer is expected to enhance the THz power and conversion efficiency significantly.

4.5 High Power THz Sources with Epilayer-Down Mounting

Despite the higher conversion efficiency, the THz power of the Čerenkov phase-matched device was lower than the modal phase-matched device because of a poorer heat removal mechanism due to the epilayer-up mounting and 340 μm thick substrate, and a less efficient current injection scheme due to the single-side current injection through a thin layer of bottom contact. Figure 4.5a, b show the temperature and electrical potential distributions of an epilayer-up mounted QCL with an asymmetric contact pattern (single-sided current injection) near the front facet. The x, y, and z directions are the lateral, longitudinal, and growth directions, respectively. The polished angle of the facet is set to be 30°. The thermal conductivities of the materials are from Lops et al. [30] and the heat source density is set to be 2.5×10^6 W/cm^3 with the average input power at 1 % duty cycle.

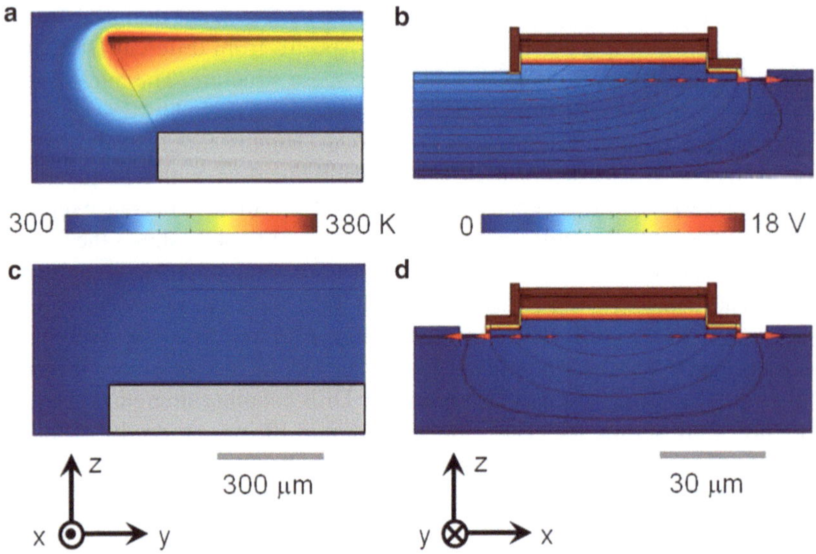

Fig. 4.5 Temperature and electrical potential distributions of devices with epilayer-up (**a**, **b**) and epilayer-down mounting schemes (**c**, **d**). The *lines* and *arrows* in (**b**) and (**d**) denote the contour lines of the electrical potential and the current flow. Current injection is single-sided in (**b**) and double-sided in (**d**)

The electrical conductivities of bulk materials are from Becker and Sirtori [31] and the electrical conductivity of the active region in the growth direction is fitted from the *I-V* curve from a reference device. Clearly, the Čerenkov device suffers severely from both thermal and electrical issues, as the heat near the front facet is not effectively dissipated and the non-uniform electrical distribution across the active region induces non-uniform current injection with the electrical field varying over 2–4 kV/cm. To address these two issues, we design and demonstrate double-sided current injection and epilayer-down mounting on a submount. Figure 4.5c, d show the temperature and electrical potential distributions of a device with epilayer-down mounting to a diamond submount. The heat dissipation is improved significantly and the electrical distribution is much more uniform with electrical field variation less than 0.5 kV/cm in the x-z plane of the active region.

In the experiment, the waveguide structure is similar to that in Lu et al. [15], except that the thickness of the InP bottom contact layer is reduced from 1 to 0.25 μm, the InP buffer layer from 5 to 3 μm, and the InP top cladding layer from 3.5 to 3.2 μm. The reduced top cladding layer helps increase the dual-period DFB coupling strength, which is important for the stable dual wavelength operation with a wide THz frequency spacing. The sample is processed into double-channel geometries with a ridge width of 20 μm. The 1.5-mm long DFB section is patterned with a dual-period grating within a 3-mm long laser cavity. The frequency spacing of the grating targets 3.5 THz. Submounts with patterns corresponding to the top

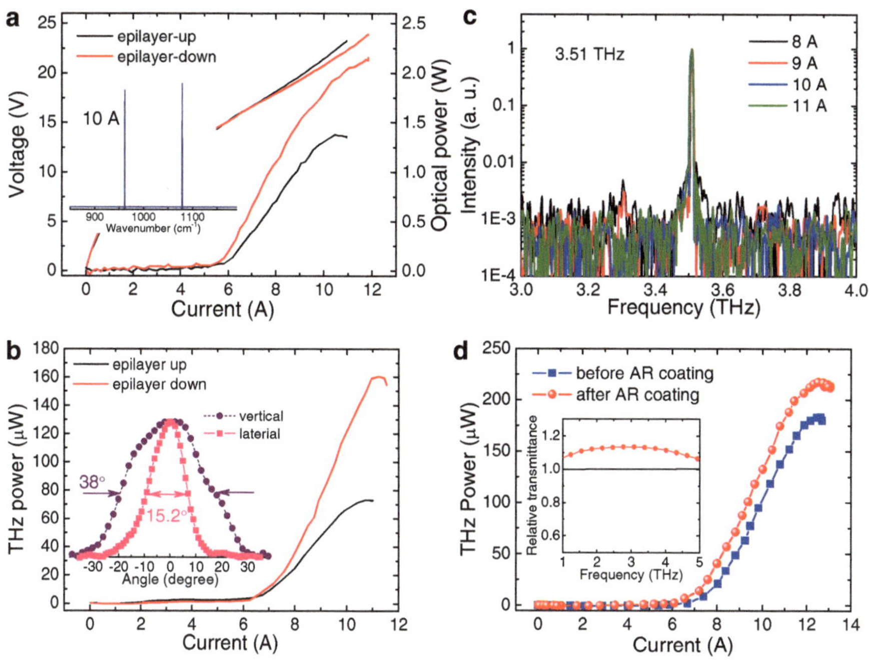

Fig. 4.6 (**a**) Mid-IR *P-I-V* and for the epilayer-up and -down mounted devices. *Inset*: mid-IR spectrum at 10 A. (**a**) THz power as a function of current for the devices with two mounting schemes. The *inset* is the THz far fields of the epilayer-down mounted device at 10 A in vertical and lateral directions. (**c**) THz spectra of the epilayer-down mounted device. (**d**) THz powers characterizations of an epilayer-down mounted device with polished mid-IR facet (*blue dotted line*) and after applying SU-8 coating (*red dotted line*). The *inset* is the relative transmission of a 15-μm thick SU-8 coating on a high-resistivity Si wafer

and bottom contact are prepared for the epilayer-down mounting. For comparison, part of the sample is processed with one bottom contact for testing with epilayer-up mounting. The substrate is polished without lapping to keep the maximum thickness around 340 μm.

Both epilayer-up and epilayer-down mounted devices were characterised at room temperature for comparison. The substrate near the front facet has been polished at 25° without damaging the mid-IR facet in order to collect the THz signal which satisfies the Čerenkov phase-matching condition. The *P-I-V* characterizations of two devices with different mounting schemes are shown in Fig. 4.6a. The epilayer-down mounted device exhibits reduced threshold current density and voltage with ~35 % higher mid-IR power compared with its epilayer-up counterpart. The mid-IR spectra of the epilayer-down device in the inset of Fig. 4.6a exhibits stable dual-wavelength emissions at $\lambda_1 = 9.26$ μm and $\lambda_2 = 10.4$ μm in the working current range. Similar spectral behaviour is also observed from the epilayer-up device. Figure 4.6b shows that THz powers up to 160 and 70 μW with power conversion efficiencies of $\eta = 160$ and 148 μW/W^2 at high currents are obtained for the epilayer-down and epilayer-up mounted devices, respectively. The higher THz power and the improved conversion efficiency of the epilayer-down mounted device is attributed to the improved thermal and electrical managements, the higher THz frequency operation, and increased THz transmission through the reduced thickness of the bottom contact layer. THz far fields in the lateral and vertical directions for the epilayer-down mounted device at a working current of 10 A are shown in inset of Fig. 4.6b. The device exhibits single-lobed beam distributions in both directions with a full width at half maximum (FWHM) of 15.2° and 38°, respectively, The THz beam peaks at 0° in both directions. The epilayer-down mounted device exhibits a stable single mode operation around 3.51 THz ($\lambda \sim 85.5$ μm), as shown in Fig. 4.6c. The SMSR of the normalized spectra is as high as 30 dB at high currents. This is compared with the results in Refs. 15 and 18 with a SMSR as high as 40 dB obtained with the liquid-helium cooled silicon bolometer. The lower SMSR here is due to the lower sensitivity of the uncooled far-IR DTGS detector used in THz spectral measurement.

All previous Cerenkov implementations share a common feature of preserving the mid-IR facets when forming the inclined facet. For our THz device with distributed feedback mechanism, since the grating feedback ($\kappa L \sim 4.5$–6 for the HR coated cavity with 1.5-mm long grating section) is much stronger than the mirror feedback, the mid-IR threshold is mainly determined by the grating feedback rather than the mirror feedback. We attempted to extend the polished area through the mid-IR mirror facet. As such, the threshold current increases by about 0.5 A, and the mid-IR output power is reduced dramatically to 0.5 W. Moreover, the THz power increases to 183 μW, as shown in Fig. 4.6d (blue curve). The nominal THz conversion efficiency is above 2 mW/W^2. This behaviour can be explained as follows. The THz signal is generated inside the cavity, so the mid-IR pumping power that is relevant to the nonlinear generation is the mid-IR power flux within the cavity rather than the output power. In this case, even though the mid-IR output power is reduced due to the facet polishing, the forward travelling power flux inside the cavity actually increases due to the reduced in-plane reflectivity.

In the epilayer-down mounting scheme, the InP substrate is exposed. Therefore, THz power can be further enhanced by THz anti-reflection (AR) coating. Several materials have been employed as the THz AR coating, such as SiO_2, Parylene, and low-density Polyethylene [32, 33]. Here SU-8 photoresist is used for its simplicity. The inset in Fig. 4.6d shows the relative transmission of a 15-μm SU-8 layer on a high-resistivity Si wafer. Given a refractive index of ~1.7, the absorption loss at 3.5 THz is estimated to be ~100 cm^{-1}, and the optimal coating thickness on InP substrate is around 11 μm with a maximum transmission of about 87 %. The SU-8 photoresist is then spin coated on the mounted device, and baked and cured. The coating thickness is controlled by the spin rate. The THz powers after the SU-8 coating (red curve) is shown in Fig. 4.6d. The power is enhanced by 17.5 % to 215 μW. Ideally, with the low-loss coatings, such as SiO_2, or low-density Polyethylene, power enhancement above 40 % can be achieved, which will be our future work.

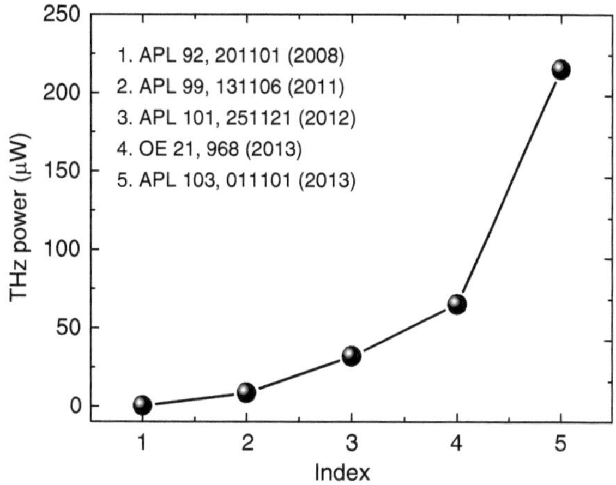

Fig. 4.7 Steady improvement of THz output power of THz QCL sources at room temperature

4.6 Conclusions

Several important breakthroughs have been demonstrated for room temperature THz QCL sources based on DFG in the recent years, which includes single mode operation with high SMSR and narrow linewidth, wide frequency coverage of 1–4.6 THz, and THz power up to 215 μW. Figure 4.7 shows a comparison of the THz maximum power in the recent publications, from which a steady upward trend is apparent. We believe that the persistent effect in this direction will eventually lead to room temperature continuous wave operation with high THz power output. This type of THz sources will enable many interesting applications in the astronomical and medical areas.

This work is partially supported by the National Science Foundation under grants ECCS-1231289 and ECCS-1306397. The authors would also like to acknowledge the encouragement and support of Dr. K.K. Law from the Naval Air Warfare Center, Dr. T. Manzur from the Naval Undersea Warfare Center, and Dr. N. Dhar from the Defense Advanced Research Projects Agency.

References

1. Tonouchi M (2007) Cutting edge terahertz technology. Nat Photonic 1:97–105
2. Woolard DL, Brown ER, Pepper M, Kemp M (2005) Terahertz frequency sensing and imaging: a time of reckoning future applications? Proc IEEE 93:1722
3. Eisele H, Haddad GI (1998) Two terminal millimeter-wave sources. IEEE Trans Microw Theory Tech 46:739–746
4. Maestrini A, Ward J, Gill J et al (2004) A 1.7–1.9 THz local oscillator source. IEEE Microw Wirel Compon Lett 14:253–255
5. Momeni O, Afshari E (2011) High power terahertz and sub-millimeter-wave oscillator design: a systematic approach. IEEE J Solid State Circ 46:583–597

6. Köhler R, Tredicucci A, Beltram F et al (2002) Terahertz semiconductor-heterostructure laser. Nature 417:156
7. Williams BS (2007) Terahertz quantum cascade lasers. Nat Photon 1:517
8. Fathololoumi S, Dupont E, Chan CWI et al (2012) Terahertz quantum cascade lasers operating up to 200 K with optimized oscillator strength and improved injection tunnelling. Opt Express 20:3866
9. Razeghi M (2009) High-performance InP-based mid-IR quantum cascade lasers. IEEE J Quantum Electron 15:941
10. Dey D, Wu W, Memis OG, Mohseni H (2009) Injectorless quantum cascade laser with low voltage defect and improved thermal performance grown by metal-organic chemical-vapor deposition. Appl Phys Lett 94:081109
11. Bai Y, Bandyopadhyay N, Tsao S, Slivken S, Razeghi M (2011) Room temperature quantum cascade lasers with 27% wall plug efficiency. Appl Phys Lett 98:181102
12. Lyakh A, Maulini R, Tsekoun A, Go R, Patel CKN (2012) Multiwatt long wavelength quantum cascade lasers based on high strain composition with 70% injection efficiency. Opt Express 22:24272
13. Belkin MA et al (2007) Terahertz quantum cascade laser source based on intracavity difference-frequency generation. Nat Photonic 1:288–292
14. Belkin MA et al (2008) Room temperature terahertz quantum cascade laser source based on intracavity difference-frequency generation. Appl Phys Lett 92:201101
15. Lu QY, Bandyopadhyay N, Slivken S, Bai Y, Razeghi M (2012) Widely tuned room temperature terahertz quantum cascade laser sources based on difference-frequency generation. Appl Phys Lett 101:251121
16. Vijayraghavan K, Adams RW et al (2012) Terahertz sources based on Čerenkov difference-frequency generation in quantum cascade lasers. Appl Phys Lett 100:251104
17. Lu QY, Bandyopadhyay N, Slivken S, Bai Y, Razeghi M (2011) Room temperature single-mode terahertz sources based on intracavity difference- frequency generation in quantum cascade lasers. Appl Phys Lett 99:131106
18. Lu QY, Bandyopadhyay N, Slivken S, Bai Y, Razeghi M (2013) High performance terahertz quantum cascade laser sources based on intracavity difference frequency generation. Opt Express 21:968
19. Lu QY, Bandyopadhyay N, Slivken S, Bai Y, and Razeghi M (2013) Room temperature terahertz quantum cascade laser sources with 215 μW output power through epilayer-down mounting. Appl Phys Lett 103:011101
20. Sirtori C, Capasso F et al (1994) Far-infrared generation by doubly resonant difference frequency mixing in a coupled quantum well two-dimensional electron gas system. Appl Phys Lett 65:445–447
21. Dupont E, Wasilewski ZR, Liu HC (2006) Terahertz emission in asymmetric quantum wells by frequency mixing of midinfrared waves. IEEE J Quantum Electron 42:1157–1174
22. Lyakh A, Maulini R et al (2009) 3 W continuous-wave room temperature single-facet emission from quantum cascade lasers based on nonresonant extraction design approach. Appl Phys Lett 95:141113
23. Liu PQ, Hoffman AJ et al (2010) Highly power-efficient quantum cascade lasers. Nat Photonic 4:95
24. Bai Y, Bandyopadhyay N et al (2010) Highly temperature insensitive quantum cascade lasers. Appl Phys Lett 97:251104
25. Lu QY, Zhang W et al (2009) Holographic fabricated photonic-crystal distributed-feedback quantum cascade laser with near-diffraction-limited beam quality. Opt Express 17:18900
26. Lu QY, Bai Y, Bandyopadhyay N, Slivken S, Razeghi M (2011) 2.4 W room temperature continuous wave operation of distributed feedback quantum cascade lasers. Appl Phys Lett 98:181106
27. Tien PK, Ulrich R, Martin RJ (1970) Optical second harmonic generation in form of coherent Cerenkov radiation from a thin-film waveguide. Appl Phys Lett 17:447

28. Auston DH, Cheung KP, Valdmanis JA, Kleinman DA (1984) Cherenkov radiation from femtosecond optical pulses in electro-optic media. Phys Rev Lett 53:1555–1558
29. Austerer M, Detz H et al (2008) Čerenkov-type phase-matched second-harmonic emission from GaAs/AlGaAs quantum-cascade lasers. Appl Phys Lett 92:111114
30. Lops A, Spagnolo V, Scamarcio G (2006) Thermal modelling of GaInAs/AlInAs quantum cascade lasers. J Appl Phys 100:043109
31. Becker C, Sirtori C (2001) Lateral current spreading in unipolar semiconductor lasers. J Appl Phys 90:1688
32. Xu J, Hensley JM et al (2007) Tunable terahertz quantum cascade lasers with an external cavity. Appl Phys Lett 91:121104
33. Lee AWM, Qin Q et al (2010) Tunable terahertz quantum cascade lasers with external gratings. Opt Lett 35:910

Chapter 5
Terahertz Photonic Devices

Miriam S. Vitiello

Abstract The paper provides an overview on recent developments of Terahertz photonics devices and components with a special focus on metrological-grade quantum cascade lasers, novel room temperature nano-detectors and low-loss waveguides, paving the way to strategic sensing and security applications across the far-infrared.

5.1 Introduction

Terahertz (THz) radiation lies in the region of the electromagnetic spectrum that is often called "THz gap". Loosely defined as the 30–300 μm wavelength region, the name "gap" originates from the lack of compact, solid-state, coherent radiation sources. However, this problem has started to be seriously addressed with the development of quantum cascade (QC) lasers operating in the THz range [1]. These semiconductor-based sources – though still limited in operating temperatures – hold the promise of being a core component of practical THz systems devoted to spectroscopy, sensing and more recently imaging in a wide variety of application areas (biomedical diagnostics, security, cultural heritage, quality and process controls, etc.), mainly due to the fact that many materials such as paper, plastics, and ceramics which are opaque at visible frequencies, are highly transmissive in the THz and micro-wave ranges [2]. The other "THz gap" which needs to be overcome for imaging oriented applications is the one concerning the detectors. Commercially

M.S. Vitiello (✉)
NEST, CNR – Istituto Nanoscienze and Scuola Normale Superiore, Piazza San Silvestro 12, 56127 Pisa, Italy

CNR – Istituto Nazionale di Ottica and LENS (European Laboratory for Non-linear Spectroscopy), Via Carrara 1, 50019 Sesto Fiorentino (FI), Italy
e-mail: miriam.vitiello@sns.it

C. Corsi and F. Sizov (eds.), *THz and Security Applications*, NATO Science for Peace and Security Series B: Physics and Biophysics, DOI 10.1007/978-94-017-8828-1_5,

available THz detectors are based on thermal sensing elements that are either very slow or require deep cryogenic cooling, while those exploiting fast nonlinear electronics are usually limited to sub-THz frequencies for best performances [3]. Despite the progressive advances in the development of compact and efficient THz sources and detectors, the realization of optical components to control THz waves in this intermediate spectral region still remains a challenge. Specifically, to move THz technology toward component integration, more effort is required to improve the mechanisms to confine light in specified areas (resonant cavities), filter certain frequencies (optical filters) and, in particular, to direct the propagation of terahertz light (optical waveguides).

5.2 Terahertz Quantum Cascade Lasers

Quantum cascade (QC) lasers are unipolar devices exploiting optical transitions between electronic states (conduction subbands) created by spatial confinement in semiconductor multi-quantum-wells, with electronic wavefunctions carefully engineering on a nanometre scale. As far as the macroscopic properties of materials are defined by their electronic structure, the QC laser is based on an artificial nano-material. The extreme precision of the material growth, necessary to obtain the peculiar characteristics required by the device design, together with the large number of layers and the complexity of the structure, makes this laser the most impressive demonstration of the capabilities offered by bandgap engineering.

After almost two decades from their invention, QCLs operating in the mid-IR have reached high performance levels. Multi-watt output power, continuous wave, room-temperature devices operating across the mid-IR, with wall-plug efficiencies larger than 50 %, have been recently reported [4], making the QCL technology the most efficient one, in the (4.3–24 μm) spectral region. However, despite the large application- oriented interests, QCL technology is not still completely mature in the $\lambda > 100$ m region, i.e. in the so called far-infrared or Terahertz range.

The first report on QCLs operating at frequencies below the optical phonon energies in the THz range is relatively recent [1]. Design and fabrication of such long wavelength QC lasers has traditionally been challenging: the emission at energies lower than the optical phonon one, makes the *e*-*e* scattering non negligible as energy relaxation channel; the huge free-carrier absorption ($\approx\lambda^2$) becomes the dominant factor of the waveguide losses; anomalies in semiconductor dielectric constants complicate laser action and must be used for the design of optical waveguides that overcome the high losses and the practical limitations induced by the prohibitive layer thickness; the non-radiative lifetimes ($\sim$0.5 ps) of intersubband transitions becomes very short and carriers tend to go everywhere enhancing leakage channels. Also, it is important to stress that at terahertz frequencies the spontaneous emission lifetime is typically much longer (microseconds) compared to the non-radiative lifetimes (picoseconds), so radiative relaxation plays no role in transport below threshold.

Many progresses have been done in the last few years and, despite the still low operating temperatures (≤199.5 K) [5], Terahertz QCLs have now a realistic chance to deeply impact technological applications, thanks to the high output power (>100 mW) [6], the quite broad operating frequency range (1.2–4.7 THz), the tunability (10 % of the emission frequency) [7], the unmatched compactness together with the possibility to be stabilized in both frequency, phase and amplitude. Frequency and phase stabilized, high power and reliable solid-state terahertz sources [8] can indeed find application in a large number of fields, spanning from far-infrared astronomy [9] and high-precision molecular gas spectroscopy [10, 11] to high resolution coherent imaging and telecommunications [2, 12]. To address such application requirements, high frequency stability sources are almost mandatory.

5.2.1 Metrological-Graded THz QCLs

It is well known that, in a QCL, environmental effects such as temperature, bias-current fluctuations or mechanical oscillations are expected to significantly affect the emission linewidth (LW). This means that any experimental LW measurement is mainly dominated by extrinsic noise. Few experimental studies of the spectral purity of THz QCLs have indeed been reported so far, giving upper limits of 30 kHz, 20 kHz and 6.3 kHz for the instantaneous LW [13–15]. Environmental effects can be here minimized by using frequency-stabilization or phase locking techniques, resulting in narrower LWs, only limited by the loop bandwidth of the specific experimental system [16].

A clear understanding of the physical effects playing a role for the QCL intrinsic LW values and their dependence on the gain media architecture/waveguide geometry is still missing so far for THz QCLs.

To address the above issue, we developed an experimental technique focused on the analysis of the spectral purity of THz QCLs via the measurement of their frequency-noise power spectral density (FNPSD). The latter provides, for each frequency, the amount of noise contributing to the spectral width of the laser emission, by deriving the laser emission spectrum over any accessible time scale [18]. It also allows calculating the LW reduction achievable using a frequency-locking loop, once its gain/bandwidth characteristics are known, while enabling to single out spurious noise sources. Intensity measurements were performed to retrieve information in the frequency domain by converting the laser frequency fluctuations into detectable intensity (amplitude) variations [17]. As a discriminator, we used the side of a Doppler- broadened methanol molecular transition. Specifically, the rovibrational molecular transition line of CH_3OH, centered at $\upsilon_o = 2.5227816$ THz, was used as a discriminator. This line is the strongest absorption transition available within the 4 GHz tunability range of the employed QCL.

Two distinctive THz detectors have been employed in order to cover a frequency range spanning from DC to 100 MHz: (i) a helium cooled Silicon Bolometer (*Infrared Laboratories Inc.*) having a nominal noise equivalent power

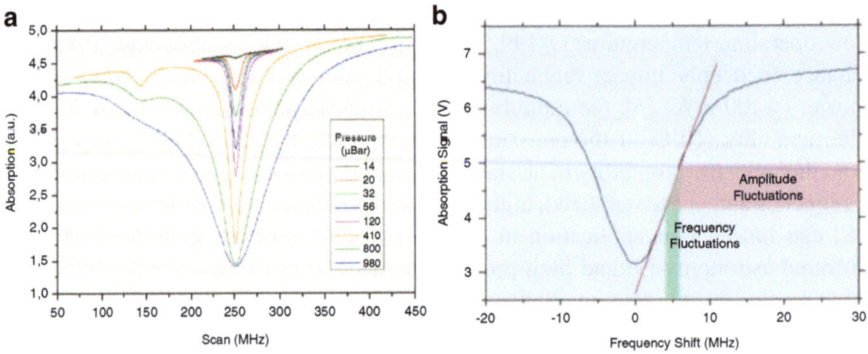

Fig. 5.1 (**a**) CH_3OH absorption line measured as a function of the gas pressure (**b**) Discriminator function of the absorption molecular line, evidencing how frequency fluctuations are converted into detectable intensity fluctuations

$NEP = 30\ \text{fW}/\sqrt{\text{Hz}}$ within its 0–10 kHz bandwidth; (ii) a hot electron bolometer (HEB) (*Scontel*) working at 4 K, with a $NEP = 70\ \text{fW}/\sqrt{\text{Hz}}$ within a 10 kHz–100 MHz bandwidth. The intensity spectrum transmitted by the discriminator reproduces the spectrum of the laser frequency fluctuations, "amplified" by the slope of the absorption profile. The absorption signal detected by means of direct-absorption spectroscopy inside a 20 cm-long cell, with a CH_3OH gas pressure of 2 Pa is reported in Fig. 5.1, as a function of the CH_3OH gas pressure (2a) and with a CH_3OH gas pressure of 2 Pa (2b).

Given the intrinsic low-noise nature of the measurement, the converter (or discriminator) must introduce a negligible noise providing, at the same time, a gain factor suitable for a good detection. By studying the pressure broadening of the selected line, we also measured a current tuning coefficient of about 8.4 ± 0.5 MHz/mA that is used to calibrate the discriminator slope. The overall transfer function of the discriminator was also retrieved from the experimental absorption profile [17] and used for correcting the frequency noise acquisitions. A schematic diagram of the employed experimental set-up is shown in Fig. 5.2.

The collimated THz QCL beam is sent to the gas cell for spectroscopy experiments. It is then split by a wire grid polarizer: the reflected beam is chopped and sent to a pyroelectric detector for the acquisition of the line profile and the frequency stabilization; the transmitted beam is acquired by means of two distinctive detectors (depending on the required bandwidth) and used for the frequency-noise measurement. It is worth noticing that the gas cell window has been properly tilted with the specific purpose to avoid any optical feedback effect on the measured frequency noise [18].

During the frequency-noise measurement the QCL frequency needs to be locked at the half-height position of the absorption line, in order to keep the conversion factor constant at its maximum value. The latter procedure was done by implementing a software PI loop on the QCL current and by using the line around the locking point as a feedback signal. The portion of the beam reflected, after the gas

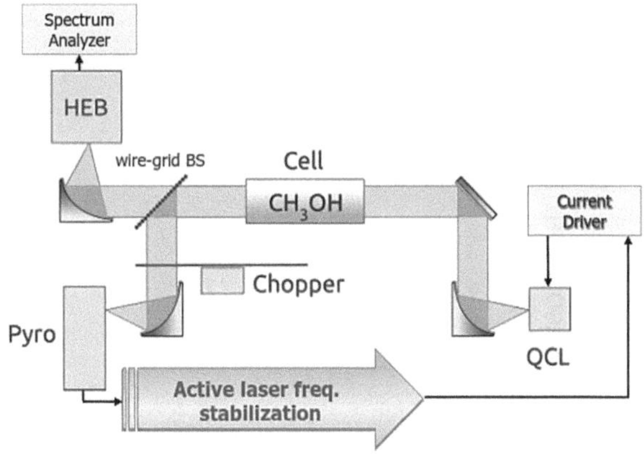

Fig. 5.2 Schematic diagram of the experimental setup

cell, by the wire-grid polarizer was devoted to this purpose: a pyroelectric detector (*Gentec SPH-62 THz*) detects the chopped radiation, its signal is demodulated by a Lock-In amplifier, then acquired by a digital acquisition board and processed by a software Proportional-Integral controller, written in LabView. Its output drives an analog voltage signal that controls the QCL current driver for the stabilization. The chopping frequency was set to 130 Hz, leading to a demodulated signal bandwidth of ≈10 Hz. This allows for an efficient stabilization of the mean QCL frequency at the right point without affecting the QCL frequency noise above 10 Hz.

The frequency noise was then measured (experimental set-up shown in Fig. 5.2) while the QCL frequency is stabilized at the center of the discriminator slope, following the procedure described above. In combination with fast detection and low-noise fast-Fourier-transform acquisition, this technique enables spectral measurements spanning over seven frequency decades (from 10 Hz to 100 MHz) and ten amplitude decades. A measurement of the residual amplitude noise is also performed, by shifting the QCL frequency out of the discriminator side. The latter is then subtracted from the former, in order to retrieve the correct frequency noise power spectral density (FNPSD). The full spectrum of the QCL was obtained by sticking together several acquisitions on smaller spectral windows, in order to ensure a high overall resolution. The resulting FNPSD spectrum is plotted in Fig. 5.3, together with the current-noise power spectral density (CNPSD) of the current driver, converted to the same units by using the current tuning coefficient. Residual external noise gives rise to the sharp peaks visible throughout the trace. Three distinct domains can be clearly recognized: (i) in the $f = 10$ Hz–10 kHz range the FNPSD is dominated by a noise not arising from the current driver and therefore ascribed to the QCL itself [19]. It is worth noticing that, in a QCL, electric field fluctuations lead to a non-negligible frequency noise contribution through the gain Stark shift and the cavity mode pulling effect [20]. The main frequency noise with respect to the CNPSD, absent at larger frequencies, can likely be attributed to this mechanism, as confirmed by the similar dependence observed in the amplitude noise. Additional spurious contributions of low-frequency background radiation

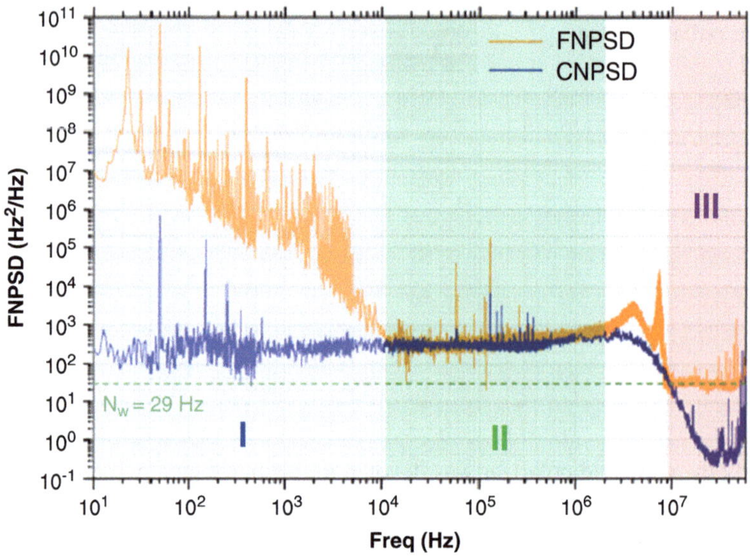

Fig. 5.3 Experimental FNPSD of the THz QCL (*orange trace*), compared with the contribution to the frequency-noise of the CNPSD of the current driver (*blue trace*). The *dashed line* marks the white noise level

signals and/or electronic noise or mechanical vibrations may probably also play a role; (ii) in the 10 kHz–5 MHz range the FNPSD is fully dominated by the contribution of the current driver, as confirmed by the perfect overlap between the two noise traces; (iii) above 8 MHz, an asymptotic flattening is observed in the FNPSD, with a significant deviation from CNPSD, thus suggesting a flattening to a white noise level (N_w).

According to frequency-noise theory, the laser power spectrum corresponding to the white component N_w of the frequency noise is purely Lorentzian, with a FWHM $\delta\upsilon = \pi N_w$, from which we extracted an intrinsic LW $\delta\upsilon = 90 \pm 30$ Hz [17].

5.2.2 *Phase Locking*

In THz QCLs, a scaling of the tuning coefficient with wavelength is in part responsible for both the lower measured intrinsic LWs [17] and the narrower free-running LW, falling in the sub-MHz range.

Phase-locking to THz coherent sources through Optical Frequency Comb synthesizers (OFCS) was achieved only recently [16] by locking a 2.5 THz QCL to a specific harmonic n-th of the repetition rate of a mode-locked erbium-doped fiber laser. The comb generated by the mode-locked laser was then mixed in a nonlinear crystal with a continuous wave THz QCL, thus generating THz sidebands around

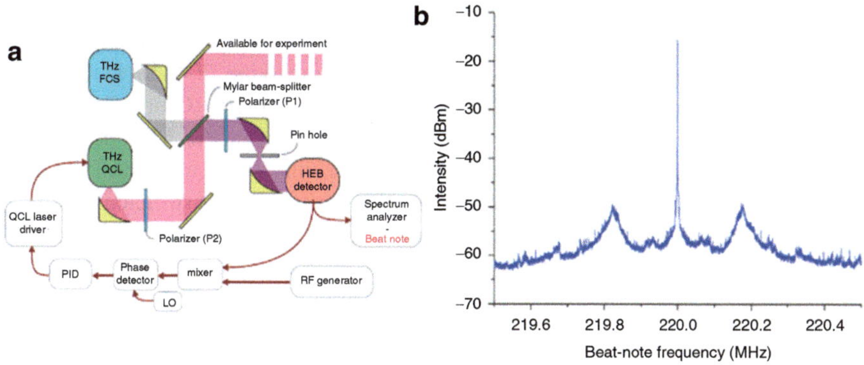

Fig. 5.4 (**a**) Beat-note detection and phase-locking setup. The beating between the QCL and the OFCS is detected on a HEB. The two beams are superimposed by means of a highly asymmetrical beam splitter, so that more than 99.5 % of the QCL radiation is available for the experiment beam. The P1 polarizer ensures polarization matching between the beams, whereas P2 selects the amount of QCL power to be sent to the HEB. The beat note is mixed with a synthesized RF signal, so that it is down-converted to the 21.4 MHz frequency of the phase-lock electronics local oscillator (*LO*). The output of the phase detector is processed by a standard proportional-integral-derivative (*PID*) unit and then sent to the QCL current driver, to close the locking loop. (**b**) Phase-locked beat-note signal acquired with a 1 MHz span and 100 Hz RBW. The two *sidebands* indicate a phase-lock electronic bandwidth of about 200 kHz

the near-infrared carrier. The beating between the original comb and its shifted replicas provide the signal for closing the phase-lock loop. Alternative approaches employing photoconductive antennas have been also proposed [21].

Despite the clear advantage of a room-temperature detection of the beat-note, the latter approaches involve low-efficiency up-conversion processes, requiring CW THz power in the mW range. One possible solution can be provided by THz detection approaches, involving beat-note acquisitions provided by a square-law high efficiency detector, via an air-propagating THz comb. Although the intrinsic comb nature of pulsed THz sources used in time domain spectroscopy has been recently demonstrated [22], no direct use of such sources as frequency "ruler" for a THz QCL had ever been reported. A further strong motivation for generation of a real THz frequency comb is its possible use as direct source for sensing, imaging or hyperspectral imaging.

A first step to address the above issue is the generation of sufficient power from each comb-tooth to allow phase locking of THz coherent sources. We recently demonstrate phase-locking of a 2.5 THz QCL to a free- space comb, generated in a LiNbO3 waveguide and covering the 0.1–6 THz frequency range. A schematic diagram of the employed experimental system is shown in Fig. 5.4a.

The principle of the THz comb generation is optical rectification, in Cherenkov configuration [23], of a femtosecond mode-locked Ti:Sa laser in a single-mode waveguide fabricated on a MgO-doped $LiNbO_3$ crystal plate [24]. The generated radiation is a train of THz pulses, each consisting of a single electric field cycle carrying a very large spectral content (from 100 GHz up to 6 THz, centered at 1.6 THz). Since the pulses are identical, the comb-like spectrum of the infinite train has a perfectly-zero offset, and a spacing corresponding to the 77.47 MHz repetition rate of the pump laser, that has been stabilized against a Rb-GPS (Global Positioning System) disciplined 10-MHz quartz oscillator. Stability in the MHz range was obtained for the repetition rate, thus ensuring a stability of each tooth of the THz comb at the 100 Hz level (Fig. 5.4b). The generation efficiency is sufficiently high to allow using a very simple setup and a commercial hot- electron-bolometer detector, with a 250 MHz bandwidth, to directly observe the beating between single teeth of the comb and a small power fraction (100 nW) from the THz QCL.

As a consequence, most of the QCL output power is thus made available for any specific experiment. Exploitations of such a system for THz-comb-assisted spectroscopy are already providing new precise results for rotational molecular transitions. The spectroscopy combination of an OFCS able to cover a broad spectrum (up to 5–6 THz) with QCL sources that can cover, point-by-point, most part of the same range with unprecedented power-levels, is a very promising perspective for a metrological-grade investigation of the THz region. Among the several possible applications, it is worth citing the development of comb-assisted THz sub-Doppler spectrometers and of absolutely-referenced local oscillators for heterodyne THz spectrometers. Regarding the latter topic, in particular, not only the narrow linewidth of the local oscillator but also, and mostly, the stability over long time periods of its absolute frequency can represent a real breakthrough. A number of demanding new applications will be possible to afford, e.g. spectroscopic interrogation of cold molecules [25] or precise measurement of the long-term variation of fundamental physical constants from astronomical observation [26].

5.3 Terahertz Detectors

Commercially available THz detectors are based on thermal sensing elements that are either very slow (10–400 Hz modulation frequency for Golay cells or pyroelectric elements, with noise equivalent powers (NEP) in the $10^{-10}\ \mathrm{W} \cdot \mathrm{Hz}^{-1/2}$ range) or require deep cryogenic cooling (4 K for superconducting hot-electron bolometers), while those exploiting fast nonlinear electronics (Schottky diodes) are usually limited to sub-THz frequencies for best performances [3, 27]. More recently, electronic devices based on the gate-modulation of the conductance channel by the incoming radiation have been realized in high-electron- mobility transistors (HEMT), field effect transistors (FET), and Si-MOSFET architectures, showing fast response times and high responsivity (R_v) [28], as well as the possibility of implementing multi-pixel focal-plane arrays [29].

The operating mechanism of a FET detector is not trivial [30], but can intuitively be interpreted as deriving from the non-linear dependence of the FET channel current on the gate voltage near the pinch-off point. These devices have the advantage that the can be maximized with the gate bias V_G, while measuring the output at the drain with no source-drain bias applied, thus dramatically improving the signal-to-noise ratio. THz detection in FETs is mediated by the excitation of plasma waves in the transistor channel. On one hand, a strong resonant photoresponse is predicted in materials having plasma damping rates lower than both the frequency ν of the incoming radiation and the inverse of the wave transit time τ in the channel. This requires mobilities of at least several thousand cm^2/Vs at frequencies >1 THz. Under these conditions, stationary states arising from the quantization of plasma waves over the gate width are excited whenever V_g is such that $n\pi s/(2L_g) = 2\pi\nu$, where n is an odd integer, s the plasma-wave velocity, and L_g the gate length. On the other hand, when plasma oscillations are overdamped, i.e. decay on a distance smaller than the channel length, broadband THz detection is predicted. In this case, the oscillating electric field of the incoming radiation applied between source and gate electrodes produces a modulation of both charge density and carrier drift velocity. Carriers travelling towards the drain generate a continuous source-drain voltage, Δu, controlled by the carrier density in the channel that can be in turn maximized by varying V_g. High mobility at room temperature (RT) is therefore crucial to take full advantage of resonant detection.

During the last year, the described approach has been extended to either semiconductor nanowires (NWs) or graphene FETs that has been shown to be very promising for highly-sensitive room temperature THz detection [31, 32].

5.3.1 *Nanowire FETs*

One (1D) nanowires are at the forefront of studies on future electronics, although issues like massive parallelization, doping control, surface effects, and compatibility with silicon industrial requirements are still open challenges. On the other hand, significant progress is being recently made in atomic to nanometer scale control of materials morphology, size, and composition, including the growth of axial, radial, and branched nanowire-based heterostructures [33].

Semiconductor nanowires are conventionally grown from metallic seed particles, meaning that mechanisms for impurity incorporation are different from those of other growth techniques. The most significant difference is related with the evidence that in nanowire growth the surfaces of already-grown segments are exposed to impurities for the duration of the growth, which can affect the physical and chemical properties of such segments. In a "*bottom-up*" vision, semiconductor nanowires can be considered peculiar "building blocks", easily synthesizable as active elements for high performance electronic devices, whose specific characteristics are highly dependent on the nanowire physical properties. In this perspective, InAs can be identified as one of the most successful candidates for such nanoscale integration. InAs nanowires have indeed a small electron effective mass and correspondingly high bulk electron mobility and can be grown epitaxially on silicon without the use of gold seeding, thus making the process viable also for low-cost silicon technology integration where deep Au-levels in the Si bandgap must be avoided.

In addition, nanowires can be easily removed from the host substrate and placed on top of a new functional one for individual contacting, even in relatively large numbers, with a simple planar technology very suitable for low capacitance circuits. Therefore, they represent, in principle, an ideal building block for implementing rectifying diodes or plasma-wave detectors that could be well operated into the Terahertz, thanks to their typical attofarad–order capacitance. Surprisingly, despite the strong effort in developing these nanostructures for a new generation of complementary metal-oxide semiconductors (CMOS), memory and/or photonic devices, their potential as Terahertz radiation sensors has not been explored so far.

We recently developed InAs NW-FETs acting as THz detectors. Field effect transistors based on InAs NWs as active element can act as rectifying diodes when the electromagnetic radiation is fed onto the source and gate terminals through a resonant or a broadband antenna. The choice of InAs NWs, with narrow bandgap and degenerate Fermi-level pinning, revealed to be successful in preserving good detecting performances while scaling down the dimension of the device to increase the detected frequency up to the 1.5–3 THz range, accessible with QCL sources [34].

The electrical characterization of the investigated samples was performed by employing the DACs of a SR830 amplifier to drive independently the source-to-drain (V_{sd}) and the gate (V_g) voltages [34]. The drain contact was connected to a current amplifier converting the current into a voltage signal with an amplification factor of 10^4V/A. The latter signal was then measured with an Agilent 34401A voltmeter reader. The threshold voltages V_{th} of each device have been determined by the intercepts with the horizontal axis of the linear fit of the $I_{sd} - V_g$ characteristics in the region of maximum transconductance (g_m). Our results show that V_{th} progressively decreases as a function of the precursor pressure, i.e. by increasing the carrier density. However, the analysis of the extracted slopes shows that the peak transconductance, normalized to the gate length, varies from 10 to 100 mS/m reaching a maximum when a precursor pressure of 0.1 Torr is employed. Under the same experimental conditions a minimum inverse subthreshold slope of ≈ 11 V/dec has been found [35].

A systematic characterization of transconductance and resistivity values measured for five prototype FET devices realized under different growth conditions [35] showed that to optimize the transport properties of the FETs, a compromise should be found between high transconductances/low threshold voltages and sufficiently low nanowire resistivities. By using the measured transconductances we can extract the carrier mobility (μ) in our nanowires by means of the Wunnicke metrics. The latter correlates the gate capacitance C with the carrier mobility through the equation $\mu = g_m l^2/(CV_{ds})$. By using a 3D finite element simulation of the electrostatic potential a gate capacitance value $C = 2.8$ aF has been estimated at $V_g = 0V$, while keeping $V_{ds} = 0.025$ V. Mobility values in the range $[10^3 - 10^4]$ cm^2/(V $\times$ s) have been extracted in NW FETs with an increasing ditertiarybutyl selenide (DtBSe) line pressure. The latter values are substantially lower than bulk ones, indicating that scattering processes at the nanowire surface play a non negligible role. From the mobility and resistivity data we extracted the carrier concentration, n, at $V_g = 0V$ from the relation $n = (\rho e \mu)^{-1}$. This allows correlating the precursor line pressure during Se doping with the effective carrier density through the nanowire.

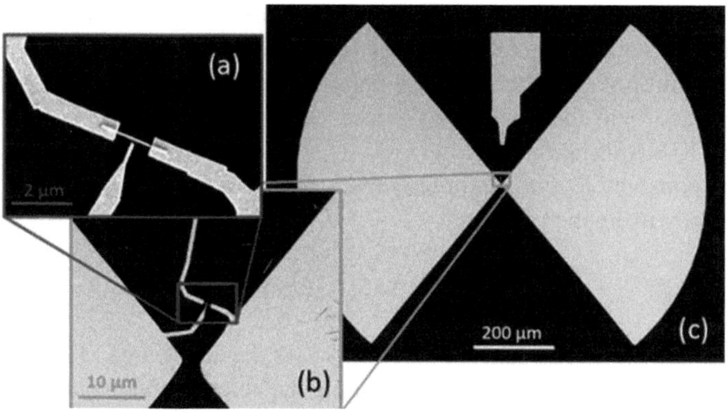

Fig. 5.5 Scanning electron micrograph of a NW-based FET broadband THz single pixel detector. (**a**) Source and drain ohmic contacts are located at a nominal distance of 1.2 m at the opposite ends of the InAs NW. A 100-nm-wide lateral gate electrode is located at 85 nm from the NW, at the same distance (550 nm) from the two contacts. (**b**) Image of the drain terminal connected to a bonding pad, with the source and gate electrodes directly connected to the two lobes of a bow-tie antenna (**c**)

By keeping fixed the ditertiarybutyl selenide (DtBSe) precursor line pressures above 0.1 Torr we then realized a set of different NW FETs (Fig. 5.5a, b) by varying the geometry of the contact patterns and/or the mounting configuration. Specifically, to increase the asymmetry and therefore the responsivity, we designed as source and gate electrodes low shunt-capacitance antenna structures to funnel the radiation into the strongly sub-wavelength detecting element. A set of devices exploiting different antenna geometries have been realized: (i) broad band bow-tie antennas (Fig. 5.5c) with arm lengths in the range 100–250 µm, (ii) wide band log-periodic circular-toothed structures, with an outer diameter of 650 µm, (iii) control structures with the antenna arms between source and drain.

InAs nanowire grown bottom-up by Chemical Beam Epitaxy (CBE) and n-doped with Se at levels between 10^{16}cm^{-3} and $5 \times 10^{18}\text{cm}^{-3}$ have been employed to fabricate FETs having a broad-band bow-tie dipole antenna or a log periodic circular toothed antenna patterned between the source and the gate contact.

In all cases the gate electrode was lithographically designed to be 80–100 nm away from the InAs nanowire, with a channel length in the range 100–150 nm. Each sample was finally glued on a dual-inline package and wire bonded. In a few devices a Silicon hyper-hemispherical lens having a 6 mm diameter was mounted on the back of the Si substrate so that the GHz beam is properly focused on the nanowire after crossing the Si/SiO_2 layers. This makes the bow antenna perfectly resonant with the 292 GHz source available for the measurements. The device was then inserted in a compact package and wire bonded, taking care that the source contact of the bow antenna was grounded on the metal package itself [34].

Photoresponse experiments were performed by using either a 0.3 THz electronic source [6] or a set of THz QCLs operating between 1.5 [34] and 2.8 THz cooled at T = 10 K and modulated at frequencies in the range 33 Hz–300 KHz. The radiation was collimated and focused by a set of two $f/1$ off-axis parabolic mirrors and the photo-induced source-drain voltage was measured by using a lock-in without any pre-amplification stage. The vertically polarized incoming radiation impinges from the free space onto the nanowire devices while the detector was moved with a motorized X-Y translation stage.

The responsivity R_v can be directly extracted from the measured photo-induced current Δu by using the relation:

$$R_v = (\Delta u S_t) \,/\, (P_t S_a)\,,$$

where S_t is the radiation beam spot area and S_a is the active area [31–34]. Since the total area of our nanowire transistor including the antenna and the contact pads was smaller than the diffraction limited area $S_\lambda = \lambda^2/4$, the active area was taken equal to S_λ. The induced drain current Δu has been extracted from the signal *LIA* measured using the lock-in amplifier, without taking account the loading effect by using the following equation:

$$\Delta u \left(V_g\right) = \frac{2\sqrt{2}\frac{\pi}{4}LIA}{G},$$

where 2 is due to peak-to-peak magnitude, $\sqrt{2}$ originates from the lock-in amplifier rms amplitude, and $\pi/4$ is the fundamental sine wave Fourier component of the square wave produced by the chopper and (G) is the preamplifier gain.

The responsivity value is highly dependent from:

(i) NW diameter: the thicker the NW the highest the responsivity;
(ii) NW doping concentration: trade off condition reached at $n = 5 \cdot 10^{17}$ cm^{-3};
(iii) Polarization condition: Maximum responsivity reached for incoming beam polarization parallel to antenna axis;
(iv) Antenna design: log-periodic antennas in general yielded better performance than bow-tie, and devices mounted with substrate-side Si lens showed an improved collection efficiency of more than one order of magnitude;
(v) NW resistivity: lower resistivity values can be reached after a few hours vacuum and increases the expected responsivity considerably;
(vi) Impedance matching: the responsivity dependence from the gate bias is largely influenced by the transistor load impedance matching;

By playing with the above parameters [35, 36], responsivity values up to 100 V/W have been reached with impressive noise equivalent powers ($NEP \sim 6 \times 10^{-11}$ W/$\sqrt{\text{Hz}}$) [34] and a wide modulation bandwidth (>300 kHz). Since the responsivity is directly linked to the electric transconductance, the photoresponse peak is reached in the proximity of the gate voltage for which the FET channel closure is achieved. For these reasons, we observed best performances with the devices having low doping concentration, since channel closure was easier to obtain.

Large area fast imaging applications of the detectors have been performed by using as test object a poppy flower [36]. Figure 5.6 show the THz transmission

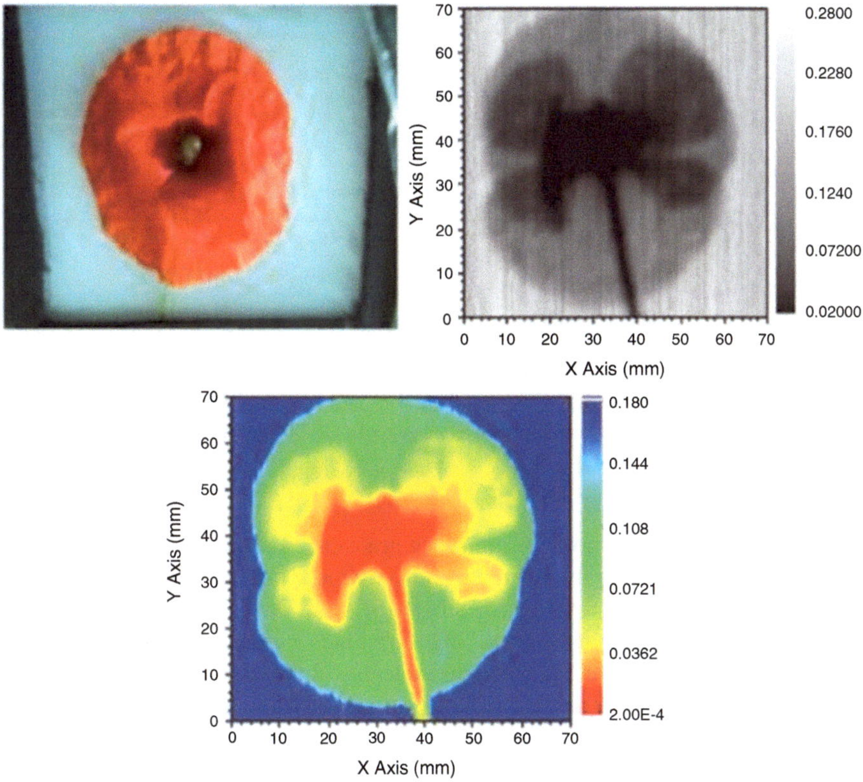

Fig. 5.6 Transmission THz image of a poppy flower; linear intensity color scale is given in relative units. The radiation absorbed by the four petals increases in correspondence of the overlaps between pairs of them. The whorl of stamens at the center of the flower and the stem absorb a higher amount of THz radiation, possibly due to thicker structures and/or higher water content

image, consisting of 200×550 scanned points, collected by raster-scanning the object in the beam focus, with an integration time of 20 ms/point. This shows that our devices are beyond proof-of-concept, and can already be used in a realistic setting, enabling large area, fast imaging of macroscopic samples.

5.3.2 *Graphene FETs*

The naturally occurring 2DEG in a doped graphene sheet has a very high mobility even at RT [37]. Furthermore, it supports plasma waves that are weakly damped in high-quality samples [38, 39]. Thus, single-layer and bilayer graphene FET (GFET) plasma-based photodetectors could easily outperform other terahertz detection technologies.

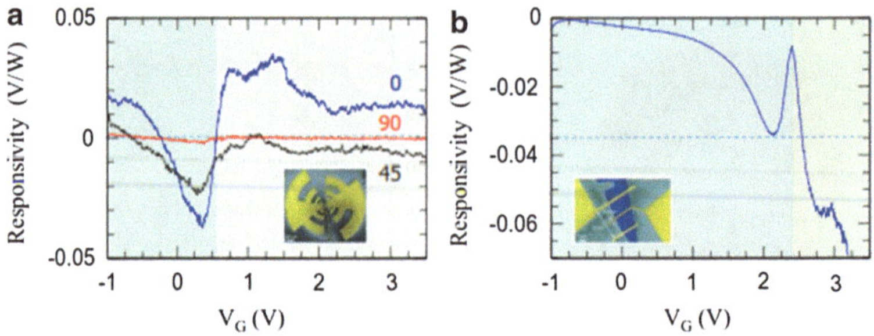

Fig. 5.7 Room temperature responsivity plotted as a function of V_G for detectors based on SL-FET (with data for different angles between the beam polarization axis and the antenna axis) (**a**) and BL-FET (**b**). Different *background colors* identify regions below and above the Dirac point. *Insets*: scanning electron microscope micrographs. The detector consists of a log-periodic circular-toothed antenna patterned between the source and gate of a GFET. The drain is a metal line running to the bonding pad [32]

Recently a novel technology employing graphene in a simple top-gate antenna-coupled configuration for the excitation of overdamped plasma waves in the channel of a FET [32] have been proposed. Single-(SL) and double-layer (BL) graphene flakes mechanically exfoliated on Si/SiO_2 substrates were used to fabricate the top gate FETs. Log-periodic circular-toothed antennas at the source and gate electrodes were used to couple the 0.3 THz radiation of an electronic source. A 35-nm-thick HfO_2 layer was used as the gate dielectric. The whole channel length was 7–10 μm, while the gate length was 200–300 nm.

Figure 5.7a, b plots R_v measured in SL and BL graphene FETs at RT, while sweeping V_G from -1 V to $+3.5$ V and modulating the THz source at 500 Hz. Individual SL responsivity curve corresponds to a different relative orientation between the source electric-field polarization and the antenna axis. In the case of the SL device the photoresponse drops rapidly with angle until it becomes almost zero when the incoming polarization is orthogonal to the antenna axis, confirming the efficacy of the dipole antenna. The dependence of Δu from $\sigma^{-1}\partial\sigma/\partial V_g$ (where σ is the extracted conductivity) is in qualitative agreement with the prediction of a diffusive theoretical model [30], thereby proving that the detectors operate in the so-called broadband overdamped regime. Together with the expected photovoltage change in the vicinity of the Dirac point, a further sign switch around gate voltages $V_G = 0$ is observed in all cases, suggesting a possible contribution of thermoelectric origin. This arises from the presence of the ungated p-doped graphene regions, and subsequent formation of p-p-p or p-n-p junctions, depending on V_G.

In the case of the BL device the responsivity curve is in excellent agreement with that predicted by the diffusive plasma-wave detection model up to the Dirac point (2.5 V). However, no change of sign is here visible, with a strongly enhanced response above the Dirac point with respect to that predicted. This suggests an additional contribution to the photovoltage, this time of constant sign. Its magnitude grows rapidly with V_G, eventually dominating in the regime in which a p-n-p junction is present. A possible role of interband transitions driven by the THz field at the p-n junction, with the resulting generation-recombination noise could here play a not negligible role. Maximum responsivity values of 150 mV/W and minimum $NEP_S = 30$ nWz$^{1/2}$ have been reached [32]. Further advances are expected by achieving the resonant detection regime, and also by understanding and exploiting the new physics emerging in BL samples, which may lead to even better device concepts.

Similar results were obtained at higher frequencies up to 3.11 THz for back-gated graphene transistors [40], reaching maximum photovoltage signals of a few μV.

5.4 Terahertz Hollow Waveguides

Low-loss waveguides with good mode selectivity in the far-infrared are quite difficult to realize due to the high ohmic losses in metals and the relatively high absorption coefficient of most dielectric materials in this spectral range. To date, THz waveguides have been realized by using single or double metal wires, [41, 42] metal tubes, solid-core [43] or porous polymer [44] fibers, photonic crystal fibers [45] and hollow core waveguides with inner metallic [46] or metallic/dielectric thin film coating [47].

In a hollow waveguide the wave energy is mostly distributed in the air region and only a small fraction propagates inside the absorbing medium. Because the terahertz waves are strongly confined within their core, hollow waveguides can be inserted in cables and this will be an advantage especially in medical endoscopic applications.

Losses in the hollow metallic waveguides arise from the non-vanishing electric field component at the metallic surface that penetrates and is absorbed in the waveguide wall. As a result, transmission losses of pure metallic waveguides are in general limited to 8–10 dB/m in the THz range [48]. In a hollow cylindrical metallic waveguide two modes have low loss characteristics, i.e. the linearly polarized TE_{11} mode and the lowest losses azimuthally polarized TE_{01} mode that however is not easily excited by linearly polarized THz sources. The losses for the linearly polarized mode can be reduced by adding a thin ($\sim\lambda/10$) dielectric coating to the inner core via liquid flow coating processes. The waveguide dominant mode structure in the dielectric-lined waveguide will change from TE_{11} to the hybrid HE_{11} mode that exhibits minimal penetration into the absorbing metallic wall, assuring transmission losses lower than 1 dB/m [4]. For the pure metallic waveguide however propagation in the TE_{01} mode offers a significant loss reduction compared to the TE_{11} mode, especially at high frequencies ($\nu > 2$ THz).

The selection of a preferential waveguide mode is a crucial issue for a complete loss and dispersion analysis of waveguides. The conventional procedure employed for the optical characterization of a THz waveguide is based on the measurement of the waveguide transmission spectra by THz time domain spectroscopy (TDS). THz TDS, however, doesn't facilitate mode selection and, as a result, experimental spectra often contain periodic patterns caused by waveguide mode interference.

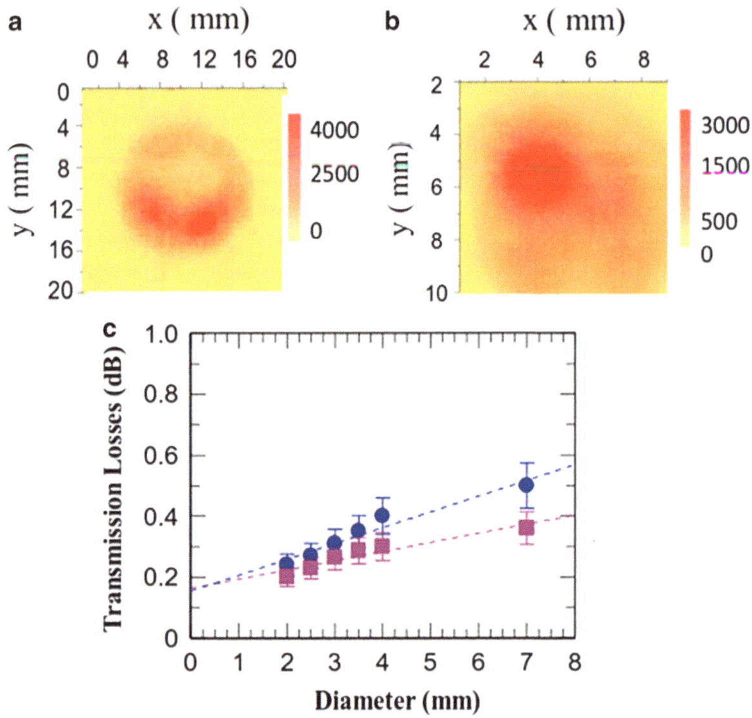

Fig. 5.8 Far field spatial intensity distribution of sample a) upon exiting a 2 cm long, 2 mm bore diameter aluminum waveguide in a back-to-back configuration (**a**) or while focusing the QCL beam on the waveguide entrance by means of a Picarin lens (**b**); (**c**) Total losses measured from the ratio between the output and input power while coupling the micro-ring QCL beam in a 2 mm bore diameter aluminum waveguide preferentially with the TE_{11} mode (●) or mostly with the TE_{01} mode (■)

Recently we demonstrate a promising new experimental approach to efficiently couple QCL sources fabricated in both surface emitting micro-ring resonator, or standard edge emitting geometry with THz hollow waveguides. We coupled the azimuthally polarized beam of a micro-ring QCL operating at 3.2 THz [49] (sample a) into hollow aluminum waveguides either by focusing the laser beam into a ≈ 1 mm spot at the waveguide input with a 5 cm focal length Picarin lens or by sending the unfocused QCL beam through a pinhole into the waveguide at $\approx$4 mm distance from the QCL (back-to-back configuration). In both cases, a calibrated pyroelectric detector has been mounted on a XY translational stage and driven by stepper motors having a 0.2 μm spatial resolution to image the mode profile at the output ($\approx$2 cm from the edge) and record the total transmitted power at the waveguide end. Figure 5.8 show the far field spatial intensity distribution of sample a) measured upon exiting a hollow waveguide having a bore diameter *d* twice the QCL ring diameter, in the back-to-back configuration (Fig. 5.8a) or while focusing the incoming THz beam on the waveguide entrance by means of the Picarin lens (Fig. 5.8b).

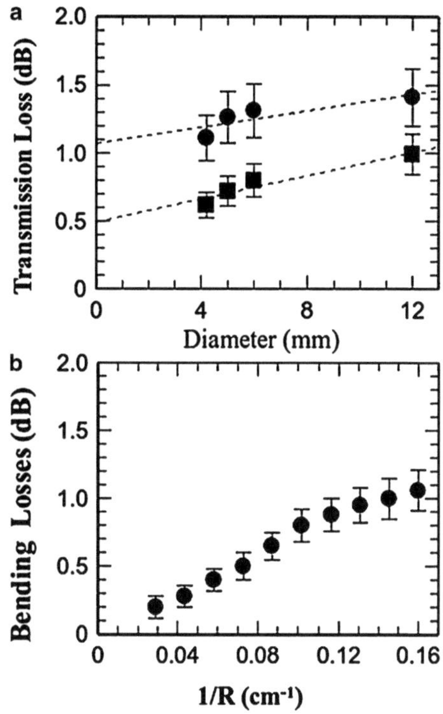

Fig. 5.9 (**a**) Total losses measured from the ratio between the output and input power while coupling the QCL beam of sample (**b**) preferentially with the TE_{01} mode (●) or mostly with the TE_{11} mode; (**b**) Measured bending losses as a function of the inverse of the bending radius

The doughnut-shape of the intensity distribution in Fig. 5.8a shows the laser mode is perfectly matched to the TE_{01} waveguide mode. In contrast, in the case of Fig. 5.8b the wave is mostly coupled to the TE_{11} mode, however showing a weak "ring" likely caused by the TE_{01} mode.

Despite the presence of this low intensity ring, the mode profile guarantees good coupling efficiency to and from free space. Transmission losses of 5.1 and 2.7 dB/m have been measured when the THz beam of samples (a) is either focused on the waveguide by using the lens or in the back-to-back configuration (Fig. 5.8c). From the intercept of the linear fit to the data we extracted coupling efficiency values as high as 98 %.

Figure 5.9a shows the total transmission losses measured while focusing a linearly polarized surface-plasmon edge emitting QCL (sample b) operating at 3.28 THz at the centre of a polycarbonate flexible pipe tube [50] having a thin Ag coating and a 1.8 mm bore diameter or by shifting the input pinhole toward the waveguide wall.

In the latter case, the position of the input end has been adjusted while monitoring the mode profile. By centering the position of the input beam on the waveguide axis the launched TE_{11} laser mode, after propagation through the waveguide, fully develops at the output. On the contrary, when the input beam is coupled close to the waveguide wall, the QCL is exciting mostly the TE_{01} mode, without any significant coupling to the TE_{11} mode. Total transmission losses of 3.0 and 4.2 dB/m have been measured for the TE_{01} and the TE_{11} mode, respectively, with coupling efficiencies as high as 81 and 89 %.

In the mid-IR region it has been shown that there are additional losses when bending hollow waveguides, which vary as 1/R, where R is the radius of curvature. This loss dependence has been verified for many different metallic and single-layer dielectric coated hollow guides in the range 2–12 μm, but no data has been reported so far for flexible metallic waveguides in the Terahertz region. To measure the bending losses of the investigated waveguides, we bent the central part of a 12 cm long hollow waveguide in a uniform curvature. Figure 5.9b shows the measured bending losses as a function of 1/R while keeping constant the amount of waveguide under bend. In the experiment, the polarization of the incoming beam of sample (b) has been kept perpendicular to the plane of bending. This condition should give in principle lower losses than polarization parallel to the bending plane. However, for TE modes, the difference is theoretically small as shown experimentally by applying complex bending shapes at the waveguides. The data plotted in Fig. 5.9b show that the losses increase as the curvature increases, remaining lower than 1.2 dB even at bending angles of 50°. It is worth noticing that the bending loss was obtained by comparing the output power of straight waveguides with that of the bent waveguides, thus implying that the coupling efficiency did not affect the bending loss calculation.

5.5 Conclusions

Terahertz photonics devices as metrological-grade THz QCLs are actually the most powerful source across the far-infrared due to the inherently high spectral purity, the quite broad operating frequency range, the high optical power, the compactness and reliability. Especially if combined with room-temperature, low noise, fast, high detectivity and ultra-compact nano-detectors based on high mobility materials (as semiconductor nanowires or graphene) QCLs have the potential to deeply impact technological applications allowing unprecedented sensitivity and resolution levels in sensing, security, process and quality control, cultural heritage, medicine, biology and space science fields.

Acknowledgements This work was partly supported by the Italian Ministry of Education, University, and Research (MIUR) through the program "FIRB-Futuro in Ricerca 2010" RBFR10LULP "Fundamental research on terahertz photonic devices". A sincere acknowledgment is addressed to the people contributing extensively to the results reported in the present work: S. Bartalini, L. Consolino, P. De Natale, A. Tredicucci, L. Viti, L. Romeo, M. Pea, D. Ercolani, L. Sorba, L. Vicarelli, M. Polini, V. Pellegrini, M. Inguscio, A. Lombardo, A. Ferrari, D. Coquillat, W. Knap, H.E Beere and D.A. Ritchie.

References

1. Köhler R, Tredicucci A, Beltram F, Beere HE, Linfield EH, Davies AG, Ritchie DA, Iotti RC, Rossi F (2002) Terahertz semiconductor-heterostructure laser. Nature 417:156–159
2. Tonouchi M (2007) Cutting-edge terahertz technology. Nat Photonic 1:97–105
3. Sizov F, Rogalski A (2010) THz detectors. Prog Quantum Electron 34:278–347
4. Bai Y, Slivken S, Kuboya S, Darvish SR, Razeghi M (2010) Quantum cascade lasers that emit more light than heat. Nat Photonic 4:99–102
5. Fathololoumi S, Dupont E, Chan CWI, Wasilewski ZR, Laframboise SR, Ban D, Mátyás A, Jirauschek C, Hu Q, Liu HC (2012) Terahertz quantum cascade lasers operating up to ~200 K with optimized oscillator strength and improved injection tunneling. Opt Express 20:3886
6. Williams BS, Kumar S, Hu Q, Reno JL (2006) High-power terahertz quantum-cascade lasers. Electron Lett 42:89–90
7. Vitiello MS, Tredicucci A (2011) Tunable emission in a THz quantum cascade lasers. IEEE Trans Terahertz Sci Tech 1:76–84
8. Williams BS (2007) Terahertz quantum-cascade lasers. Nat Photonic 1:517–525
9. Reix J-M et al (2009) The Hershel/Planck programme, technical challenges for two science missions, successfully launched. Acta Astron 34:130–148
10. Mittleman DM (2003) Sensing with THz radiation. Springer, New York
11. De Natale P, Lorini L, Inguscio M, Nolt IG, Park JH, Di Lonardo G, Fusina L, Ade PAR, Murray AG (1997) Accurate frequency measurements for H_2O and $^{16}O_3$ in the 119-cm^{-1}OH atmospheric window. Appl Optics 36:8526–8532
12. Capasso F et al (2002) Quantum cascade lasers: ultrahigh-speed operation, optical wireless communication, narrow linewidth and far-infrared emission. IEEE J Quantum Electron 38:511–532
13. Barkan A et al (2004) Linewidth and tuning characteristics of terahertz quantum cascade lasers. Opt Lett 29:575–577
14. Hubers H et al (2005) Terahertz quantum cascade laser as local oscillator in a heterodyne receiver. Opt Express 13:5890–5896
15. Baryshev A et al (2006) Phase locking and spectral linewidth of a two-mode terahertz quantum cascade laser. Appl Phys Lett 89:031115
16. Barbieri S, Gellie P, Santarelli G, Ding L, Maineult W, Sirtori C, Colombelli R, Beere H, Ritchie D (2010) Phase-locking of a 2.7-THz quantum cascade laser to a mode-locked erbium-doped fibre laser. Nat Photonic 4:636–640
17. Vitiello MS et al (2012) Quantum-limited frequency fluctuations in a terahertz laser. Nat Photonic 6:525–528
18. Ravaro M, Barbieri S, Santarelli G, Jagtap V, Manquest C, Sirtori C, Khanna SP, Linfield EH (2012) Measurement of the intrinsic linewidth of terahertz quantum cascade lasers using a near-infrared frequency comb. Opt Express 20:25654
19. Bartalini S et al (2010) Observing the intrinsic linewidth of a quantum-cascade laser: beyond the Schawlow-Townes limit. Phys Rev Lett 104:083904
20. Walther C, Scalari G, Beck M, Faist J (2011) Purcell effect in the inductor-capacitor laser. Opt Lett 36:2623–2625
21. Ravaro M, Manquest C, Sirtori C, Barbieri S, Santarelli G, Blary K, Lampin J-F, Khanna SP, Linfield EH (2011) Phase-locking of a 2.5 THz quantum cascade laser to a frequency comb using a GaAs photomixer. Opt Lett 36:3969–3971
22. Yasui T, Yokoyama S, Inaba H, Minoshima K, Nagatsuma T, Araki T (2011) Terahertz frequency metrology based on frequency comb. IEEE J Sel Top Quantum Electron 17: 191–201
23. Askaryan GA (1962) Cherenkov and transition radiation from electromagnetic waves. ZhETF 42:1360

24. Bodrov SB, Stepanov AN, Bakunov MI, Shishkin BV, Ilyakov IE, Akhmedzhanov RA (2009) Highly efficient optical-to-terahertz conversion in a sandwich structure with $LiNbO_3$ core. Opt Express 17:1871
25. Carr LD, De Mille D, Krems RV, Ye J (2009) Cold and ultracold molecules: science, technology and applications. New J Phys 11:055049
26. Steinmetz T, Wilken T, Araujo-Hauck C, Holzwarth R, Hänsch TW, Pasquini L, Manescau A, D'Odorico S, Murphy MT, Kentischer T, Schmidt W, Udem T (2008) Laser frequency combs for astronomical observations. Science 321:1335–1337
27. Siegel PH (2002) Terahertz technology. IEEE Trans Microw Theory Tech 50:910
28. Knap W, Dyakonov M, Coquillat D, Teppe F, Dyakonova N, Lausakowski J, Karpierz K, Sakowicz M, Valusis G, Seliuta D, Kasalynas I, El Fatimy A, Meziani YM, Otsuji T (2009) Field effect transistors for terahertz detection: physics and first imaging applications. J Infrared Millim Terahertz Waves 30:1319–1337
29. Öjefors E, Pfeiffer UR, Lisauskas A, Roskos HG (2009) A 0.65 THz focal-plane array in a quarter-micron CMOS process technology. IEEE J Solid State Circ 44:1968–1976
30. Dyakonov M, Shur M (1993) Shallow water analogy for a ballistic field effect transistor: new mechanism of plasma wave generation by dc current. Phys Rev Lett 71:2465–2468
31. Vitiello MS, Coquillat D, Viti L, Ercolani D, Teppe F, Pitanti A, Beltram F, Sorba L, Knap W, Tredicucci A (2012) Room-temperature terahertz detectors based on semiconductor nanowire field-effect transistors. Nano Lett 12:96–101
32. Vicarelli L, Vitiello MS, Coquillat D, Lombardo A, Ferrari AC, Knap W, Polini M, Pellegrini V, Tredicucci A (2012) Graphene field-effect transistors as room-temperature terahertz detectors. Nat Mater 11:865–871
33. Li Y, Qian F, Xiang J, Lieber CM (2006) Nanowire electronic and optoelectronic devices. Mater Today 9:18
34. Vitiello MS, Viti L, Romeo L, Ercolani D, Scalari G, Faist J, Beltram F, Sorba L, Tredicucci A (2012) Semiconductor nanowires for highly sensitive, room-temperature detection of terahertz quantum cascade laser emission. Appl Phys Lett 100:241101
35. Viti L, Vitiello MS, Ercolani D, Sorba L, Tredicucci A (2012) Se-doping dependence of the transport properties in CBE-grown InAs nanowire field effect transistors. Nanoscale Res Lett 7:159
36. Romeo L, Coquillat D, Pea M, Ercolani D, Beltram F, Sorba L, Knap W, Tredicucci A, Vitiello MS (2013) Nanowire-based field effect transistors for terahertz detection and imaging systems. Nanotechnology 24:214005
37. Geim AK, Novoselov KS (2007) The rise of graphene. Nat Mater 6:183–191
38. Grigorenko AN, Polini M, Novoselov KS (2012) Graphene plasmonics. Nat Photonic 6: 749–758
39. Jablan M, Buljan H, Soljac M (2009) Plasmonics in graphene at infrared frequencies. Phys Rev B 80:245435
40. Knap W, Rumyantsev S, Vitiello MS, Coquillat D, Blin S, Dyakonova N, Shur M, Teppe F, Tredicucci A, Nagatsuma T (2013) Nanometer size field effect transistors for THz detectors. Nanotechnology 24:2104002
41. Wang K, Mittleman DM (2004) Metal wires for terahertz waveguiding. Nature 432:376–379
42. Mbonye M, Mendis R, Mittleman DM (2009) A terahertz two-wire waveguide with low bending loss. Appl Phys Lett 95:233506
43. Chen LJ, Chen HW, Kao TF, Lu HY, Sun CK (2006) Low-loss subwavelength plastic fiber for terahertz waveguiding. Opt Lett 31:308–310
44. Atakaramians S, Afshar S, Fischer BM, Abbott D, Monro TM (2008) Porous fibers: a novel approach to low loss THz waveguides. Opt Express 16:845–8854
45. Lu J-K, Yu C-P, Chang H-C, Chen H-W, Li Y-T (2008) Terahertz air-core microstructure fiber. Appl Phys Lett 92:064105
46. Harrington JA, George R, Pedersen P, Mueller E (2004) Hollow polycarbonate waveguides with inner Cu coatings for delivery of terahertz radiation. Opt Express 12:5263–5268

47. Bowden B, Harrington JA, Mitrofanov O (2008) Low-loss modes in hollow metallic terahertz waveguides with dielectric coatings. Appl Phys Lett 93:181104
48. Ito T, Matsuura Y, Miyagi M, Minamide H, Ito H (2007) Flexible terahertz fiber optics with low bending-induced losses. J Opt Soc Am B 24:1230–1235
49. Vitiello MS, Xu J-H, Kumar M, Beltram F, Tredicucci A, Mitrofanov O, Beere HE, Ritchie DA (2011) High efficiency coupling of terahertz micro-ring quantum cascade lasers to the low-loss optical modes of hollow metallic waveguides. Opt Express 19:1122–1130
50. Vitiello MS, Xu J-H, Beltram F, Tredicucci A, Mitrofanov O, Harrington JA, Beere HE, Ritchie DA (2011) Guiding a terahertz quantum cascade laser into a flexible silver-coated waveguide. J Appl Phys 110:063112

Chapter 6
Terahertz Imaging System Based on Superconducting Heterodyne Integrated Receiver

R.V. Ozhegov, K.N. Gorshkov, Yu B. Vachtomin, K.V. Smirnov, M.I. Finkel, G.N. Goltsman, O.S. Kiselev, N.V. Kinev, L.V. Filippenko, and V.P. Koshelets

Abstract The development of terahertz imaging instruments for security systems is on the cutting edge of terahertz technology. We are developing a THz imaging system based on a superconducting integrated receiver (SIR). An SIR is a new type of heterodyne receiver based on an SIS mixer integrated with a flux-flow oscillator (FFO) and a harmonic mixer which is used for phase-locking the FFO. Employing an SIR in an imaging system means building an entirely new instrument with many advantages compared to traditional systems.

In this project we propose a prototype THz imaging system using an 1 pixel SIR and 2D scanner. At a local oscillator frequency of 500 GHz the best noise equivalent temperature difference (NETD) of the SIR is 10 mK at an integration time of 1 s and a detection bandwidth of 4 GHz. The scanner consists of two rotating flat mirrors placed in front of the antenna consisting of a spherical primary reflector and an aspherical secondary reflector. The diameter of the primary reflector is 0.3 m. The

---

R.V. Ozhegov • Y.B. Vachtomin • K.V. Smirnov
Moscow State Pedagogical University, 1 Malaya Pirogovskaya str., Moscow, Russia

CJSC "Superconducting nanotechnology", 5/22 Rossolimo str., Moscow, Russia

K.N. Gorshkov • M.I. Finkel
Moscow State Pedagogical University, 1 Malaya Pirogovskaya str., Moscow, Russia

G.N. Goltsman (✉)
Moscow State Pedagogical University, 1 Malaya Pirogovskaya str., Moscow, Russia

CJSC "Superconducting nanotechnology", 5/22 Rossolimo str., Moscow, Russia

National Research University Higher School of Economics, 20 Myasnitskaya Ulitsa, Moscow 101000, Russia
e-mail: goltsman@mspu-phys.ru

O.S. Kiselev • N.V. Kinev • L.V. Filippenko • V.P. Koshelets
Kotel'nikov Institute of Radio Engineering and Electronics, 125009 Moscow, Russia

C. Corsi and F. Sizov (eds.), *THz and Security Applications*, NATO Science for Peace and Security Series B: Physics and Biophysics, DOI 10.1007/978-94-017-8828-1_6,

operating frequency of the imaging system is 600 GHz, the frame rate is 0.1 FPS, the scanning area is $0.5 \times 0.5\ m^2$, the image resolution is 50×50 pixels, the distance from an object to the scanner was 3 m. We have obtained THz images with a spatial resolution of 8 mm and a NETD of less than 2 K.

6.1 Introduction

Modern terahertz heterodyne receivers with ultimate sensitivity were developed primarily for astronomical missions as a part of big international projects. As a result, we have such instruments as the balloon-borne telescope TELIS, the plane-borne telescope SOFIA, the space-borne telescope HERSHEL, and a project of future space-borne telescope Millimetron. Recently there have been great advances in the development of both receivers and radiation sources of the terahertz range combined with a decrease of their prime cost so that these devices have become more accessible for a wide range of tasks.

Obtaining an image of an object is one of the most important tasks in the terahertz region. Such imaging systems can be used in many areas from medicine and security systems to navigation under poor visibility. However, this frequency region is rather difficult to work in owing to the so-called terahertz gap [1]: the power level of radiation sources falls off as one of approaches this region either from the radio or from the infrared frequencies. At the same time terahertz imaging systems have certain advantages over analogous systems operating at different frequencies. X-ray systems offer excellent spatial resolution of hidden objects but do not allow analysis of the composition of the object, and besides X-rays are potentially harmful to humans. IR systems provide good spatial resolution but most materials used in daily life are not transparent in the IR region. Radio frequencies have a good penetrating capability but its relatively low frequency does not afford good spatial resolution. Falling in the region between radio and IR waves, terahertz radiation has good penetrating capability and offers high spatial resolution, which makes it very attractive to security systems.

Terahertz imaging systems also allow performing spectral analysis of the substance of the object in hand [2] (see Fig. 6.1) and obtaining pictures of slices of this object with the use of mathematical processing [3] (see Fig. 6.2).

Passive systems receive thermal radiation of an object and thus an image is constructed. Such systems are called thermal imaging systems and require very sensitive receivers, which are usually direct detection receivers. Employment of a heterodyne receiver in a thermal imaging system allows one to use spectral information present in the thermal radiation of the object to deduce its chemical composition. This makes heterodyne systems attractive to medical and security systems. An example of a passive direct-detection imaging system based on antenna-coupled superconducting microbolometers technology is presented in [4].

Heterodyne systems employ either Schottky diode mixers [5], or superconductor-insulator-superconductor (SIS) mixers [6], or HEB mixers [7]. SIS mixers offer

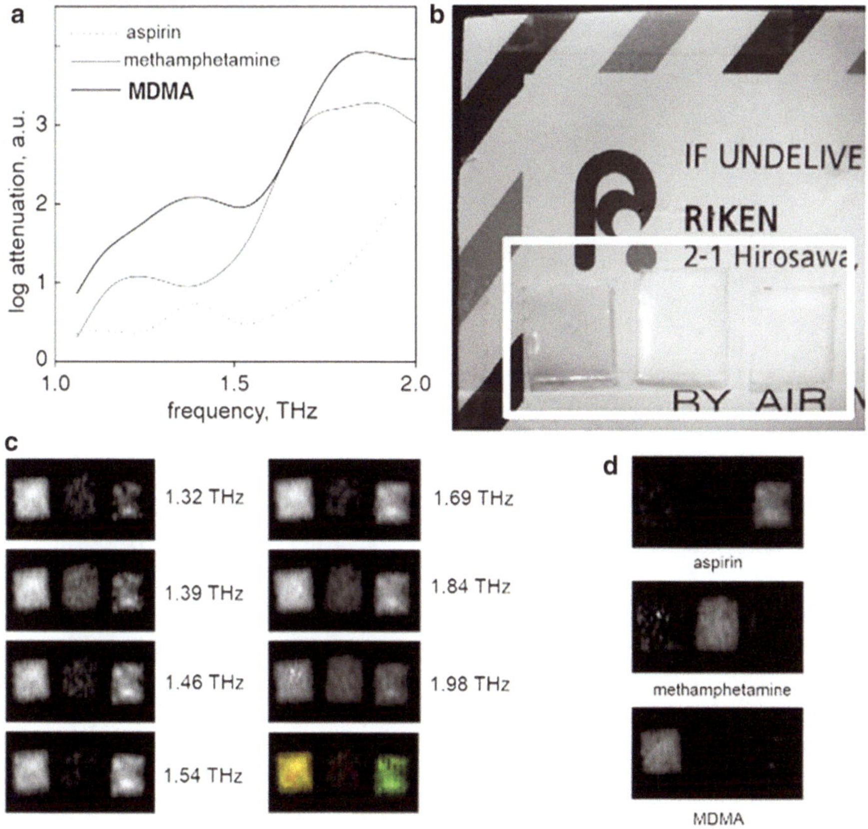

Fig. 6.1 Demonstration of capabilities of terahertz imaging systems to determine chemical composition of an object. (**a**) Spectral transmission lines of certain illegal drugs (MDMA, methamphetamine) and aspirin. (**b**) Observed objects, *left* to *right*: MDMA, aspirin, methamphetamine. (**c**) Terahertz images of objects taken at different frequencies. *Colour images* demonstrate possibility of detecting objects with the use of "colour" terahertz imaging. (**d**) Results of image processing [2]

low noise temperature and theoretically unlimited intermediate frequency (IF) bandwidth, but the RF range is limited by the properties of the superconductor. HEB mixers also offer a low noise temperature, but at frequencies below 1.3 THz SIS mixers have better noise performance. Schottky-diode mixers have worse noise performance than SIS and HEB mixers but do not require to be cooled down to the liquid He temperature. The noise performance of these three types of mixers is compared in Fig. 6.3.

Currently SIS mixers are the most sensitive heterodyne receivers in the frequency range 0.1–1.3 THz (see Fig. 6.3). Besides Josephson effects [8], one can observe in an SIS junction very strong nonlinearity of its current-voltage curve, which allows this junction to be used in a heterodyne receiver with a conversion gain and quantum-limited noise temperature. Superconducting junctions are also used

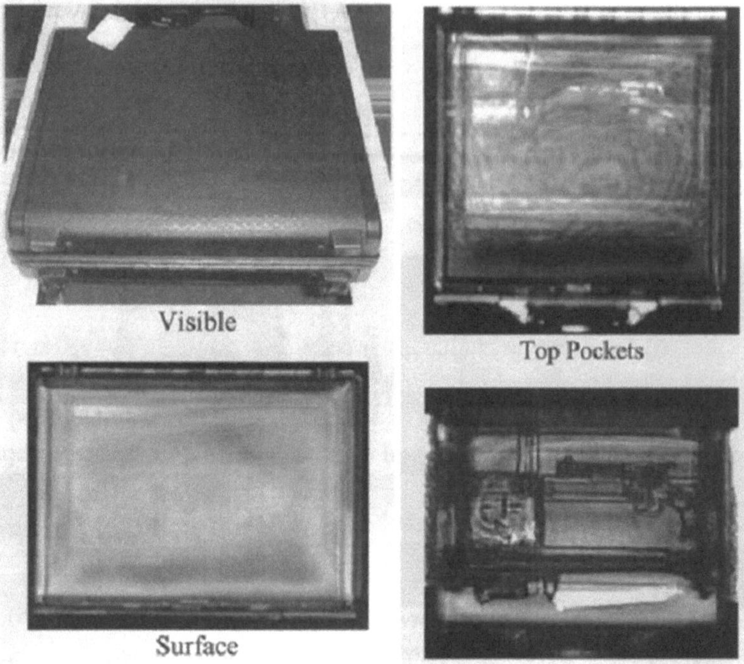

Fig. 6.2 An example of the performance of an active pulse imaging system receiving radiation reflected off the object [3]. By compensating the delay in the signal path it is possible to obtain images of various layers within the object. *Top left*: a photograph of the object (suitcase); *bottom left*: a terahertz image of the surface of the suitcase; *top right*: terahertz image of a plastic bag inside the suitcase; *bottom right*: terahertz image of the contents of the bag. One can clearly observe a knife and a gun inside the bag

as sources of high frequency current. An example of such a source is a flux flow oscillator (FFO) [9, 10].

Within the TELIS (TErahertz Limb Sounder) project, the research groups from the Institute of Radio Engineering and Electronics, the Russian Academy of Sciences (IREE), and the Space Research Organisation of the Netherlands (SRON) have built a superconducting integrated receiver (SIR)$4 \times 4 \times 0.5^3$ in size. The SIR carries an SIS mixer coupled with a planar quasioptical antenna and an FFO-based local oscillator (LO), and a harmonic mixer used to phase-lock the LO. The harmonic mixer technology is the same as that of the SIS mixer. A schematic of the SIR is presented in Fig. 6.4, and a photograph of the SIR chip is shown in Fig. 6.5. The SIR developed for TELIS [11, 12] has demonstrated successful performance under severe temperature, pressure and altitude conditions, and allowed collecting a lot of valuable information about the Earth's atmosphere. Despite the fact that the noise performance of SIRs is slightly worse compared to single SIS mixers, they become more and more attractive as candidates for compact spectrometers and imaging systems.

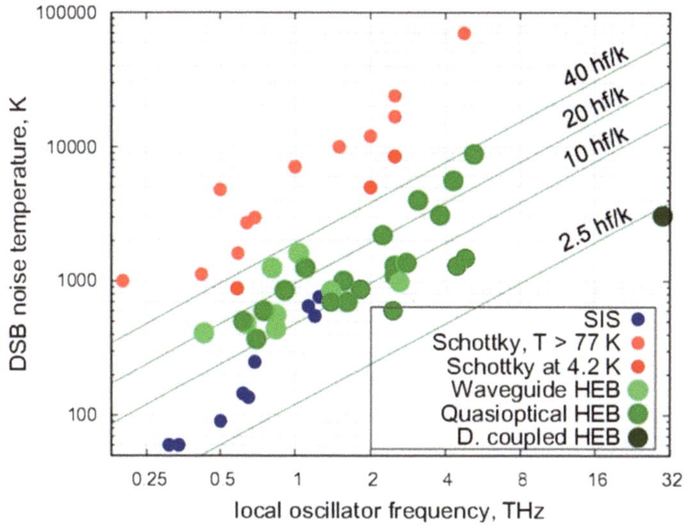

Fig. 6.3 Comparison of the noise performance of most widely used terahertz heterodyne receivers. For HEB receivers the data are given for both kinds of HEB mixers: waveguide and quasioptical ones

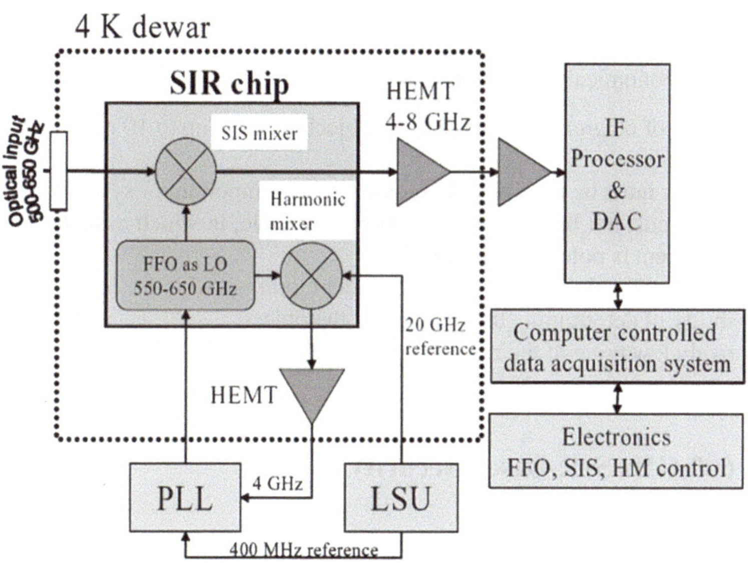

Fig. 6.4 A schematic of the SIR. To make the full use of the SIR one also needs a frequency synthesiser (LSU) and a PLL

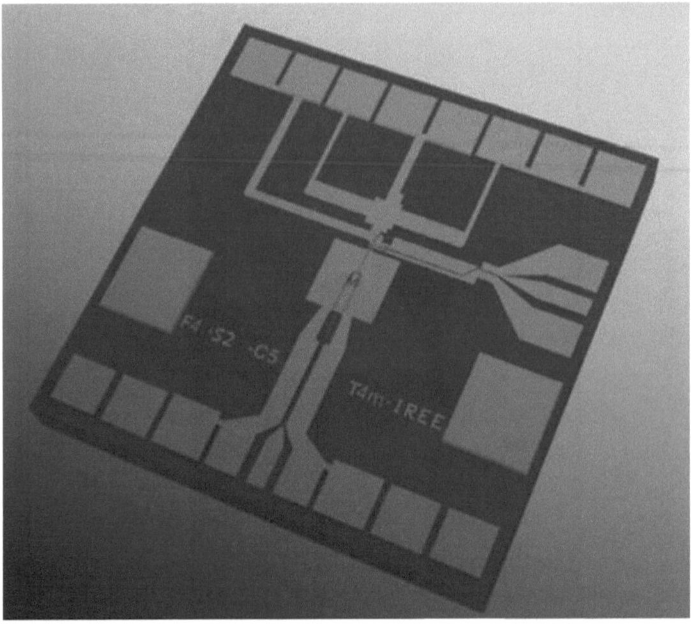

Fig. 6.5 A photograph of an SIR

Thus, one can put forward the requirements that a security system must meet in order to be economically efficient:

- capability of obtaining an image of an object which is up to 10 m away from the system;
- the system must be passive (this requirement is important in systems where the observer must not be seen and that monitor people, in which case the use of an active system is potentially harmful);
- capability of "seeing" objects hidden under clothing and in the luggage;
- capability of determining the material of the object;
- relatively low price of the ready-to-use system.

6.2 600 GHz SIR-Based Security System

The passive heterodyne thermal imaging system that we offer is currently unique. It potentially can be used for security purposes to detect plastic or metal weapon and explosives hidden under clothing, to check the post for illegal drugs, or in medicine to diagnose cancer at an early stage. The system has certain other advantages that makes it potentially the system of choice on the market of terahertz technology. The heart of the system is the SIR developed at IREE RAS. A series of tests of the SIR receiver performance was performed, and a model of an SIR-based imaging system was built without the phase-lock loop (PLL).

6.2.1 Receiver Performance

When an imaging system is used to obtain pictures of objects with temperatures of about 300 K, it may so happen that the system noise temperature will either be comparable with or less than the object temperature. In this case the system sensitivity is determined chiefly by the object temperature, and in the limiting case of a low-noise receiver the temperature resolution of the system is given by [13]:

$$\Delta T = 0.612\alpha \frac{T_S}{\sqrt{B_\tau}}. \tag{6.1}$$

In order to obtain the lowest possible temperature resolution we used a receiver with a large IF bandwidth, and also employed a PLL. The measurements were performed with the SIR T4m-093#6m offering a noise temperature of about 90 K at an LO frequency of 507 GHz. The main characteristics of this receiver are presented in Fig. 6.6.

We determined the temperature resolution of the receiver by measuring the rms voltage of the lock-in amplifier. Figure 6.7 shows time dependence of the lock-in read- out expressed in temperature units. As the magnitude of the input signal was lowered the temperature resolution improved. For the receiver T4m-093#6m looking at a temperature difference of about 173.5 K the temperature resolution was found to be 140 ± 15 mK; when the temperature difference was decreased to about 2.95 K, the temperature resolution reached 10 ± 1 mK. The error is determined by the integration time and is proportional to $(\tau/T)^{0.5}$, where τ is the time constant of the lock-in amplifier and T is the integration time.

Besides temperature resolution, another important parameter of a receiver operating as an imaging system is its temporal stability. In a system with a mechanical scanner, the time required to make one scan (frame time) can be a few seconds. This means that the time during which the system is considered stable (see below) must be much longer than the frame time, otherwise the temperature resolution of the system will be rather poor and the resulting image blurred.

To characterise the receiver stability one plots the so-called Allan variance as a function of the integration time [14]. The plot typically consists of three regions: the variance first decreases with time, then reaches a plateau, and finally increases as one makes the integration time longer. The Allan time is the integration time at which the Allan variance plot starts to depart from the $\sigma^2 \sim 1/\tau$ dependence.

In order to increase the Allan time one usually tries to achieve better temperature stability of amplifiers and bias sources. Alternatively, one can decrease the receiver input bandwidth [11].

The method of thermal stabilisation is rather challenging, expensive, and highly undesirable for mass production of receivers. The alternative method leads to degradation of the receiver sensitivity. That is why one has to look for different methods to improve the receiver stability. This is discussed below.

Various instabilities inherent in a receiver may manifest themselves either as fluctuations of the overall conversion gain, including the gain of the IF chain, or as fluctuations of the receiver noise temperature. As experience of working with HEB

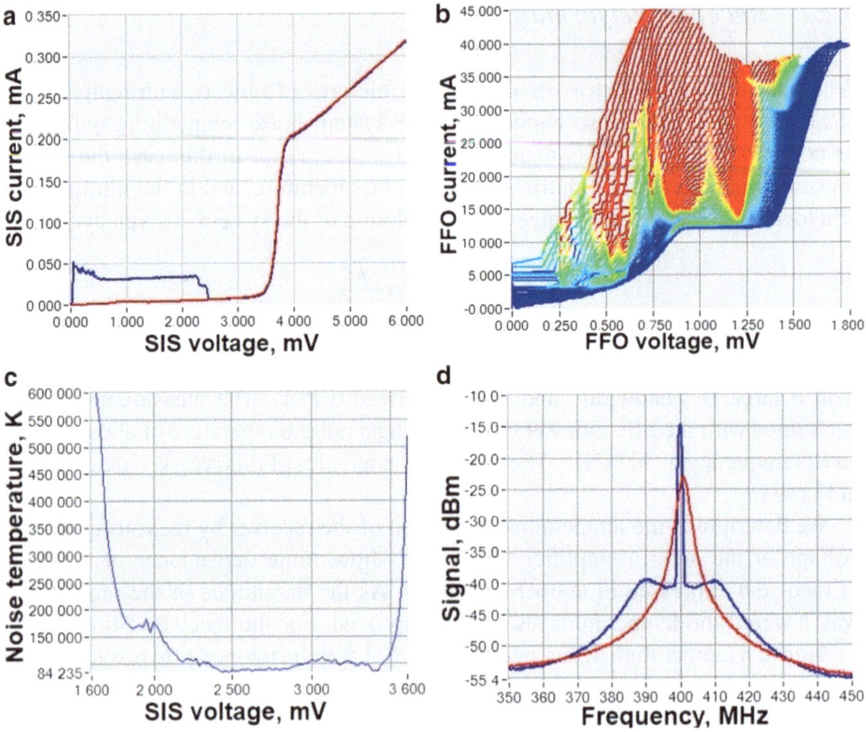

Fig. 6.6 The main characteristics of receiver T4m-093#6m. (**a**) Current-voltage characteristics (*CVC*) of the SIS mixer: the *blue* CVC corresponds to the unsuppressed DC Josephson effect; the *red* one corresponds to the optimal current of the control line suppressing the critical current of the tunnel junction. (**b**) CVCs of the FFO for different levels of magnetic field penetrating the junction; *colours* show the local oscillator (*LO*) drive level of the SIS mixer. (**c**) The receiver noise temperature as a function of the bias voltage. (**d**) The LO line with the PLL (*blue*) and without the PLL (*red*); the LO line is taken from the output port of the harmonic mixer integrated on the same chip with the SIS mixer at an IF of about 400 MHz

receivers of SIRs shows, gain fluctuations have a more dramatic effect on the system stability. That is why, as the first approximation, the receiver output is determined by the object brightness temperature and the instantaneous gain:

$$P_{OUT} = k_B\,(T_R + T_S)\,GB_H, \tag{6.2}$$

where k_B is the Boltzmann constant; B_H is operating frequency range; G is the total receiver IF gain, including the mixer conversion loss; T_R is the receiver noise temperature; T_S is the object brightness temperature.

To correct for gain fluctuations we used a wobbling mirror that could be switched between two positions: in one position the receiver was looking towards the reference load (77 K), in the other it was looking towards the signal load (300 K). The switching time was 1.8 ms [15].

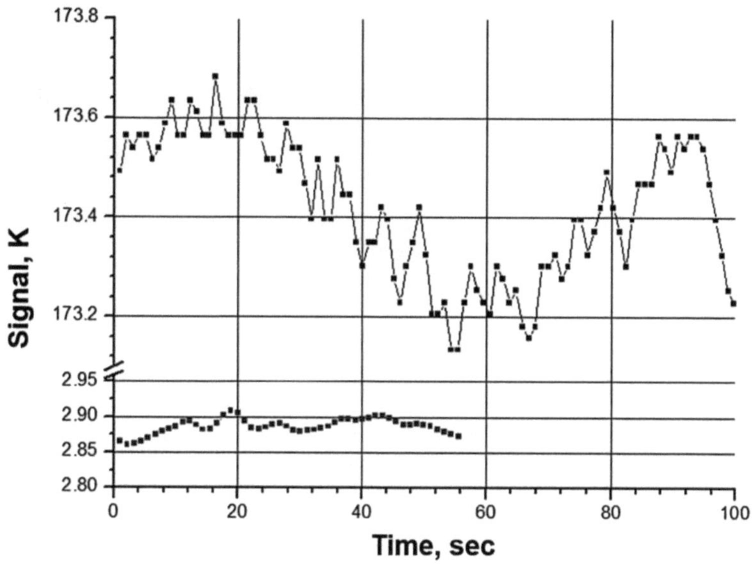

Fig. 6.7 Time dependence of the lock-in output for different levels of the receiver input signal for receiver T4m-093#6m. The input signal is the difference between the temperatures of the chopper blades and the load. In the case of a large difference (*top line*) NETD = 143 ± 14 mK, and for a small difference (*bottom line*) NETD = 10 ± 1 mK. In both cases the integration time of the lock-in amplifier was 1 s

The IF output was fed to the data acquisition system which also controlled the mirror and recorded its position. The data acquisition system computed the magnitude of the detected signals coming from the reference and source loads and adjusted them with the use of the expression:

$$S'_{OUT_i} = \frac{S_{OUT_i}}{S_{REF_i}} S_{REF_0}, \tag{6.3}$$

Here S_{REF_0} is the initial value of the output signal of the receiver looking towards the reference load; S_{OUT_i} and S_{REF_i} are instantaneous values of the output signal of the receiver looking towards the reference source load and reference load respectively, and S'_{OUT_i} is the corrected value of the output signal.

Figure 6.8 shows the Allan plots for the reference signal, the source signal, and the signal corrected according to 1.3. Fluctuations of the corrected signal at short integration times are shorter than fluctuation of each of the measured signals, in accordance with the expression $\sqrt{\sigma_{REF}^2 + \sigma_{OUT}^2}$. As can be seen from the plot, corrections allow one to reduce the contribution of the 1/*f* noise and drift noise by more than an order of magnitude, and increase the Allan time from fractions of a second up to 5 s, with an IF bandwidth of 4 GHz. The inset in Fig. 6.8 shows the result of correcting the output signal with a narrow IF bandwidth. In this case with

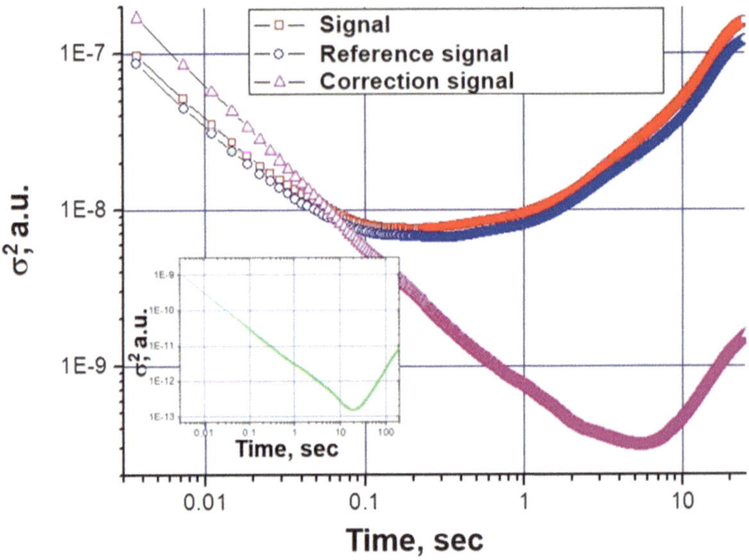

Fig. 6.8 The Allan plots for reference, source and corrected signals. The IF bandwidth is 4 GHz. The *inset* shows the corrected signal obtained with an IF bandwidth of 40 MHz

an IF bandwidth of 40 MHz it is possible to achieve an Allan time of 20 s. The data of the temperature resolution and the Allan time are unprecedented for heterodyne receivers of the terahertz range and are comparable with the best results obtained for direct detection receivers.

6.2.2 *Model of the Imaging System*

The imaging system consists of two main parts: SIR and front-end.

The SIR chip is glued onto a Si elliptic lens with an anti-reflection coating. The lens is installed into the mixer block mounted onto the cold plate of the He cryostat and enclosed with a cryomagnetic shield.

A photograph of the model of the terahertz imaging system is shown in Fig. 6.9. The front-end comprises a primary spherical dish 0.3 m in diameter with a secondary mirror 40 mm in diameter, and a mechanical scanner in front of the primary.

The scanner consists of two flat mirrors wobbling in mutually perpendicular directions. One is responsible for frame scans, the other for column scan. The sizes of the mirrors are 0.405×0.275 m^2 and 0.28×0.38 m^2, respectively. All the mirrors were made from a bulk piece of aluminium with the use of a milling machine. The mirrors are positioned with step motors. The scanning system allows performing observation of a relatively large spatial area. This however is possible at the expense of spatial resolution owing to decrease of the effective aperture. We

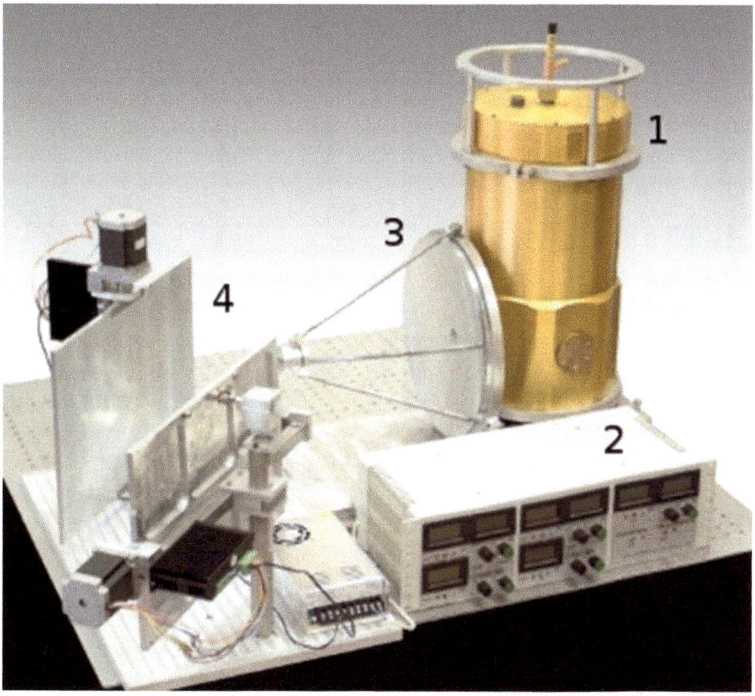

Fig. 6.9 A photograph of the model of the terahertz imaging system. *1* He cryostat with an SIR, *2* bias source of the SIR, *3* primary mirror, *4* mechanical scanner

achieved the following performance of system: time required to capture an image is 10 s, scanning area is 0.5×0.5 m^2, distance from the system to the object is 3 m, spatial resolution is better than 10 mm.

Some of the images taken with the model imaging system are presented in Fig. 6.10. The objects hidden under clothing were: a piece of chipboard 1 cm thick and measuring 10×3 cm^2 (top), and a packet of cigarettes in a breast pocket (bottom). The temperature resolution was less than 2 K.

6.3 Summary and Outlook

We have built a laboratory model of the terahertz imaging system with characteristics that will allow it to be developed into a commercial instrument for security applications. In order to develop such an instrument it is necessary to improve the scanner so that the frame time will be 1 s at most. This can be accomplished by replacing the step motors with more powerful AC servo motors, and by optimising the shape of the mirrors to reduce their moments of inertia. Replacing the He cryostat with a closed-cycle refrigerator will increase the system operation time

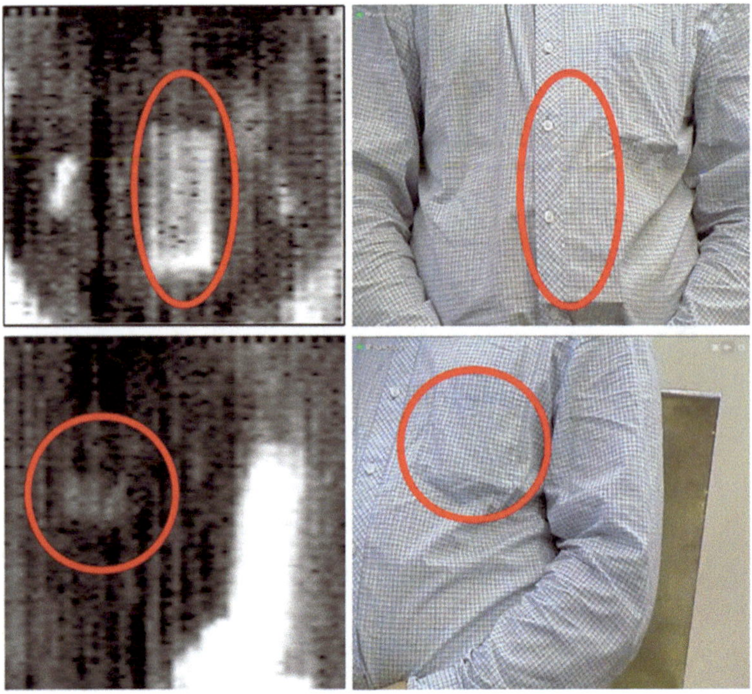

Fig. 6.10 Pictures taken with the model terahertz imaging system. *Left*: thermal images; *right*: optical images

by several orders of magnitude. Besides, it is necessary to improve the optics of the front-end of the system by adding elements that will allow switching between the object and the reference load. This will considerably improve the temperature resolution and stability of the system.

The passive heterodyne security imaging system has certain advantages over analogous active heterodyne and direct-detection systems, which will make it more attractive to the potential user. This research is partly supported by the Ministry of Education and Science of the Russian Federation, contract # 14.B25.31.0007.

References

1. Sisov F (2010) THz radiation sensors. Opto Electron Rev 18:10–36
2. Kodo Kawase, Yuichi Ogawa, Hiroaki Minamide, Hiromasa Ito (2005) Terahertz parametric sources and imaging applications. Semicond Sci Technol 20:258–265
3. Zimdars D, White J, Stuck G et al (2007) Time domain terahertz imaging of threats in luggage and personnel. Int J High Speed Electron Syst 17(2):271–281
4. Luukanen A, Gronberg L, Helisto P, Penttil JS, Seppa H, Sipola H, Dietlein CR, Grossman EN (2006) An array of antenna-coupled superconducting microbolometers for passive indoors real-time THz imaging. Proc SPIE 6212:270–27

5. Crowe TW, Mattauch RJ, Roser HP, Bishop WL, Peatman WCB, Liu X (1992) GaAs Schottky diodes for THz mixing applications. Proc IEEE 80:1827–1841
6. Richards PL et al (1979) Quasiparticle heterodyne mixing in SIS tunnel junctions. Appl Phys Lett 34:345–347
7. Gershenzon EM, Gol'tsman GN, Gogidze IG, Gousev YP, Elant'ev AI, Karasik BS, Semenov AD (1990) Millimeter and submillimeter range mixer based on electronic heating of superconducting films in the resistive state. Sov Phys Supercond 3:1582–1597
8. Josephson BD (1962) Possible new effects in superconducting tunneling. Phys Rev B 1:251
9. Nagatsuma T, Enpuku K, Irie F, Yoshida K (1983) Flux-flow type Josephson oscillator for millimeter and submillimeter wave region. J Appl Phys 54:3302. doi:10.1063/1.332443
10. Qin J, Enpuku K, Yoshida K (1988) Flux-flow-type Josephson oscillator for millimeter and submillimeter wave region. IV. Thin-film coupling. J Appl Phys 63:1130. doi:10.1063/1.340019
11. de Lange G, Boersma D, Dercksen J, Dmitriev P, Ermakov AB, Filippenko LV, Golstein H, Hoogeveen RWM, de Jong L, Khudchenko AV, Kinev NV, Kiselev OS, van Kuik B, de Lange A, Rantwijk J, Sobolev AS, Mikhail Y, Torgashin EV, Yagoubov PA, Koshelets VP (2010) Development and characterization of the superconducting integrated receiver channel of the TELIS atmospheric sounder. Supercond Sci Technol 23:45–61
12. Koshelets VP, Dmitriev PN, Ermakov AB, Filippenko LV, Khudchenko AV, Kinev NV, Kiselev OS, Sobolev AS, Torgashin MY (2009) On-board Integrated submm spectrometer for atmosphere monitoring and radio astronomy. ISTC thematic workshop on perspective materials, devices and structures for space applications, Yerevan, Armenia, 26–28 May
13. Esepkina NA, Korolkov DV, Pariyskiy YN (1973) Radiotelescopes and radiometers (in Russian)
14. Allan D (1966) Statistics of atomic frequency standards. Proc IEEE 54:221–230
15. Ozhegov RV, Gorshkov KN, Gol'tsman GN, Kinev NV, Koshelets VP (2011) Stability of terahertz receiver based on superconducting integrated receiver. Supercond Sci Technol 24:035–038

Chapter 7
Application of Surface Waves for THz Spectroscopy

Oleg Mitrofanov

Abstract Weak coupling of THz waves to sub-wavelength size objects makes detection and spectroscopic analysis of small amounts of illicit materials challenging at THz frequencies. THz surface waves can be used to enhance the interaction of the THz wave with small objects and mitigate the weak coupling. We will discuss excitation THz surface waves at metallic edges as well as detection and imaging of THz surface waves using the integrated sub-wavelength aperture THz near-field microscopy method. This near-field probe technology can improve the sensitivity of THz spectroscopy analysis and therefore relax power and sensitivity requirements for THz sources and detectors.

7.1 Introduction

Spectroscopic analysis at terahertz (THz) frequencies can provide a number of unique capabilities for industrial applications, security and scientific research. One of the major challenges for applications of THz spectroscopy in security and sensing is the difficulty of providing sufficient sensitivity for detection of small amounts of illicit materials. A similar challenge limits scientific research applications, where sensitivity of THz spectroscopy is presently insufficient for the analysis of single nano-scale systems, such nano-wires and quantum dots. This challenge is largely due to weak coupling of THz waves to small (sub-wavelength size) objects.

THz surface waves on metallic surfaces can be used to enhance the electromagnetic wave interaction with a sub-wavelength size object and, as a result, improve

O. Mitrofanov (✉)
Electronic and Electrical Engineering, University College London, London, UK
e-mail: o.mitrofanov@ucl.ac.uk

C. Corsi and F. Sizov (eds.), *THz and Security Applications*, NATO Science for Peace and Security Series B: Physics and Biophysics, DOI 10.1007/978-94-017-8828-1_7,

the sensitivity of THz spectroscopy on the sub-wavelength scale. Surface waves however are non-radiative and special techniques for excitation and detection of THz surface waves must be developed.

It was demonstrated recently that THz surface waves can be excited at the edges of metallic patterns deposited on dielectric substrates by a plane THz wave incident from the substrate side [1]. They can be efficiently detected by a near-field probe, which incorporates a sub-wavelength size aperture [2–4]. A system consisting of a surface wave supporting structure and the near-field probe, which also enables high spatial resolution imaging, provides a sensitive platform where THz time-domain spectroscopy can be applied to a sub-wavelength size object positioned on the metallic surface. An example of such a system with a bow-tie antenna as a surface wave supporting structure will be discussed.

We will also discuss the use of surface waves for confining THz waves to extremely small volumes ($<\lambda^3/10^4$), surpassing the diffraction limit experienced by the free space propagating beams. A configuration with two metallic needles can couple a THz beam into surface wave propagating on cylindrical surfaces of the needles to guide the electromagnetic energy to the tips of the needles, where the energy is confined between two metallic surfaces to volume smaller than 10 μm across [5]. This twin-needle probe configuration enables THz spectroscopic analysis for single small particles only several microns in size.

Utilizing THz surface waves and the near-field probe technology provides a step toward applying THz spectroscopy to individual micron-size objects. Through enhancing the interaction between the THz wave and individual sub-wavelength objects, sensitivity of the THz spectroscopy analysis can be improved and, as a result, the challenging requirements for efficient THz sources and detectors can be relaxed. This chapter is focused on recently developed near-field probe methods that can help mitigating the weak coupling and enable application of THz spectroscopy to a wider range of sub-wavelength scale systems.

7.2 Application of THz Spectroscopy to Sub-wavelength Size Systems

Spectroscopic analysis of nano-scale objects has been enabled across the electromagnetic spectrum through advances in scanning probe near-field microscopy. THz spectroscopy still however remains limited predominantly to large objects due to the lack of sensitive high-spatial resolution THz near-field probes. The challenge of applying THz spectroscopic analysis to a single 100 nm object becomes clear if one considers the classical approach of near-field microscopy with a sub-wavelength aperture probe incorporating a 100 nm aperture. The aperture transmission coefficient, governed by the Bethe's law of electromagnetic wave transmission through a sub-wavelength hole, reduces quickly as the sixth power of the aperture size a ($T \sim a^6$) [6]. As a result, the gain in spatial selectivity is achieved

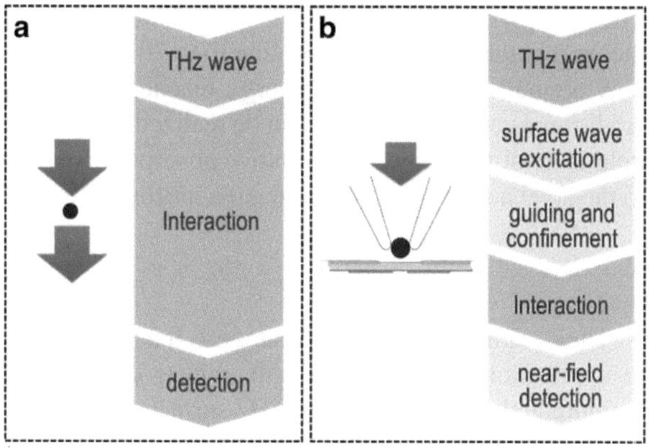

Fig. 7.1 Schematic representation of the conventional (far-field) THz spectroscopy (**a**) and near-field spectroscopy involving THz surface waves (**b**)

at a considerable loss in sensitivity [7]. For this reason, optical near-field microscopy fails to employ apertures smaller than ~1/50 of the wavelength (λ). A 100 nm aperture at 1 THz is smaller than $\lambda/1{,}000$, making the transmission coefficient $T \sim 10^{-12}$, impracticable for applications. THz spectroscopy on the 100 nm–1 μm scales therefore requires an alternative experimental configuration.

First we consider the level of interaction between a THz beam and an object. For the entire THz beam to interact with the object, the object is required to have dimensions larger than the THz beam size in a typical THz spectroscopy experiment. Determined by the diffraction limit, the beam size is comparable to the wavelength and therefore for nano-scale objects only a fraction of even a tightly focused beam interacts with a sub-wavelength object (Fig. 7.1a). A large part of the beam remains unaffected by the object. If the detector is located in the far-field zone of the object, it cannot differentiate between the unaffected part and the part that has interacted with the object. The strength of the absorption signal can be estimated as relative absorption of the THz beam caused by the object. It can be expressed (neglecting geometrical resonances) as

$$S_f = \frac{I_{ob}}{I_0} = \frac{d^2 e^{-(\alpha d)}}{\lambda^2}, \tag{7.1}$$

where α and d represent the absorption coefficient and the object size. For a micron-size object the signal reduces significantly just due to the geometrical cross-section of the object. It suggests that one of the ways for improving the technique sensitivity is to enhance the interaction with the sub-wavelength object through concentration of the incident field to an area of sub-wavelength size.

Focusing of the electromagnetic waves beyond the diffraction limit is possible with surface waves. A potential configuration for the THz spectroscopy system is schematically illustrated in Fig. 7.1b [5]. The incident wave must be coupled to the surface wave, which guides the energy to the object and enhances the interaction with the object. It is also important to collect the absorption signal efficiently.

The region of interaction of the wave and the particle is now much smaller than the wavelength. Therefore coupling to the free-space propagating wave is weak and only a small fraction of the energy reaches a detector positioned in the far-field. Additional gain in sensitivity can be realized if the detector is introduced in the near-field zone of the object. The relative absorption signal in this case can be increased to the level achievable with large area uniform films of thickness d:

$$S_n = e^{-(\alpha d)} \tag{7.2}$$

These simple estimations suggest that the sensitivity of THz spectroscopy in application to sub-wavelength objects can be improved by several orders of magnitude through employing THz surface waves and near-field detection.

7.3 Sub-wavelength Aperture THz Near-Field Probe

As pointed out earlier, a single sub-wavelength aperture exhibits very weak transmission, limiting application of apertures in standard near-field microscopy to $\sim\lambda/20$. Significantly smaller apertures, as small as $\lambda/100$, can be realized by employing the integrated near-near probe concept [8]. In an integrated probe, a detector of THz waves, typically a photoconductive antenna, is incorporated in the near-field zone of the aperture [9]. The advantage of integrating a detector into the THz near-field probe is the gain in sensitivity that is achieved by detecting the evanescent components of the wave transmitted through the aperture [9].

The field transmitted on the other side of the aperture contains both the propagating and the evanescent components. For a sub-wavelength aperture the evanescent components give the largest contribution. This field however remains confined in the near-field zone of the aperture, not reaching a detector located in the far-field region. Therefore, it is possible to improve sensitivity of the near-field probe by integrating the detector with the aperture in one device. The first demonstrations of the integrated probe with improved sensitivity allowed reducing the aperture size to 5 μm with sufficient sensitivity to THz pulses with a broad spectrum of frequencies (0.2–2.5 THz) [8].

A schematic diagram of the integrated sub-wavelength aperture probe is shown in Fig. 7.2. A photoconductive antenna is used as the THz detector. The probe sensitivity depends on the distance between the aperture and the THz antenna and ideally is should be smaller than the aperture size in order to detect the evanescent components of the transmitted field [8]. In this device the antenna is positioned within several microns from an aperture in the metallic screen.

A useful property of this probe for applications in THz spectroscopy is its spectral response. It was found experimentally that the size of the sub-wavelength aperture does not affect the waveform of the detected pulse in the regime of the aperture being much smaller than the wavelength [10]. The response was described analytically both in the frequency and time domains. For the incident wave polarized in the

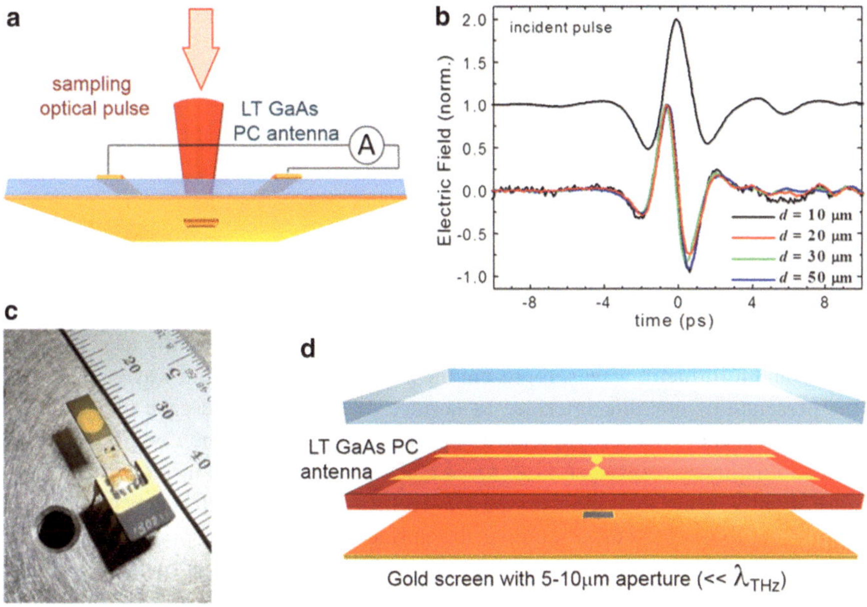

Fig. 7.2 Schematic diagram of the integrated sub-wavelength aperture probe (**a**) and normalized waveforms THz pulses detected by probes with apertures of 10, 20, 30 and 50 μm (**b**). A photograph of mounted probe (**c**) and a schematic diagram of the probe structure (**d**)

plane of the aperture screen the detected photocurrent is proportional to the temporal derivative of the incident field $E_{inc}(t)$:

$$i_t(t) = g_t \frac{dE_{inc}(t)}{dt}. \tag{7.3}$$

g_t is a constant representing the aperture transmission properties for the transverse incident field and the photoconductive antenna response. In the frequency domain, the amplitude response is proportional to a product of the incident field spectrum with frequency, $\omega E(\omega)$, and the phase between the incident and the detected field is $\pi/2$, independent of frequency. The analytical relationship between the incident field and the induced photocurrent allows direct reconstruction of the temporal evolution of the field.

Application of the integrated aperture probe for near-field imaging of metallic structures showed that the probe is sensitive to THz surface waves, which propagate on metallic surfaces [1, 10–12]. Remarkably, the response to surface waves is also described analytically. The detected photocurrent is proportional to the gradient of the field normal to the surface of the probe $E_z(x,y,t)$:

$$i_n(t) = g_n \nabla E_z(x, y, z) \cdot \vec{A}, \tag{7.4}$$

where g_n is a constant representing the aperture transmission properties for the surface wave and the photoconductive antenna response, and vector $\overrightarrow{A}$ represents the orientation of the dipole antenna in the detector.

7.4 THz Surface Waves

Two isolated metallic surfaces can confine the electromagnetic waves between them. Wave propagation is this case is supported by the metallic surfaces and it can be described in terms of surface waves. Surface waves, the term typically used for radio waves and millimetre waves, represent the same solution of to the wave equation in the presence of metallic interface as surface plasmons. In the case of an infinite flat surface, the wave propagation is described by the in-plane complex k-vector and an imaginary transverse k-vector. At THz frequencies the in-plane k-vector is only slightly larger than the free-space k-vector making THz surface waves weakly confined to a single surface unlike the surface plasmons at optical frequencies. Therefore at THz frequencies, confinement of THz wave is better realized with two surfaces.

Unlike the free-space propagating electromagnetic waves, surface waves are not subject to the diffraction limit of focusing and they allow confinement of the electromagnetic energy to volumes significantly smaller than the wavelength. Sub-wavelength confinement in one plane for example is realized in a parallel plate waveguide, where the electromagnetic wave propagation (the TEM mode) is supported by the two plates of the waveguide. THz spectroscopic applications have been realised with the plate separation of only 50 μm [13]. Confinement in two dimensions can be realized with tapered parallel plate waveguides [14], two metallic needles [5], transmission lines [15] and antennas [16].

The non-radiative nature of surface waves prevents their direct coupling to free-space propagating waves. To utilize THz surface waves for microscopy and local spectroscopy at THz frequencies, it is essential to develop a reliable method for surface wave excitation. Although the non-radiative surface waves couple weakly to the free-space waves, surface wave excitation can occur at metallic edges. If a metallic pattern on a dielectric substrate is illuminated by a plane wave with the electric field perpendicular to edge of metal, the incident wave induces oscillating change density at the edge. The corresponding electric field distribution has a strong concentration of the field, including a component perpendicular to the surface, E_z. Since the E_z component is localized only near the edge, its angular spectrum contains a broad distribution of wavevectors and includes components that match the k-vector of the surface wave. Therefore illumination of the metallic pattern causes excitation of the surface waves at the pattern edges, which begin propagating along the metallic surface (Fig. 7.3).

Formation and propagation of THz surface waves at the edges of a bow-tie antenna pattern was studied in Mueckstein et al. [1]. Figure 7.3b illustrates the process of excitation at the edges for two metallic triangles joint at their corners.

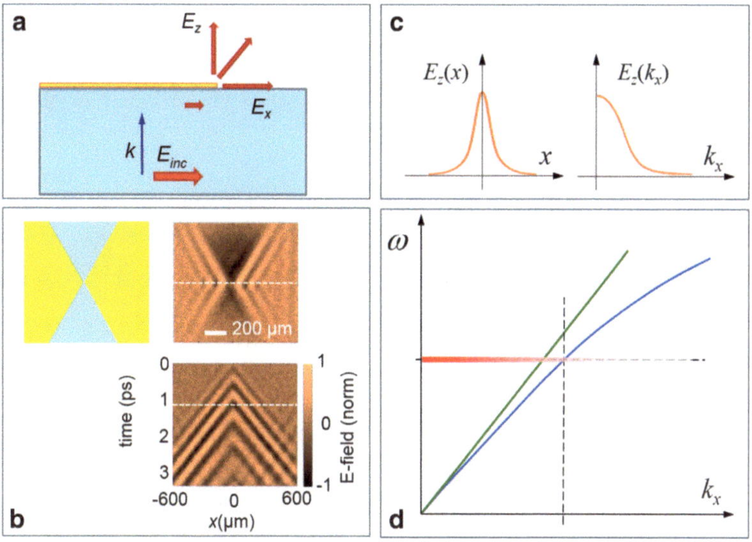

Fig. 7.3 THz surface wave excitation at the edges of metallic patterns. (**a**) Schematic diagram of the excitation process; (**b**) experimental observation of THz surface waves: spatial distribution of the instantaneous field distribution and a space-time map showing propagation of the waves away from the metallic edges [1]; (**c**) Schematic diagrams of the induced field E_z at the edge and its angular spectrum; (**d**) dispersion diagram for the surface waves showing the spectrum of the induced field (*red*) overlapping with the surface wave dispersion curve (*blue*)

The incident THz pulse in directed into the region of the corners and it is polarized horizontally to form a component of the electric field perpendicular to the edges of the triangles. The surface wave is detected by scanning the THz near-field probe with respect to the pattern within several microns from the surface. An instantaneous image of the electric field shows fringes formed along the edges of the triangles. The pattern was shown to evolve over time with the fringes moving away from the edges as it is illustrated in the space-time map recorded for a linear horizontal scan across the edges of the triangles.

7.5 Near-Field Imaging with THz Surface Waves

Any metallic edge illuminated by THz waves produces a localized distribution of the E_z component. If a sample contains a metallic feature, for example an antenna patch or a transmission line, surface waves are excited at every edge and the field distribution on the patch is a superposition of all excited waves [1]. Detection of surface waves requires a probe placed in the near-field region of the surface. In Fig. 7.3, the surface wave is detected by the integrated sub-wavelength aperture near-field probe. The same probe type was also used to study THz surface waves on metallic antennas as well as on the metallic surface of the probe itself. A previously

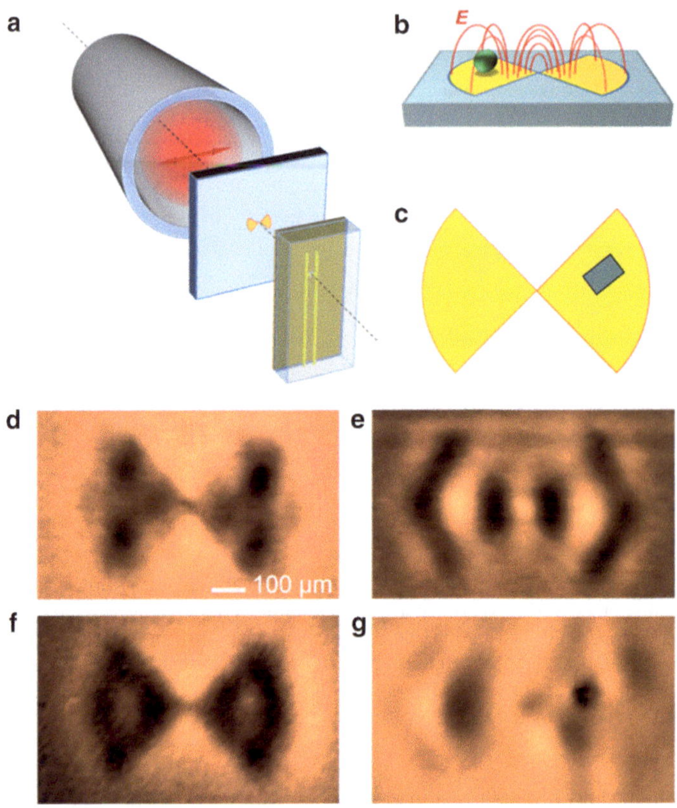

Fig. 7.4 (**a**) Schematic diagram of the experimental system for excitation and detection of surface waves on a bow-tie antenna, and (**b**) schematic representation of the surface wave field. (**c**) A map of the bow-tie antennas sample with a rectangular particle placed on the antenna surface. (**d–g**) THz near field images of the bow-tie without (**d, e**) and with (**f, g**) the particle [1, 17]

undetected phenomenon, a surface wave formed on a metallic surface by a tightly focused THz beam, was observed using this probe [10].

The introduction of the integrated sub-wavelength aperture probe near the sample surface is equivalent to the introduction of a metallic plane parallel to the surface of the sample. Although the distance between the sample surface and the surface of the probe is only several microns, much smaller than the wavelength, surface waves are nevertheless excited at the edges of the metallic pattern. Numerical analysis of surface waves on a bow-tie antenna showed that the electric distribution (field patterns) on the antenna is not changed significantly by the metallic surface of the probe. However the amplitude of the field can be increased considerably making the system of the antenna and the near-field probe a possible configuration for probing sub-wavelength objects.

Figure 7.4 shows the distribution of the surface electric field detected on the surface of a bow-tie antenna at two moments, at the beginning and the end of the

THz pulse. These field distributions illustrate the surface wave behaviour. As the incident pulse, which uniformly illuminates the antenna, excites surface waves at the antenna edges, an interference pattern is formed on the bow-tie. As time progresses the interference pattern becomes steady corresponding to a standing wave, i.e. one of the fundamental modes of the bowtie. The second row of images shows the field distribution for the same antenna with a sub-wavelength particle placed on the right hand side of the antenna. At the beginning of the THz pulse the electric field pattern is hardly perturbed by the object because the surface waves have not reached the object, which is positioned about 50 μm from the nearest edge. At the end of the pulse, however the surface wave is strongly affected by the object. First of all the pattern of the antenna mode is destroyed and, secondly, the object perturbs the field in its vicinity. It is interesting to note that there is a standing wave inside the particle due to the large value of its dielectric constant.

The experimental configuration in Fig. 7.4a, b allows probing single dielectric particles of sub-wavelength size at THz frequencies. A dielectric sphere only 30 μm in diameter was detected in a similar experimental configuration [17].

To perform quantitative analysis or to improve sensitivity further different metallic patterns can be explored, for example patterns with a high Q-factor or, alternatively, patterns that exhibit no resonant behavior, such as a tapered metallic strip.

7.6 Sub-wavelength Confinement Using THz Surface Waves

To enable efficient interaction of THz waves with small objects, the wave must be confined to a volume similar to the size of the object. Figure 7.5 shows a twin-needle probe configuration, designed to couple an incident THz pulse to surface waves, which are guided between the needles toward the needle tips. If a sub-wavelength object is placed in the region between the needles, where the field is the strongest, the interaction of the wave with the object is increased and the signal strength is described by Eq. 7.2. Since the field is distributed in a region smaller than the wavelength, the confined wave is evanescent and it remains near the tips. To detect this field, the integrated near-field THz probe must be placed near the tips (Fig. 7.5a).

The experimental results in Fig. 7.5 show that the THz wave can be confined between two isolated metallic surfaces to a volume significantly smaller than the wavelength by means of surface plasmon waves [5]. The twin-needle method avoids the frequency cut-off effect while achieving a high degree of confinement, required for single micron- and nano-scale objects. Unlike resonant structures, which increase the interaction between the wave and the object within narrow frequency bands, this approach is not frequency-selective and it will allow accessing the full spectroscopic signature of the object.

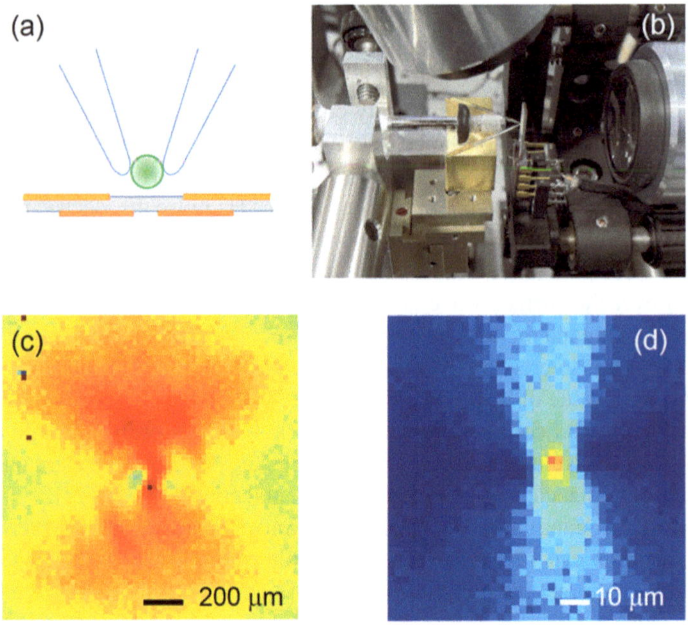

Fig. 7.5 Confinement of the THz pulses to a region of ~5–10 μm. using a twin-needle probe [5]. A schematic diagram (**a**) and photograph (**b**) of the detection configuration showing the twin needle probe and the sub-wavelength aperture near-field detector (the *green circle* indicates a particle that can be placed between the needle tips for THz analysis. Experimental maps of the electric field distribution in the plane just under the needles showing the surface waves that emanate from the tips (**c**) and the field concentration between the tips (**d**)

7.7 Conclusions

Spectroscopic analysis of single sub-wavelength size objects at THz frequencies is restricted by weak coupling of free-space propagating THz waves to the object. The interaction however can be enhanced substantially with help of THz surface waves on metallic surfaces. Excitation of the THz surface waves occurs at the edges of metallic patterns illuminated by free-space THz waves. The edge excitation is broadband and therefore it is ideal for THz timed-domain spectroscopy employing broadband THz pulses. Detection of the surface waves has been demonstrated with the integrated aperture probe. It provides local broadband (0.2–3 THz) THz spectroscopic analysis that can be used for probing small quantities of substances or for high spatial resolution imaging. Two experimental configurations described here can be applied for analysis of illicit substances in trace amounts (security and forensic applications) and for scientific investigations of small-scale systems. THz surface waves and near-field probe technology, including other near-field probes, provides a step toward enabling THz spectroscopy on the sub-wavelength scale.

Acknowledgements I gratefully acknowledge the contribution of many researchers with whom I have been working over the years on the development and applications of THz near-field microscopy and near-field spectroscopy, including R. Mueckstein, M. Navarro-Cia, C. Graham, M. Natrella, C.C. Renaud, A.J. Seeds, J.A. Harrington, Z. Jiang, W. Pan, I. Brener, J. Reno, P. Kuzel, P. Mounaix and J-C. Delagnes. I also would like to acknowledge the support of the Royal Society and the EPSRC.

References

1. Mueckstein R, Graham C, Renaud CC, Seeds AJ, Harrington JA, Mitrofanov O (2011) Imaging and analysis of THz surface plasmon polariton waves with the integrated sub-wavelength aperture probe. J Infrared Millim Terahertz Waves 32:1031
2. Natrella M, Mitrofanov O, Mueckstein R, Graham C, Renaud CC, Seeds AJ (2012) Modelling of surface waves on a THz antenna detected by a near-field probe. Opt Express 20:16023
3. Mueckstein R, Mitrofanov O (2011) Imaging of terahertz surface plasmon waves excited on a gold surface by a focused beam. Opt Express 19:3212–3217
4. Misra M, Andrews SR, Maier SA (2012) Waveguide artefacts in THz near field imaging. Appl Phys Lett 100:191109
5. Mitrofanov O, Renaud CC, Seeds AJ (2012) Terahertz probe for spectroscopy of sub-wavelength objects. Opt Express 20:6197
6. Bethe HA (1944) Theory of diffraction by small holes. Phys Rev 66:163; Bouwkamp CJ (1950) On Bethe's theory of diffraction by small holes. Philips Res Rep 5:321
7. Mitrofanov O, Lee M, Hsu JWP, Pfeiffer LN, West KW, Wynn JD, Federici JF (2001) Terahertz pulse propagation through small apertures. Appl Phys Lett 79:907
8. Mitrofanov O, Lee M, Hsu JWP, Brener I, Harel R, Federici JF, Wynn JD, Pfeiffer LN, West KW (2001) Collection-mode near-field imaging with 0.5-THz pulses. IEEE J Sel Top Quantum Electron 7:600
9. Mitrofanov O, Harel R, Lee M, Pfeiffer LN, West KW, Wynn JD, Federici J (2001) Study of single-cycle pulse propagation inside a terahertz near-field probe. Appl Phys Lett 78:252
10. Mitrofanov O, Pfeiffer LN, West KW (2002) Generation of low-frequency components due to phase-amplitude modulation of subcycle far-infrared pulses in near-field diffraction. Appl Phys Lett 81:1579
11. Misra M, Pan Y, Williams CR, Maier SA, Andrews SR (2013) Characterization of a hollow core fibre-coupled near field terahertz probe. J Appl Phys 113:193104
12. Mitrofanov O, Mueckstein R (2011) Near-field imaging of terahertz plasmon waves with a subwavelength aperture probe. Proc SPIE 8096:809605
13. Laman N, SreeHarsha S, Grischkowsky D, Melinger JS (2008) 7 GHz resolution waveguide THz spectroscopy of explosives related solids showing new features. Opt Express 16:4094
14. Zhan H, Mendis R, Mittleman DM (2010) Superfocusing terahertz waves below λ/250 using plasmonic parallel-plate waveguides. Opt Express 18:9643
15. Schnell M, Alonso-González P, Arzubiaga L, Casanova F, Hueso LE, Chuvilin A, Hillenbrand R (2011) Nanofocusing of mid-infrared energy with tapered transmission lines. Nat Photonics 5:283
16. Grober RD, Schoelkopf RJ, Prober DE (1997) Optical antenna: towards a unity efficiency near-field optical probe. Appl Phys Lett 70:1354–1356
17. Navarro-Cía M, Natrella M, Dominec F, Delagnes J-C, Kužel P, Mounaix P, Graham C, Renaud CC, Seeds AJ, Mitrofanov O (2013) Terahertz imaging of sub-wavelength particles with Zenneck surface waves. Appl Phys Lett 103:221103

Chapter 8
Quartz Enhanced Photoacoustic Sensors for Trace Gas Detection in the IR and THz Spectral Range

Pietro Patimisco, Simone Borri, Angelo Sampaolo, Miriam S. Vitiello, Gaetano Scamarcio, and Vincenzo Spagnolo

Abstract Quartz-enhanced photo-acoustic spectroscopy (QEPAS) is one of the most robust and sensitive trace-gas detection techniques, which in the mid-IR range offers the advantage of high sensitivity, compactness and fast time-response. One of the main features of the photoacoustic techniques is that no optical detection is required. Thus, the use of the QEPAS technique in THz range would allow to avoid the use of low-noise but expensive, bulky and cryogenic bolometers. The results obtained in the development of QEPAS sensors for trace gas detection of several chemical species, employing mid-IR and THz laser sources will be reviewed. Normalized noise equivalent absorption coefficients (NNEA) down to 10^{-10} cm^{-1} $W/Hz^{½}$ and part per trillion concentration detection ranges have been attained.

P. Patimisco • G. Scamarcio • V. Spagnolo (✉)
Dipartimento Interateneo di Fisica, Università e Politecnico di Bari, Via Amendola 173, I-70126 Bari, Italy

IFN-CNR UOS Bari, via Amendola 173, 70126 Bari, Italy
e-mail: v.spagnolo@poliba.it

S. Borri
IFN-CNR UOS Bari, via Amendola 173, 70126 Bari, Italy

A. Sampaolo
Dipartimento Interateneo di Fisica, Università e Politecnico di Bari, Via Amendola 173, 70126 Bari, Italy

IFN-CNR UOS Bari, via Amendola 173, 70126 Bari, Italy

M.S. Vitiello
NEST, CNR-Istituto Nanoscienze and Scuola Normale Superiore, Piazza San Silvestro 12, I-56127 Pisa, Italy

C. Corsi and F. Sizov (eds.), *THz and Security Applications*, NATO Science for Peace and Security Series B: Physics and Biophysics, DOI 10.1007/978-94-017-8828-1_8,

8.1 Introduction

The development of compact optical sensors for trace gas sensing is of interest for a number of applications, including atmospheric chemistry, volcanic activity, industrial processes, and medical diagnostics [20]. Conventional photoacoustic spectroscopy (PAS) is a well-established trace gas detection method based on the photoacoustic effect [6]. In this method, light energy absorbed by a gas is converted to heat that generally occurs via molecular collision-induced non-radiative relaxation of excited states. This absorption causes the gas to expand. If the light is chopped or modulated, the expansion produces pressure waves that can be detected using a microphone. The common approach used to detect the acoustic signal utilizes an acoustic resonator filled with the gas. In this way the absorbed laser power is accumulated in the acoustic mode of the resonator. In order to obtain an optimal acoustic signal, the laser modulation frequency is typically selected to match the first acoustic resonance of the cell, given by the equation $f = v/2\,L$, where v is the speed of the sound and L is the length of the cell. Most often the Q factor is in the range 40–200 and $f = 1{,}000$–$4{,}000$ Hz. Ideally, PAS is a background-free technique because only the absorption of modulated laser radiation generates an acoustic signal. However, background signals can originate from nonselective absorption of the gas cell windows and external acoustic noise, which can be reduced by means of a proper isolation of the cell from any mechanical vibrations. Quartz-enhanced photoacoustic spectroscopy (QEPAS) sensors are based on a recent approach to photoacoustic detection which employs a quartz tuning fork as an acoustic transducer [6, 8]. In this way, the common approach is inverted: the absorbed energy is not accumulated in the gas but in the sensitive element, i.e. the tuning fork. These quartz tuning forks have recently become widely used for atomic force and optical near field microscopy and therefore their properties have been carefully analyzed. A standard tuning fork (S-QTF) used in electronic watches for timepiece applications has a resonant frequency close to $f_0 = 32{,}768$ (i.e. 2^{15}) Hz. The mode at this frequency corresponds to a symmetric vibration mode where the prongs move in opposite directions. S-QTF has typically $Q \approx 100{,}000$ or higher when it is encapsulated in vacuum and a $Q \approx 8{,}000$ at atmospheric pressure. At this resonant frequency, QEPAS devices result in a high immunity of to background acoustic noise as a consequence of the following behavior and the ambient acoustic noise density approximately follows a $1/f$ dependence, so it is very low above 10 KHz. In addition, sound waves emanating from a distant source tend to apply a force in the same direction upon the two tuning fork prongs, not exciting the piezoelectrically active mode in which the two prongs move in opposite directions.

8.2 Mid-IR QEPAS

To detect a trace gas, modulated laser radiation is directed between the tines of the tuning fork and absorbed by the gas which gives rise to an acoustic pressure. This pressure wave excites a resonant vibration of the tuning fork thereby generating an electrical signal via the piezoelectric effect, thanks to thin gold or silver films deposited on the quartz surfaces. The electrical signal can be measured as either a voltage or a current, depending on the electronic circuit used. The majority of reported QEAPS sensors include a spectrophone (the module for detecting laser-induced sound) consisting of a S-QTF and a microresonator composed of a pair of thin metal tubes [4, 5, 9–12, 14]. It was shown that the microresonator yields a signal gain of at least one order of magnitude. The sensor including optical pathlength is very small (a few mm in length) and can be located within a larger sample volume.

The generation of a photoacoustic wave involves the energy transfer from internal to translational molecular degrees of freedom. If a ro-vibrational state is excited, a collision-induced vibrational to translational (V-T) or rotational to translational (R-T) relaxation are followed. The V-T or R-T energy transfer times for a particular molecule is dependent on the presence of other molecules and intermolecular interactions. The QEPAS measurements that are usually performed at a detection frequency of ~32 kHz are more sensitive to the energy relaxation rate compared to the conventional PAS (which is commonly performed at <4 kHz frequency). In case of slow V-T or R-T relaxation with respect to the modulation frequency, the translational gas temperature cannot follow fast changes of the laser induced molecular vibrational excitation. Thus the generated photoacoustic wave is weaker. On the other hand, in case of fast energy relaxation processes, the amplitude of the photoacoustic signal is not affected by the relaxation time. Thus for fast-relaxing excited energy levels the photoacoustic signal S is given by:

$$S_0 \propto \frac{F\alpha PQ}{f_0} \tag{8.1}$$

where α is the absorption coefficient, P is the exciting radiation power, and F is a micro-resonator enhancement factor included in this equation to account for a particular QEPAS sensor design.

In addition, the Q factor depends on the pressure P and can be expressed as:

$$Q = \frac{Q_{vac}}{1 + Q_{vac}a\sqrt{P}} \tag{8.2}$$

where Q_{vac} is the quality factor in vacuum and a is a parameter dependent on a specific quartz tuning fork design. From Eq. 8.2, it is clear that the Q-factor decreases at higher pressure but, on the other hand, the energy transfer in energy relaxation processes is faster at higher pressure, resulting in more efficient sound excitation. In addition, merging of closely spaced absorption lines should be taken into account at higher pressures and broader linewidth than desirable can limit the gas spectral selectivity. So, the operating pressure is an important parameter to choose.

Due to this intrinsic coupling of strain and charge displacement, the tuning fork can be modeled both electrically and mechanically. Thus, the tuning fork can be described as a circuit with a capacitance C, a resistance R, and an inductance L.

A generally accepted way of observing the tuning fork electrical response is to utilize a transimpedance amplifier with a feedback resistance R_F. Feedback maintains a zero voltage between the tuning fork electrodes. Thus, the resonant frequency is related to the electrical parameters,

$$f_0 = \frac{1}{2\pi}\sqrt{\frac{1}{LC}} \tag{8.3}$$

and the Q-factor

$$Q = \frac{1}{R}\sqrt{\frac{L}{C}} \tag{8.4}$$

while the impedance at resonance was equal to its resistance.

It is verified in many experiments that the QEPAS spectrophone noise is primarily determined by the thermal noise of QTF. The QTF noise measured at the transimpedance amplifier output at the resonant frequency f_0 is equal to the thermal noise of the equivalent resistor R:

$$\frac{\sqrt{\langle v_N^2 \rangle}}{\sqrt{\Delta f}} = R_F\sqrt{\frac{4KT}{R}} \tag{8.5}$$

where $\sqrt{\langle v_N^2 \rangle}$ is the root-means-squared (rms) voltage noise at the transimpedance amplifier output, Δf is the detection bandwidth, K is the Boltzmann constant and T is the tuning fork temperature.

Different QEPAS configurations have been demonstrated, two of them exploit a longitudinal acoustic resonance in an organ pipe type micro resonators (mR). The first of them was originally proposed by Kosterev et al. [9] in which the light come through the QTF and the two co-axial adjacent resonator tubes and is used in most of the reported QEPAS sensors. The second was reported by K. Liu et al. [13] and called "Off-Beam (OB) QEPAS". In this second configuration the mR is a single tube with a small opening in the middle, and a QTF is coupled to the mR by putting it outside the mR tube near the opening. Very recently, Y. Cao et al. [2], demonstrate evanescent-wave PAS, which uses tapered optical micro/nano fibers (OMNFs) for photoacoustic signal generation. An OMNF is threaded through the gap between the two prongs of a QTF. Light is guided along the OMNF with a very small beam size, and no precise optical alignment is needed. Modulated light from a single longitudinal mode laser is transmitted to the OMNF, and the evanescent field is absorbed by the target gas, generating an acoustic pressure wave that is detected by the QTF. Finally, an innovative spectroscopic technique, the modulation cancellation method (MOCAM) that relies on the use of two QEPAS sensors have been demonstrated for isotope, gas temperature and concentration measurements [22–24].

Table 8.1 QEPAS detection of trace gases

Molecule (host)	Frequency cm^{-1}	Pressure, Torr	NNEA, cm^{-1} $W/Hz^{1/2}$	Power, mW	NEC, ppmv
H_2O (N_2)[a]	7,306.75	60	$1.9 \cdot 10^{-9}$	9.5	0.09
HCN (air:50%RH)[b]	6,539.11	60	$4.6 \cdot 10^{-9}$	50	0.16
C_2H_2 (N_2)[b]	6,523.88	720	$4.1 \cdot 10^{-9}$	57	0.03
NH_3 (N_2)[b]	6,528.76	575	$3.1 \cdot 10^{-9}$	60	0.06
C_2H_4 (N_2)[b]	6,177.07	715	$5.4 \cdot 10^{-9}$	15	1.7
CH_4 (N_2 + 1.2 % H_2O)[b]	6,057.09	760	$3.7 \cdot 10^{-9}$	16	0.24
N_2H_4	6,470.00	700	$4.1 \cdot 10^{-9}$	16	1
H_2S (N_2)[b]	6,357.63	780	$5.6 \cdot 10^{-9}$	45	5
HCl (N_2 dry)	5,739.26	760	$5.2 \cdot 10^{-8}$	15	0.7
CO_2 (N_2 + 1.5 % H_2O)[b]	4,991.26	50	$1.4 \cdot 10^{-8}$	4.4	18
CH_2O (N_2:75 % RH)[b]	2,804.90	75	$8.7 \cdot 10^{-9}$	7.2	0.12
CO (N_2 + 2.2 % H_2O)	2,176.28	100	$1.4 \cdot 10^{-7}$	71	0.002
CO (propylene)	2,196.66	50	$7.4 \cdot 10^{-8}$	6.5	0.14
N_2O (air + 5 % SF_6)	2,195.63	50	$1.5 \cdot 10^{-8}$	19	0.007
NO (N_2 + H_2O)	1,900.07	250	$7.5 \cdot 10^{-9}$	100	0.003
C_2H_5OH (N_2)[a]	1,934.20	770	$2.2 \cdot 10^{-7}$	10	90
C_2HF_5 (N_2)[c]	1,208.62	770	$7.8 \cdot 10^{-9}$	6.6	0.009
NH_3 (N_2)[b]	1,046.39	110	$1.6 \cdot 10^{-8}$	20	0.006
SF_6	948.62	75	$2.7 \cdot 10^{-10}$	18	$5 \cdot 10^{-5}$

NNEA normalized noise equivalent absorption coefficient, *NEC* noise equivalent concentration for available laser power and 1 s constant time, 18 dB/oct filter slope
[a]Improved microresonator and double pass through ADM
[b]Improved microresonator
[c]With amplitude modulation and metal microresonator

Several chemical species, such as NH_3, NO, CO_2, N_2O, CO, CH_2O, etc. have been detected with QEPAS sensors, using laser emitting in the 1–10 μm spectral range [3–5, 10–12, 14, 18, 19, 21, 23–26] and the results reported so far are summarized in Table 8.1.

The normalized noise equivalent absorption coefficient measured to date using QEPAS are better than the best conventional PAS results and experimental study of the long-term stability of a QEPAS sensors showed that the sensors exhibit low drift, which allows long term data averaging thus allowing significant improvement of the signal-to noise ratio in concentration measurements. In mid-IR range (4–12 μm), continuous wave quantum cascade laser (QCL) sources capable of thermo-electrically cooled, room-temperature operation, single mode emission with mode-hop-free frequency tuning, high power (tens to hundreds of mW), and intrinsic narrow emission linewidth are the best performing light sources and are commercially available. In combination with these laser sources, QEPAS offers the advantage of high sensitivity (part-per-billion (ppb) detection limits), large dynamic range, compact set-up, fast time-response and simple optical alignment. Very recently, a record detection limit of 50 part per trillion (ppt) has been reported for SF_6, corresponding to a normalized noise equivalent absorption (NNEA) value of $2.7 \cdot 10^{-10}$ W $\cdot$ cm$^{-1} \cdot$ Hz$^{-1/2}$, employing an external cavity QCL and a single mode mid-IR hollow-core waveguide [15, 16, 24–26] with internal core diameter of 300 μm for guiding the laser beam.

8.3 THz QEPAS

QEPAS technique is an excellent candidate for high performance gas spectroscopic techniques at long wavelength, i.e. in the THz range, mainly for two reasons: (i) no optical detection is required, allowing to avoid the use of expensive, bulky and cryogenic bolometers; (ii) QEPAS signal strongly depends from the energy relaxation rates of the absorption gas species. In gas absorption processes that occur in the THz range, R-T relaxation rates are involved, and it was demonstrated that R-T relaxation rates are up to three order of magnitude faster with respect to V-T one [7] commonly involved in the mid-IR absorption. Thus the possibility to work with fast relaxing transition levels allows to operate at low pressure, so taking advantages of the very high QTF Q-factors (see Eq. 8.2) and enhancing the selectivity of the QEPAS sensor systems. In the last few years, the rapid growth of THz QCLs is stimulating a resurgence in THz spectroscopic application. In fact, THz QCL can provide single-mode emission with considerable frequency tunability and with output powers up to 100 mW in continuous wave operation at cryogenic temperatures. In addition, THz QCLs have recently demonstrated interesting performances in high resolution molecular spectroscopy, mainly thanks to their frequency stability. Despite this, the extension of the QEPAS technique to the THz spectral region has been delayed, mainly due to the difficulty of a proper focalization of the THz laser beam between the prongs (300 μm distance for S. QTF). Indeed, one of the main criticism in QEPAS experiments is represented by the laser beam waist between the prongs of the fork. The laser beam has not to touch the prongs, otherwise an undesirable non-zero background due to the laser contribution arises. This background is often several times larger than the thermal noise level of QEPAS sensor, limiting the detection sensitivity [5, 21]. For this reason we adopted a custom-made quartz tuning fork (C-QTF), about six times bigger than the commercial ones [17]. In Fig. 8.1 a schematic of the S-QTF and C-QTF is reported:

The prong spacing is ∼800 μm. The single prong is $L = 17.7$ cm long and 1.4 cm wide, with a thickness of 0.8 mm. In order to employ C-QTF in a QEPAS sensing system we have to verify that it behaves like a S-QTF in terms of resonance frequency and quality factor. The in-plane resonance frequencies of the tuning fork can be found analytically by considering one prong of the fork as a cantilever beam. In the fundamental mode of oscillation, the tines move in opposite directions and the center-of-mass of the fork remains unchanged. The frequency of the vibration modes of a single beam are obtained by solving the classical

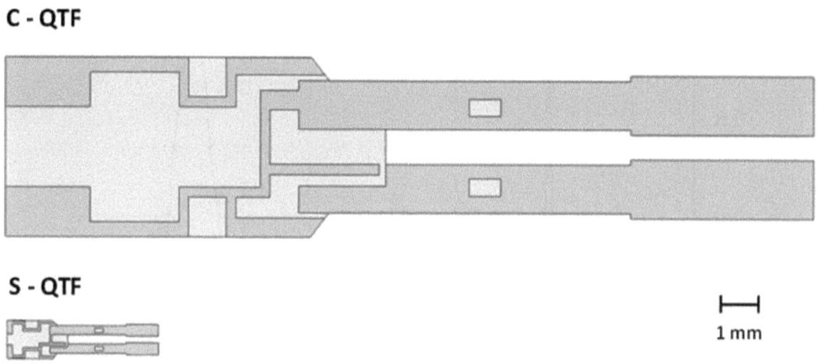

Fig. 8.1 Schematic of Custom quartz tuning fork (*C-QTF*) and standard quartz tuning fork (*S-QTF*)

Table 8.2 n values and the resonant frequencies for the S-QTF and the C-QTF fork employed in this work, calculated from Eq. 8.6

n	f_n (Hz) – S-QTF	f_n (Hz) – C-QTF
1.194	31,978	4,118
2.988	200,263	25,786
5	560,764	72,204
7	1,099,097	141,520

Euler-Bernoulli equation, including clamped-free boundary condition. Thus the resonance frequencies can be predicted by the following equation:

$$f_n = \frac{\pi G}{8L^2}\sqrt{\frac{E}{\rho}}n^2 \tag{8.6}$$

where G is the radius of gyration of the beam cross section that for a bar with rectangular cross section is equal to $1/\sqrt{12}$ times the thickness, E is the elastic Young modulus and ρ is the density of the quartz. The first four solutions of Eq. 8.6 ($E = 0.72\ 10^{11}$ N/m^2 and $\rho = 2{,}650$ kg/m^3) for C-QTF and S-QTF (3.2 mm long and 0.33 mm wide) are shown in Table 8.2.

A control electronics unit (CEU) is used to determine the equivalent electrical parameters of the quartz tuning fork and to estimate the quality factor Q, the resonant frequencies f_n and the resistance of the C-QTF. Two resonance frequencies at $f_{n=1} = 4.245$ kHz and $f_{n=3} = 25.4$ kHz have been observed at atmospheric pressure in pure N_2, as reported in Fig. 8.2. These two resonant frequencies can be easily ascribed to the first and the third in-plane resonance (see Table 8.1). The discrepancies from the theoretical values, calculated under vacuum conditions, are due to damping effects of the ambient gas, additional weight of the electrode gold layers, dependence of the elasticity modulus of quartz on the crystallographic axes orientation and deviations in geometry between the modeled and the real C-QTF.

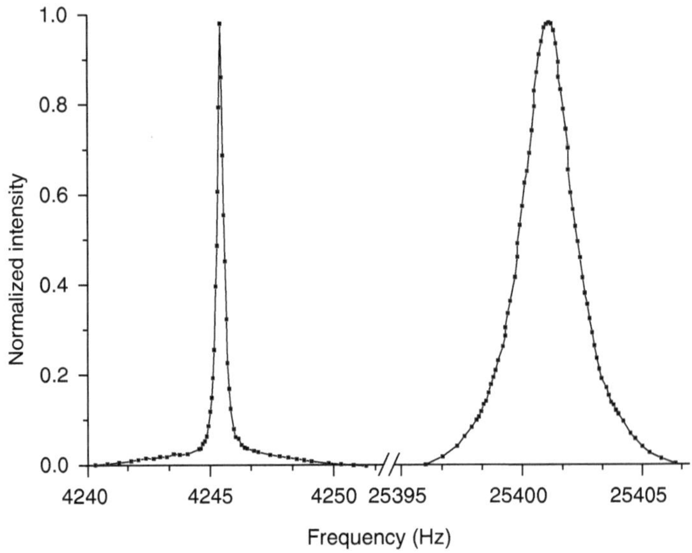

Fig. 8.2 Frequency profiles measured at atmospheric pressure in pure N_2

To study the damping effects induced by the environmental gas on the quality factor, we measured the resonant frequency f_1 and f_3 and the respective quality factor Q_1 and Q_3as a function of the N_2 pressure. In Fig. 8.3 are shown the experimental results of f_1 (f_3) versus the gas pressure is linear in the whole investigated pressure range 10–700 Torr, as theoretically predicted [17], with a slope of $-1.19 \cdot 10^{-3}$ Hz/Torr ($-4.49\ 10^{-3}$ Hz/Torr) and the intercept value is 4,246.3 Hz (25,404.5 Hz), which gives the resonant frequency in the vacuum. The Q_1 (Q_3) factor pressure dependence shows an exponential behavior, as predicted in Eq. 8.2, and rapidly decreases with the gas pressure. The best fit parameters obtained using Eq. 8.2 are: $a = 1.98 \cdot 10^{-6}$ Torr^{-1} and $Q_0 = 146{,}350$ for Q_1, and $a = 8.73\ 10^{-7}$ Torr^{-1} and $Q_0 = 13{,}180$ for Q_3.

8.4 THz QCL-Based QEPAS Sensor

A picture of the THz QCL-based QEPAS sensor is shown in Fig. 8.4. A THz QCL is used as the excitation source for generating the QEPAS signal. It is a single-mode bound-to-continuum QCL emitting at 3.93 THz (76.3 μm), driven in CW mode and mounted on the cold finger of a continuous-flow cryostat equipped with polymethylpentene (TPX) windows. 90° off-axis paraboloidal gold mirrors are used to collimate/focalize the THz beam. At 6° K, the laser optical frequency can be scanned over a 0.025 cm^{-1} by applying a low-frequency (10 mHz) voltage ramp up to 1 V to the external analog modulation input of the QCL current supply,

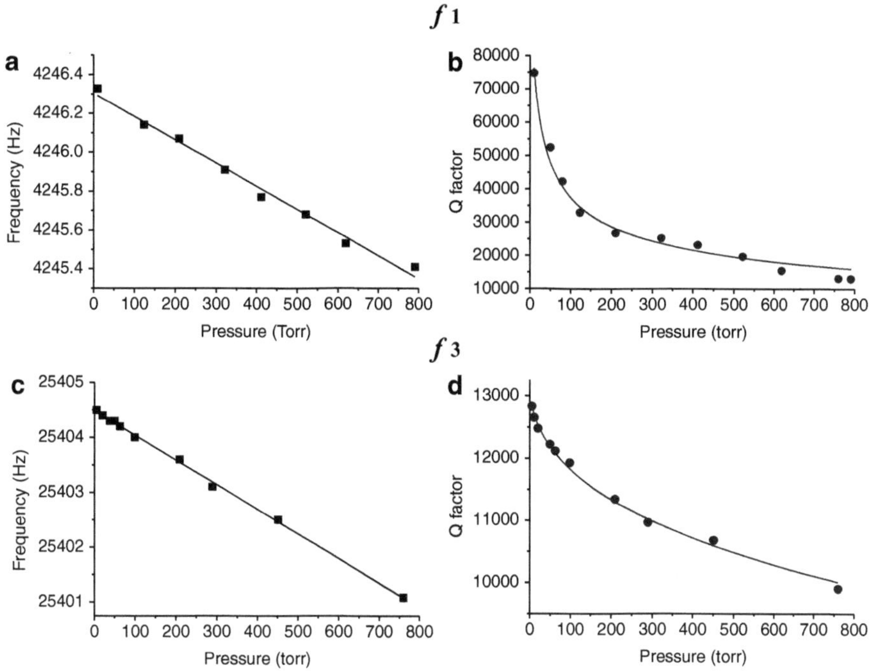

Fig. 8.3 QTF resonance frequency f_1 measured as a function of the N_2 gas pressure (**a**) f_3 measured as a function of the N_2 gas pressure (**c**) solid lines are linear fit to the data. Q_1 factor measured as a function of the N_2 gas pressure (**b**) Q_3 factor measured as a function of the N_2 gas pressure (**d**) solid lines are the best fits obtained by using Eq. 8.2

with laser output powers up to 180 μW. A sinusoidal dither at frequency $f_1$or f_3 is contemporarily added to the low-frequency voltage ramp, to obtain up to 0.01 cm^{-1}optical frequency modulation. The signal generated by the quartz tuning fork is amplified by a custom designed transimpedance amplifier ($R_f = 10$ MΩ). Subsequently the signal is demodulated by a lock-in amplifier (Stanford Research Model SR830) and digitalized by a USB data acquisition card (National Instruments DAQ-Card USB6008), which is connected to a personal computer. By means of a pyroelectric camera (mod. Spiricon Pyrocam III-C), we measured a focused beam waist of ~430 μm after the second parabolic mirror (see Fig. 8.4) well below the gap between the QTF prongs (~800 μm). As a result, in the THz QEPAS experiments almost all of the laser beams were transmitted through the C-QTF without touching it.

We selected methanol as the target gas molecule. Methanol is widely used as a solvent, detergent or even denaturant additive for industrial ethanol. The selected methanol absorption line was the (v = 1, K = 6, J = 11) → (1, 5, 10) rotational translational transition, falling at $\nu_{line} = 3.9289$ THz (131.054 cm^{-1}) with line-strength $S = 4.28 \cdot 10^{-21}$ cm/mol in HITRAN units. Gas mixtures with different

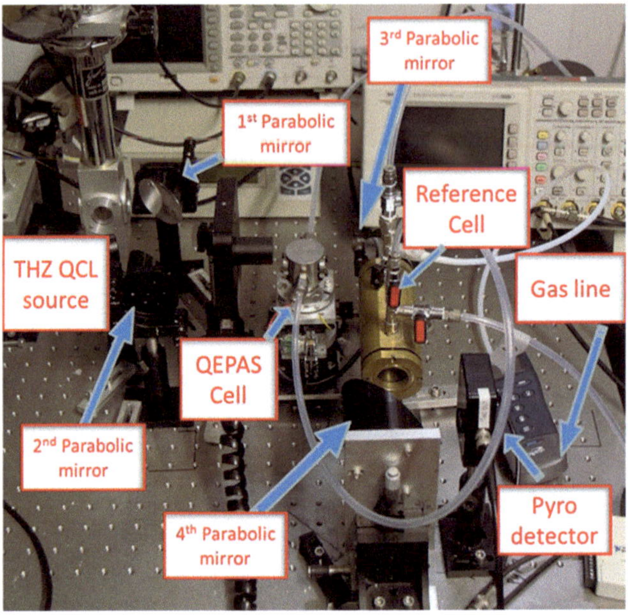

Fig. 8.4 THz QCL-based sensor

methanol concentrations in pure N_2 have been obtained by diluting methanol vapors, collected from a reservoir held at the vapor pressure (P = 120 Torr at 300 K), with pressurized N_2. Instead, for measurements at low concentrations, we used a certified 100 ppm methanol/N_2 gas mixture.

To find the optimal operating conditions in terms of QEPAS signal-to-noise ratio, we investigated the effects of gas pressure and modulation amplitude. The optimal sensor operating conditions were found to occur by using the first resonant frequency f_1 of the C-QTF (see Fig. 8.2), and at a gas pressure of 10 Torr and a modulation amplitude of 600 mV [17]. Under this operating conditions, the linearity and detection sensitivity of the methanol channel were evaluated by measuring its response to varying methanol concentration in pure N_2. In Fig. 8.5 is shown a spectral scan of the certified 100 ppm methanol sample (the lowest concentration tested) in N_2 at 10 Torr.

The data in Fig. 8.5 are collected by setting the lock-in amplifier time constant to 3 s and the filter slope to 12 dB/oct, corresponding to an equivalent noise related bandwidth of $\Delta f_{\text{lock-in}} = 0.05558$ Hz. It was verified that with such integration parameters the observed spectral lines were not distorted by the lock-in detection. The total fundamental noise also includes the feedback resistor noise and the operational amplifier noise, integrated over the full lock-in detector bandwidth. However, the power density of these noise sources is usually low and therefore expanding the detection bandwidth beyond the tuning fork response does not significantly increase the noise level. The extracted tuning fork thermal noise, by

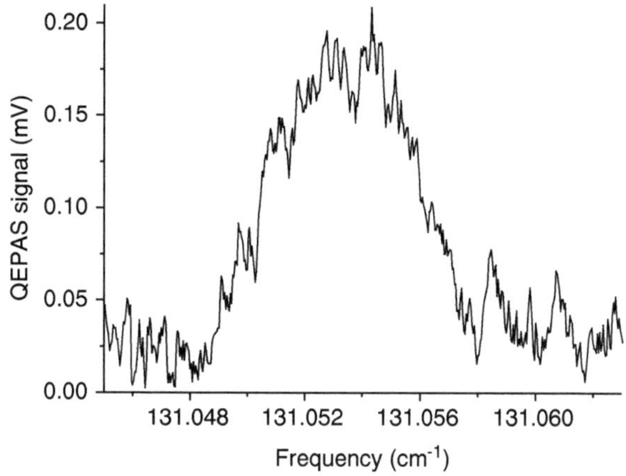

Fig. 8.5 QEPAS acquisition of a methanol/N2 sample with a certified concentration of 100 ppm and lock-in integration time of 3 s

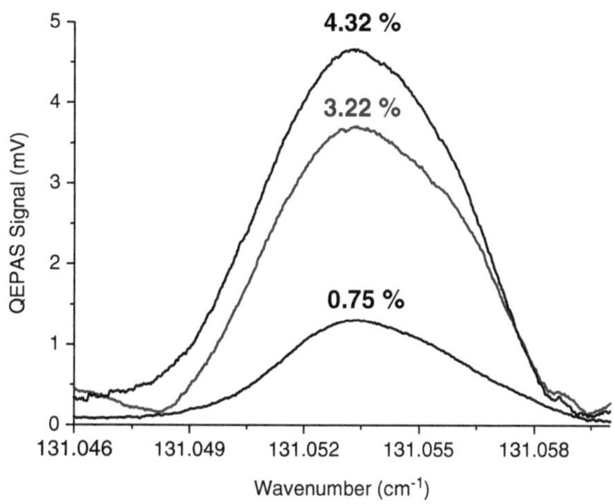

Fig. 8.6 Spectral scan of three representative methanol concentrations at 0.75, 3.22 and 4.32 % at 10 Torr using N_2 as the diluting gas, obtained by a current modulation of f_1, a peak-to-peak voltage amplitude of 600 mV and a lock-in integration time of 0.5 s. The background signal has been removed

using Eq. 8.5, is ~0.12 μV (R corresponding at f_1 is 6.5 MΩ). For 100 ppm methanol concentration the experimentally measured noise level (rms of short term point-to-point scatter) was in the ±25 μV range, several times larger than the thermal noise level. This can be mainly ascribed to the QCL power fluctuations. The QEPAS peak signal for 100 ppm methanol concentration results ~170 μV, thus the corresponding Noise Equivalent Concentration (NEC) is ~15 ppm.

High-resolution QEPAS scans of a methanol/N_2 calibrated mixtures with different methanol concentrations with lock-in integration time of 500 ms are shown in Fig. 8.6.

We found a clear not-zero background signal: we verified that it is stable over at least 1 h and this permit an efficient background subtraction by post processing of the spectra, thus improving the resolution, but also increasing the overall measuring time.

To verify the linearity of the THz QEPAS signal as a function of the methanol concentration, the sensor was operated in the locked mode, i.e. with the THz QCL frequency set to the center of the absorption line. The experimental data are easily linearly fitted, confirming the linearity of the system response to the concentration [1, 17]. In order to characterize long-term drifts and establish signal averaging limits, we performed an Allan analysis in terms of the QEPAS signal at zero methanol concentration (pure N_2) [1]. For this analysis, the laser frequency was locked to the methanol absorption line at 131.054 cm^{-1}, and pure carrier gas N_2 was introduced into the housing where the C-QTF is placed. For 4 s averaging time (and related bandwidth of 0.04169 Hz) we achieve a detection sensitivity of 7 ppm, corresponding to a normalized noise equivalent absorption coefficient (NNEA) of $2.7 \cdot 10^{-10}$ cm^{-1} W/Hz.

8.5 Conclusions

In this work we have reported an overview on the results obtained using QEPAS technique for trace gas sensing. Mid-IR quantum cascade lasers represent the ideal radiation sources for QEPAS gas sensing thanks to their excellent spectroscopic and technical properties, i.e., narrow linewidth, tunability, reliability and room-temperature operation. We described also the first THz QEPAS sensor employing a THz quantum cascade laser operating in continuous-wave and a custom-made quartz tuning fork. The normalized noise equivalent absorption coefficient obtained is comparable with the best result obtained in the mid-IR and in strong competition with the sensitivities achieved with the most performing cryogenic bolometers. THz QEPAS detection limit can be further improved by employing THz QCL with higher emission power (>100 mW has been already demonstrated [27]) and selecting molecules such as HF, H_2S, OH, NH_3, HCN, having absorption strengths larger than 10^{-19} cm/mol, potentially pushing down to the part per trillion concentration the QEPAS detectivity range.

References

1. Borri S, Patimisco P, Sampaolo A, Beere HE, Ritchie DA, Vitiello MS, Scamarcio G, Spagnolo V (2013) THz quartz enhanced photo-acoustic sensor. Appl Phys Lett 103:021105
2. Cao Y, Jin W, Ho LH, Liu Z (2012) Evanescent-wave photoacoustic spectroscopy with optical micro/nano fibers. Opt Lett 37:214–216
3. Curl RF, Capasso F, Gmachl C, Kosterev AA, McManus B, Lewicki R, Pusharsky M, Wysocki G, Tittel FK (2010) Quantum cascade lasers in chemical physics. Chem Phys Lett 487:1–18
4. Dong L, Lewicki R, Liu K, Buerki PR, Weida MJ, Tittel FK (2012) Ultra-sensitive carbon monoxide detection by using ECQCL based quartz-enhanced photoacoustic spectroscopy. Appl Phys B 107:275–283
5. Dong L, Spagnolo V, Lewicki R, Tittel FK (2011) Ppb-level detection of nitric oxide using an external cavity quantum cascade laser based QEPAS sensor. Opt Express 19:24037–24045

6. Elia A, Lugarà PM, Di Franco C, Spagnolo V (2009) Photoacoustic techniques for trace gas sensing based on semiconductor laser sources. Sensors 9:9616–9628
7. Flygare WH (1968) Molecular relaxation. Acc Chem Res 1:121–127
8. Kosterev AA, Bakhirkin YA, Curl RF, Tittel FK (2002) Quartz-enhanced photoacoustic spectroscopy. Opt Lett 27:1902–1904
9. Kosterev AA, Tittel FK, Serebryakov DV, Malinovsky AL, Morozov IV (2005) Applications of quartz tuning forks in spectroscopic gas sensing. Rev Sci Instrum 76:043105
10. Kosterev AA, Bakhirkin YA, Tittel FK (2005) Ultrasensitive gas detection by quartz-enhanced photoacoustic spectroscopy in the fundamental molecular absorption bands region. Appl Phys B 80:133–138
11. Kosterev AA, Buerki PR, Dong L, Reed M, Day T, Tittel FK (2010) QEPAS detector for rapid spectral measurements. Appl Phys B 100:173–180
12. Lewicki R, Wysocki G, Kosterev AA, Tittel FK (2007) QEPAS based detection of broadband absorbing molecules using a widely tunable, cw quantum cascade laser at 8.4 μm. Opt Express 15:7357–7366
13. Liu K, Guo XY, Yi HM, Chen WD, Zhang WJ, Gao XM (2009) Off-beam quartz-enhanced photoacoustic spectroscopy. Opt Lett 34:1594–1596
14. Liu K, Yi H, Kosterev AA, Chen WD, Dong L, Wang L, Tan T, Zhang WJ, Tittel FK, Gao XM (2010) Trace gas detection based on off-beam quartz enhanced photoacoustic spectroscopy: optimization and performance evaluation. Rev Sci Instrum 81:103103
15. Patimisco P, Spagnolo V, Vitiello MS, Tredicucci A, Scamarcio G, Bledt CM, Harrington JA (2012) Coupling external mid-IR quantum cascade lasers with low loss metallic/dielectric waveguides. Appl Phys B 108:255–260
16. Patimisco P, Spagnolo V, Vitiello MS, Scamarcio G, Bledt CM, Harrington JA (2013) Low-loss hollow waveguide fibers for mid-infrared quantum cascade lased sensing applications. Sensors 13:1329–1340
17. Patimisco P, Borri S, Sampaolo A, Beere HE, Ritchie DA, Vitiello MS, Scamarcio G, Spagnolo V (2013) Quartz enhanced photo-acoustic gas sensor based on custom tuning fork and terahertz quantum cascade laser. Analyst. First published online 04 Oct 2013. doi: 10.1039/c3an01219k
18. Phillips MC, Myers TL, Wojcik MD, Cannon BD (2007) External cavity quantum cascade laser for quartz tuning fork photoacoustic spectroscopy of broad absorption features. Opt Lett 32:1177–1179
19. Schilt S, Kosterev AA, Tittel FK (2009) Performance evaluation of a near infrared QEPAS based ethylene sensor. Appl Phys B 95:813–824
20. Sigrist W (2003) Trace gas monitoring by laser photoacoustic spectroscopy and related techniques (plenary). Rev Sci Instrum 71:486–490
21. Spagnolo V, Kosterev AA, Dong L, Lewicki R, Tittel FK (2010) NO trace gas sensor based on quartz-enhanced photoacoustic spectroscopy and external cavity quantum cascade laser. Appl Phys B 100:125–130
22. Spagnolo V, Dong L, Kosterev AA, Thomazy D, Doty JH, Tittel FK (2011) Modulation cancellation method for measurements of small temperature differences in a gas. Opt Lett 36:460–462
23. Spagnolo V, Dong L, Kosterev AA, Thomazy D, Doty JH, Tittel FK (2011) Modulation cancellation method in laser spectroscopy. Appl Phys B 103:735–742
24. Spagnolo V, Dong L, Kosterev AA, Tittel FK (2012) Modulation cancellation method for isotope $^{18}O/^{16}O$ ratio measurements in water. Opt Express 20:3401–3407
25. Spagnolo V, Patimisco P, Borri S, Scamarcio G, Bernacki BE, Kriesel J (2013) Part-per-trillion level SF6 detection using a quartz enhanced photo acoustic spectroscopy-based sensor with single-mode fiber-coupled quantum cascade laser excitation. Opt Lett 37:4461–4463
26. Spagnolo V, Patimisco P, Borri S, Scamarcio G, Bernacki BE, Kriesel J (2013) Mid-infrared fiber-coupled QCL-QEPAS sensor. Appl Phys B 112:25–33. doi:10.1007/s00340-013-5388-3
27. Williams BS, Kumar S, Hu Q, Reno JL (2006) High-power terahertz quantum-cascade lasers. Electron Lett 42:89–90

Chapter 9
THz Active Imaging Systems with Real-Time Capabilities

Fabian Friederich, Wolff von Spiegel, Maris Bauer, Fanzhen Meng, Mark D. Thomson, Sebastian Boppel, Alvydas Lisauskas, Bernd Hils, Viktor Krozer, Andreas Keil, Torsten Loffler, Ralf Henneberger, Anna Katharina Huhn, Gunnar Spickermann, Peter Haring Bolívar, and Hartmut G. Roskos

Abstract This paper presents a survey of the status of five active THz imaging modalities which we have developed and investigated during the last few years with the goal to explore their potential for real-time imaging. We start out by introducing a novel waveguide-based all-electronic imaging system which operates at 812 GHz. Its salient feature is a 32-pixel linear detector array heterodyne-operated at the 8th subharmonic. This array in combination with a telescope optics for object distances of 2–6 m reaches a data acquisition speed suited for real-time imaging. The second system described then is again an all-electronic scanner (now for around 300 GHz), designed for object distances of $\geq$8 m, which combines mechanical scanning in vertical direction, synthetic-aperture image generation in horizontal direction, and frequency-modulated continuous-wave sweeping for the depth information. The third and fourth systems follow an optoelectronic approach by relying on several- to multi-pixel parallel electro-optic detection. One imager is based on a pulsed THz-OPO and homodyne detection with a CCD camera, the other on either continuous-wave electronic or femtosecond optoelectronic THz sources and a photonic-mixing device (PMD) camera. The article concludes with a description of the state of the art of imaging with focal-plane arrays based on CMOS field-effect transistors.

F. Friederich • W. von Spiegel • M. Bauer • F. Meng • M.D. Thomson • S. Boppel • A. Lisauskas
B. Hils • V. Krozer • H.G. Roskos (✉)
Physikalische Institut, Johann Wolfgang Goethe-Universität, 60438 Frankfurt am Main, Germany
e-mail: roskos@physik.uni-frankfurt.de

A. Keil • T. Loffler
SynView GmbH, 61348 Bad Homburg, Germany

R. Henneberger
Radiometer Physics GmbH, 53340 Meckenheim, Germany

A.K. Huhn • G. Spickermann • P. Haring Bolívar
Institut für Höchstfrequenztechnik und Quantenelektronik, Universität Siegen,
57068 Siegen, Germany

C. Corsi and F. Sizov (eds.), *THz and Security Applications*, NATO Science for Peace and Security Series B: Physics and Biophysics, DOI 10.1007/978-94-017-8828-1_9,

9.1 Introduction

Imaging with electromagnetic radiation in the THz frequency regime (whose lower bounds we understand here to be 300 GHz) has made considerable progress in recent years. THz imaging and sensing is being investigated and found to be promising for a plethora of applications, both in science and beyond, such as security and safety screening, process monitoring and non-contact materials testing, biological, medical and pharmaceutical analysis, etc. [1–11].

For many application areas, it is indispensable to generate images with a high frame-rate. Imaging at video rates or close to it is certainly necessary for many tasks concerning industrial on-line process monitoring. Frame-rates on the level of at least one frame per second are required for security screening or in the medical field, if living people are scanned.

It appears fair to state that the route to active real-time imaging is via multi-pixel parallel detection of the radiation. Much work has been devoted in recent years to develop such parallel detection schemes. Progress hinges to a considerable degree on advances in the available power of emitters and the sensitivity of detectors, but not only on these. Also system integration, the optimization of the interplay of emitters, sources, and optics, plays an important role.

Outstanding examples of dedicated work toward high-frame-rate imaging include the development of all-electronic scanners for stand-off threat detection [12, 13]. But also the optoelectronic side of THz technology has seen remarkable progress, such as the realization of integrated line arrays of photoconductive and electro-optic (EO) detectors suitable for parallel read-out [14, 15]. It is hoped that stand-off imaging can be combined with stand-off spectroscopic analysis of remotely located objects [1, 16–18].

This paper describes and summarizes several approaches towards real-time imaging which we have investigated in recent years. This includes all-electronic imaging, hybrid imaging – with a microelectronic emitter and laser-based EO read-out, and all-optical imaging. We should state here, that we have focused only on actively illuminating imaging systems. We also have left approaches unconsidered which require cryogenic cooling of detectors (especially superconducting ones) and of emitters (such as THz quantum-cascade lasers), because we believe that there is only a limited acceptance of such cooling outside of laboratory environments.

9.2 Active Electronic Imaging Systems

9.2.1 Mechanically Scanning Imaging Systems

This section focuses on bringing fully electronic THz imaging with high-performance waveguide-based components towards video-rate operation. The first fully electronic THz-imaging system, which we realized, utilized a single emitter

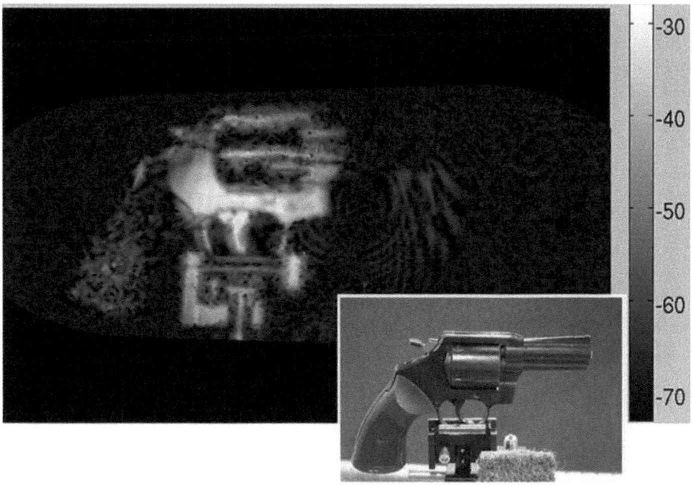

Fig. 9.1 Photograph and 620-GHz image of a handgun. The latter taken with a single-pixel scanner from a distance of about 0.5 m. The measuring time was 9 s. The *gray-scale bar* shows the IF power level detected by the lock-in-amplifier in dBm. The length of the weapon from hammer to muzzle is 15 cm. Typically, 50,000 data points are acquired in the field of view. They are binned to about 10,000 points which serve as nodes on a triangular mesh to render an image with regular pixel spacing

and single detector, operating at 620 GHz, and x-y-scanning by linear translation stages. The data acquisition time of several minutes was clearly limited by the mechanical scanning process. By switching from translational scanning to two-axes rotational scanning, we reduced the acquisition time for an image to 9 s [19–21]. With a frequency-agile emitter-detector combination [12], this system was also extended to range-resolving capabilities employing the frequency-modulated continuous-wave (FMCW) concept [22, 23]. We come back to it at the end of this section.

Figure 9.1 displays a 620-GHz image of a handgun taken with this rotational-scanning system (without range resolution) [20]. The barrel is not visible, because the specularly reflected beam returning from it misses the aperture of the optics and hence does not reach the detector. With minor changes to the software, the image generation time could be reduced further to 1.25 s, then being limited by the maximum angular velocity of the rotational stage (8°/s). Further speed improvements should be possible by employing a faster rotational stage, but the approach would reach mechanical limits, before it could get close to video rates. Apart from the high angular velocities, the integration time available for each pixel would also become a limiting factor for the system's performance.

Both the mechanical problems and those concerning signal averaging can be alleviated by employing multiple detectors (and/or multiple sources). Since waveguide-based components are expensive, especially for high frequencies (>0.3 THz), the costs set a limit for the number of elements, and 2D arrays with one detector for

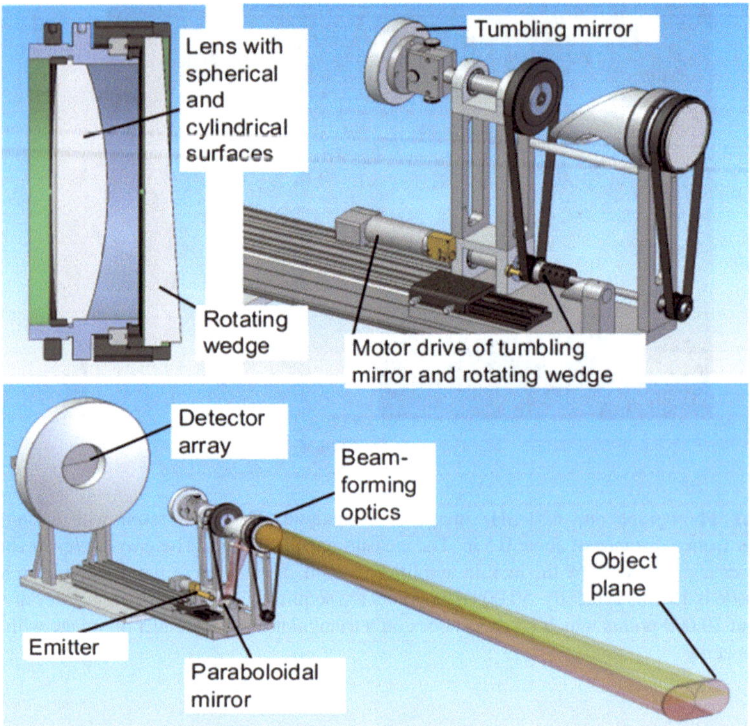

Fig. 9.2 Reflective 812-GHz real-time line scanner. *Bottom panel*: Total view, with the THz beam focused to a line and moved such that it illuminates a race-track-shaped section of the object plane. *Top right*: Frontal part of the system showing the motor-driven tumbling mirror at the back and the beam-forming optics of the illumination with the motor-driven wedged plate in the front. The emitter is placed at the bottom in the middle part of the display (*brass-colored* object, color online). *Top left*: Cross-section through beam-forming optics

each pixel are not an option with this technology today. A good alternative are line scanners. 2D images are generated either by exploiting the motion of the object (e.g., in the case of production lines or transport belts) or by active scanning.

We realized an explorative imaging system with a single emitter and 32 detectors for a radiation frequency of 812 GHz. The detectors are arranged in a line, and images are acquired by means of a single rotating mirror. Figure 9.2 displays CAD representations of the set-up. The panel at the bottom displays a view of the whole system, the salient features being the telescope optics and the illumination beam. The system contains two motor-driven units (see upper right panel of Fig. 9.2) rotating at the frame-rate of about 10 Hz: a wedged plate being part of the beam-forming optics of the illumination, and the secondary mirror of the telescope. Both motions are synchronized to each other.

The first generates a linear illumination pattern of the emitter's radiation in the object plane, 4 m away from the primary mirror. The illumination line keeps its

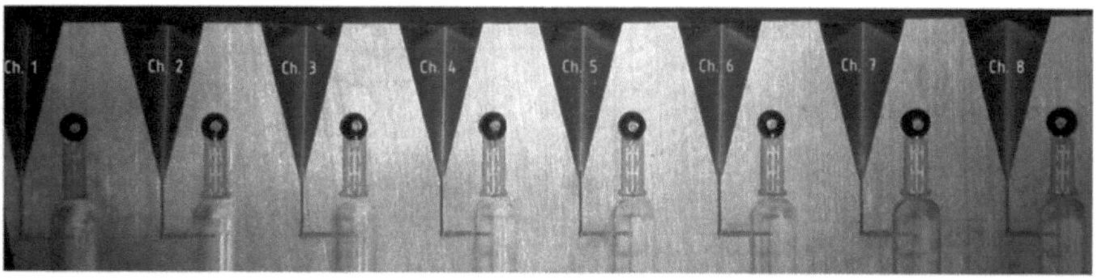

Fig. 9.3 View into one of the eight-channel 812-GHz receiver blocks. The spacing of the channels is 4 mm. THz radiation impinging from the *top* onto a horn antenna and guided into the block by a L-shaped waveguide is mixed with the local-oscillator signal arriving in the waveguide at the *bottom*. The intermediate-frequency signal is extracted through a coaxial waveguide seen next to the horn antenna

horizontal orientation while each point of the line moves along a circular path in the object plane. The radius of the circle is 7.5 cm in a distance of 4 m, which results in a race-track-shaped illuminated area covering approximately 13×26 cm^2. The synchronous motion of the secondary mirror ensures that the illuminated section of the object plane is imaged onto the linear 32-channel receiver array which is located in the central bore of the large primary mirror of the telescope.

The telescope's quasi-optics consist of two rotationally symmetric mirrors designed for diffraction-limited stand-off imaging at an adjustable distance between 2 and 6 m. For a working distance of 4 m, the numerical aperture of the optics is 0.06. The aspherical primary mirror has a diameter of 480 mm and a radius of curvature at the center of 1,373 mm, the rotating secondary mirror is also aspherical, has a diameter of 140 mm, and a radius of curvature at the center of 393 mm. The secondary mirror is tilted with an adjustable angle with respect to its rotational axis in order to produce a tumbling motion. Extra weights mounted on its shaft cancel moments of torque and ensure dynamic balancing.

Considering the limited radiation power available from the source, we refrained from illumination of the whole scene in the object plane to be imaged, but rather implemented the synchronized line illumination addressed above. The upper left panel of Fig. 9.2 illustrates the implementation of the beam-forming optics. The THz beam is shaped by a single-element lens with a convex spherical surface on one side and a concave cylindrical surface on the other. The scanning is realized by a beam-steering wedge, which rotates in synchronicity with the secondary mirror of the imaging telescope, both driven by the same motor. The angle of the wedge is selected corresponding to the desired working distance. Two different wedges for imaging distances of 2 and 4 m were realized. The lens and the rotating wedge are home-made from high-density polyethylene (HDPE) via CNC-machining.

The detector array, located in the central bore of the main mirror, consists of four brass blocks each containing eight detector channels. Figure 9.3 shows a view into one of the blocks, opened to display its inner structure. The detectors are heterodyne receivers operating at the 8th sub-harmonic.

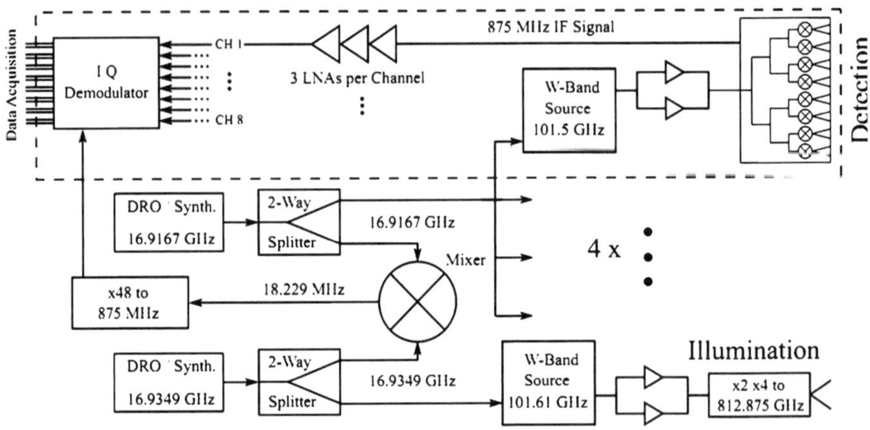

Fig. 9.4 Schematic representation of the electronics of the 812-GHz imaging system

Figure 9.4 presents the layout of the electronics of the system. The lower part represents the power generation chain. It starts with a synthesizer controlled by a dielectric resonator oscillator (DRO). The output signal is split, one part being used for the intermediate-frequency (IF) reference generation chain, the other for the radiation source. The latter part is fed to a multiplier chain, with power splitting, amplification and recombination in the W-band, followed by frequency doubling and then quadrupling to finally obtain the 812-GHz radiation used for active illumination.

The radiation frequency was chosen as a compromise between optimal spatial resolution and the limits of radiation-power generation. At 812 GHz, power levels of 1 mW appear feasible (albeit challenging). Diffraction, on the other hand, does not yet limit the resolution very severely the spatial resolution in 4 m distance being 7.5 mm, which still allows to obtain fairly detailed images of security-relevant objects. Issues of penetration through clothing and other materials certainly have to be considered also [24].

The upper part of Fig. 9.4 sketches the detector electronics. The detectors are heterodyne receivers with sub-harmonic mixers. The local-oscillator (LO) signal is provided by one W-band source per each of the four detector modules, the sources being driven by a common DRO synthesizer. The output power of each W-band source is amplified in two stages and split three times in order to provide each detector with sufficient LO power (about 12 dBm per pixel). The LO signal, at 101.5 GHz, is then mixed with the incoming THz radiation, generating an IF signal at the 8th harmonic of the LO signal minus the frequency of the THz wave. The IF is about 875 MHz, given by the difference frequency of the two DROs multiplied by the factor of 48. The DROs are not locked to each other, which results in slight IF drifts. The IF signal of each channel is amplified by three low-noise amplifiers, and then I/Q demodulated (I/Q: In-phase/Quadrature). The resulting 64 low-frequency signals for the in-phase and quadrature components are low-pass-filtered and fed into the data acquisition unit for digitization and image generation on a PC.

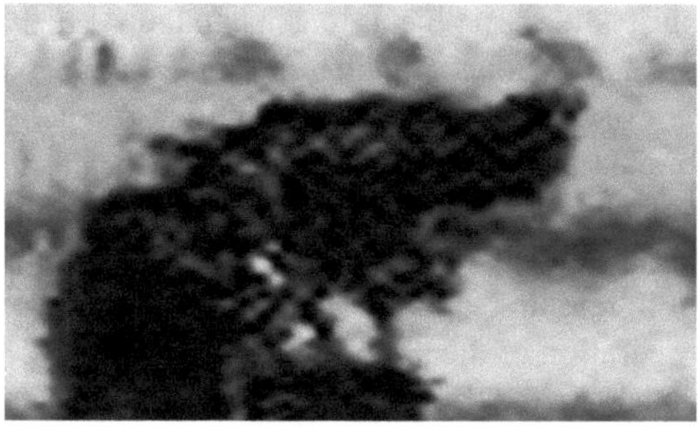

Fig. 9.5 Captured image from the real-time line scanner with a source power of less than 40 W. Reflection image of the handgun of Fig. 9.1, positioned 4 m in front of the primary mirror. The image is a composition of several measurements, since only a small group of connected detectors with full performance was available. Their data, taken at different horizontal object positions, were combined to this image

The central part of Fig. 9.4 shows that the IF reference signal for the I/Q demodulator is generated by multiplying the difference frequency of the two synthesizers by the factor of 48. Any IF drift is hence present also on the reference signal which cancels any drift effect.

For each channel, data is recorded at a total of 154 angle positions per revolution of the scanner optics. The data is further processed by software to compensate for differences in offset or gain for each of the 32 detectors. The power is then calculated from the corrected I/Q components, combined with the position information from the angular encoder of the driving motor, and used for the generation and on-screen display of a real-time or close-to-real-time image.

This is achieved by a triangulation technique which averages and interpolates the data. The software spans a triangular net over the field of view consolidating the measured 154×32 data points into approximately 2,000 effective pixels. Considering the diffraction-limited spot size, we over-sample the field of view slightly by a factor of about 2.6 ($2{,}000\ \pi(0.75\ \text{cm}/2)^2/13 \times 26\ \text{cm}^2$), which is desirable.

Figure 9.5 shows a reflection image of the handgun of Fig. 9.1. At the present time, the performance of the system is still limited by three aspects. First, the targeted power level of the source of ≥ 1 mW could not yet be reached, but so far has been limited to ≤ 105 μW. Second, the manual assembly of the detector blocks led to significant variations in detector sensitivity by up to 20 dB which could be compensated only partially by data processing in the software. And finally, we found that a lower IF, further away from mobile phone frequencies, strongly reduces crosstalk between channels and pick-up of external signals. After implementation of corrective measures, the system will be ready for field tests.

To summarize, with this portable and fully electronic system, real-time stand-off imaging at 812 GHz has been successfully demonstrated for the first time. A maximum frame-rate of 12 fps and a spatial resolution of 7.5 mm at 4 m distance has been achieved. With a sufficiently homogeneous detector performance, the identification of security-relevant objects, e.g. a handgun (Fig. 9.5), will be possible. The system described here is conceptually quite similar to the 620-GHz single-pixel imaging radar system of Cooper et al. [12]. With that second-generation machine, the authors have demonstrated the principle capability to detect objects such as pipe bombs concealed under clothes, with 0.5 mW of power for illumination at a distance of 25 m. There is no reason why our system, after optimization, should not perform in a similar way, but in real time and not with the 5-s data acquisition time dictated by the single-pixel operation of the system of Cooper et al. [12].

We now come back to the FMCW (frequency-modulated continuous wave) imaging approach. Frequency modulation of the THz wave allows time-of-flight ranging. When the frequency sweeps are performed with a high repetition rate, FMCW ranging can be readily combined with scanning imaging as described before. Three-dimensional (3D) imaging significantly improves object identification [12], which otherwise represents a major challenge for THz imaging modalities.

With the FMCW method, the range of uniqueness for the determination of the distance is much larger than with single-frequency phase measurements, but the depth resolution is lower: When we denote c as the speed of light, and $\Delta\nu$ as the chirp bandwidth, the fundamental limit for the depth resolution is given by $c/(2\Delta\nu)$. Applications like stand-off imaging can benefit from FMCW because the ranging resolution and range of uniqueness fit much better to their demands than the high-resolution, half-wavelength range of uniqueness provided by phase measurements.

Figure 9.6 provides a demonstration, how the range information provided by FMCW imaging can facilitate the recognition of an object. The example shown in this figure is a raster scan with a SynView300 transceiver working around 300 GHz with a chirp wavelength of 90 GHz (230–320 GHz) and a FMCW sweep time of 240 μs. In the power reflection data of the top panel of Fig. 9.6, one can recognize the hand, but only because the contour of the hand is clearly identified by the signal reflected from the metallic background on which the hand rests. The hand in free space would only provide strong signals from those parts of it which reflect specularly. An object recognition will, in general, be difficult. This is different with the time-of-flight data. Here (see bottom panel of Fig. 9.6), the hand is clearly identified.

9.2.2 Synthetic-Aperture Imaging

We now address the synthetic-aperture approach of imaging. Here, a number of sources (transmitters, Tx) and detectors (receivers, Rx) are arranged either in a line or in a two-dimensional (2D) configuration. The transmitters illuminate the scene, while the receivers collect the scattered energy. The coherent signals from each Rx/Tx combination are acquired and numerically transformed to the image data for the corresponding line in the object plane [25]. Multi-element arrays have been demonstrated predominantly at millimeter-wave frequencies [26–29], while most

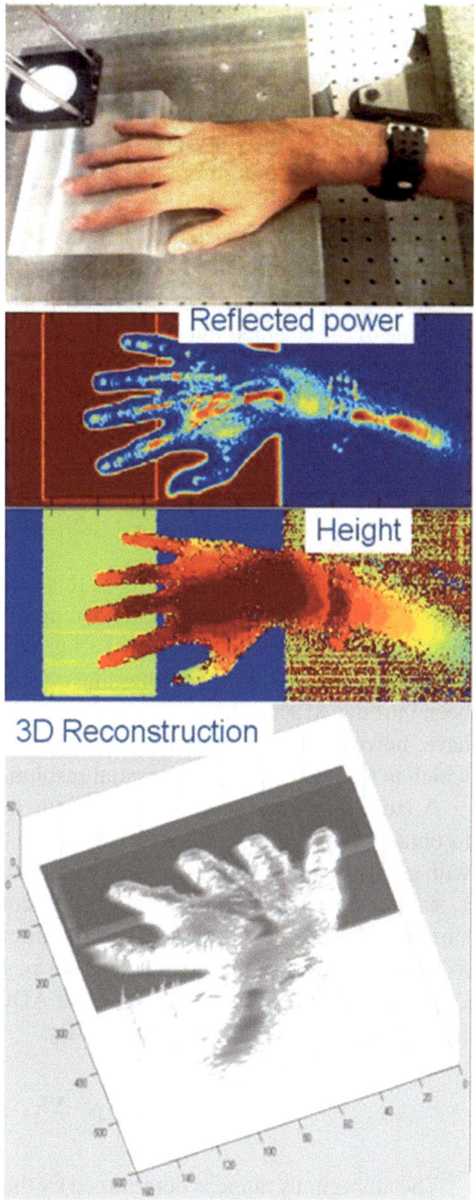

Fig. 9.6 300-GHz raster-scan images of Torsten Löffler's hand. The photograph shows the experimental situation in the visible range: The hand rests on metal blocks below the white teflon lens of the THz imaging system (the illumination and detection beam paths are equal). *Second panel*: reflected-power image; *third panel*: time-of-flight data representing the height profile; *bottom panel*: 3D reconstruction of the hand employing the THz data. While one can only recognize certain features in the reflected power image (if the reflection of the metallic background on which the hand rests is disregarded), the 3D reconstruction gives a much clearer impression of the object

higher frequency synthetic-aperture systems have been using a scanning of a single element and synthetic-aperture reconstruction.

The advantage of synthetic-aperture over conventional imaging is the need for fewer or no moving mechanical parts. This can enhance data acquisition speeds tremendously, but comes at the price, that more electronic components are needed and that the numerical effort of image reconstruction is enormous. An additional advantage of synthetic-aperture imaging is that there does not exist a focal plane as in conventional systems. The object can principally be imaged with equal resolution over a considerable range of depth.

Synthetic-aperture techniques with minimum redundancy in 2-D arrangement have been successfully presented in [30], however only for passive systems. Active imaging system configurations with low redundancy are only known for line configurations [31].

The THz camera considered here has been designed for operation at stand-off distances of ≥ 8 m, with a scene area of $0.7 \times 2\ m^2$ and producing images close to real time. Quasi-optics is employed for one dimension only (vertical direction, focusing of the scanned line) and hence consist of cylindrically-shaped mirrors. The other dimension (imaging of the line) is handled by synthetic reconstruction.

The angular resolution α is determined by the overall size of the linear antenna array $N_{Tx}\ N_{Rx}\ \Delta_{Rx}$, where Δ_{Rx} is the Rx antenna element spacing:

$$\alpha = \frac{\lambda}{N_{Tx}\, N_{Rx}\, \Delta_{Rx}}, \tag{9.1}$$

With a wavelength of $\lambda = 1$ mm, this result in $\alpha = 0.0015625$, for 8 Tx and 16 Rx elements, and an Rx element spacing of $\Delta_{Rx} = 5$ mm. At a distance of 10 m, the spatial resolution is then $\Delta_S = 15.6$ mm. The angular ambiguity separation is $\theta = N_{Tx} N_{Rx} \alpha = 0.2$. This assumes that no redundancy in the Tx and Rx elements has been introduced according to algorithms presented in [31]. The realized components have, however, finite dimensions and cannot be positioned at the exact locations, which in turn deteriorates the spatial resolution to $\Delta_S \approx 20$ mm.

A large time-bandwidth product of the transmitted signal is necessary in order to obtain good signal-to-noise performance (requiring long pulses) simultaneously with good range resolution (requiring large bandwidth). The temporal range window τ_f should be chosen shorter than the uncompressed pulse length τ_p in order to simplify the baseband system. The minimum pulse length, which must be recorded to focus the range window, is $\tau_r = \tau_p + \tau_f$. For a scene with a range extension of Δ_r, the required IF bandwidth B_{IF} (and ADC bandwidth) after analog deramp is a fraction of the modulation bandwidth B_p:

$$B_{IF} \geq N_{Tx}\ \Delta f_{Tx} \geq N_{Tx} \frac{\tau_f}{\tau_p} B_p. \tag{9.2}$$

The ambiguous range is determined by the transmitter frequency separation Δf_{Tx} between individual channels:

$$r_{amb} = \frac{c \Delta f_{Tx} \tau_p}{2B_p} \tag{9.3}$$

The image acquisition and processing of a line is performed with a refresh rate of $\tau_r < 5$ ms, for an operating frequency regime of 220–320 GHz with a modulation

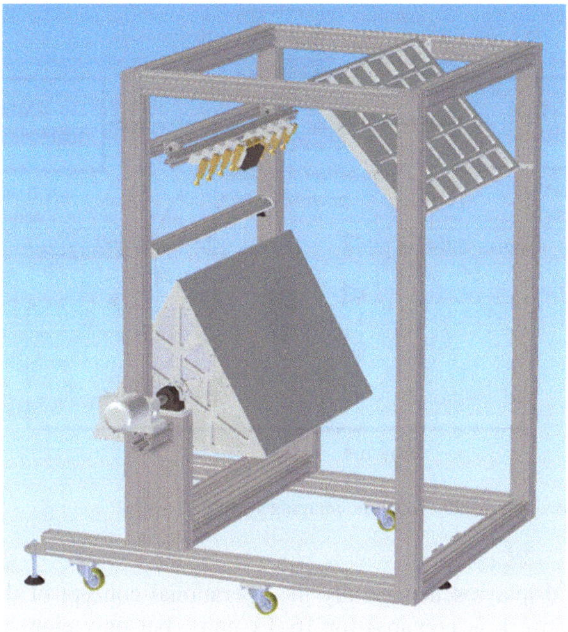

Fig. 9.7 CAD representation of the scanning synthetic-aperture imaging system. The line of emitters and detectors is visible on the *top left side* of the system (*brass-colored* components). The scanner has the dimensions: 1 × 2 × 1.5 m. The dimensions of both large mirrors are 1 × 0.7 m. The mirror is designed to rotate with 0.5 Hz. The radiation from the emitters hit the narrow cylindrical reflector located under them, which steers the beams to the large fixed mirror at the *top right edge*. From here, they impinge onto the rotating triangular deflector which illuminates the scene at the *right side* of the system. Reflected radiation returns along the same path in the opposite direction

bandwidth $B_p = 100$ GHz. For an uncompressed pulse length $\tau_p = 1$ ms and a Tx frequency separation $\Delta f_{Tx} = 1$ MHz, the ambiguous range is $r_{amb} = 1.5$ m. For every pulse and for every pixel in the output image, an interpolated sample must be computed and accumulated. Fast back-projection algorithms exist [32–34], including near-field scenarios, with improvements in speed by a factor of 100 compared with direct back-projection in 2D [33, 34]. These algorithms have the advantage of providing the flexibility and robustness of a time-domain algorithm but exhibiting a computational performance in parity with fast transformation algorithms.

The mechanical layout of the imager is illustrated in Fig. 9.7. The system combines a linear array of emitter-receiver elements, used for synthetic imaging in the horizontal direction, with classical scanning optics for the vertical direction, the latter employed in order to keep the numerical effort of image reconstruction limited for the sake of the real-time capability of the system. Currently, the system employs 16 Rx units and 8 Tx units with 8-mm receiver and 128-mm transmitter spacing. The eight transmitters are switched sequentially and each of them provides an output power of approximately 1 mW. The beam divergence is determined by the horn antenna of each emitter unit.

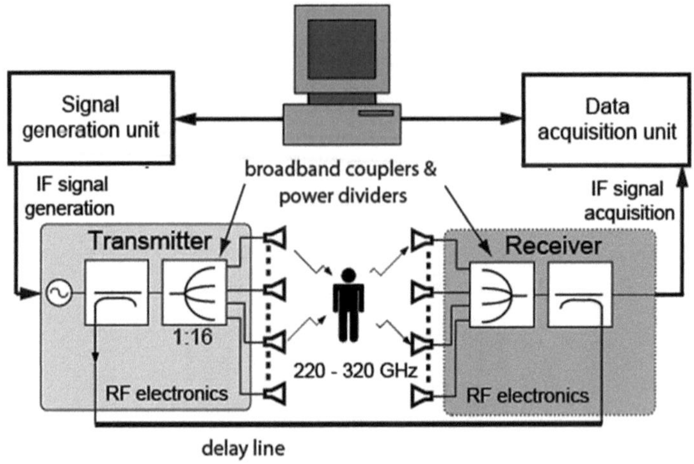

Fig. 9.8 Block diagram of the electronic imaging system

Figure 9.8 displays schematically the operational concept of the system. The distribution network is prepared for 16 Tx units, but only eight were employed at the experiments. The FMCW approach provides the depth resolution, while the emitter-detector line array generates the data for synthetic image reconstruction. The data acquisition system, which copes with the high data stream, is built around SPECTRUM ADC and DAC cards and a Cyclone Microsystems backbone. The delay line connecting the transmitter and the receiver is a fixed delay line employing a 50-semi-rigid coaxial cable.

Real-time reconstruction was demonstrated using a parallel implementation of the back-projection algorithm on an NVIDIA GTX260 graphics processing unit (GPU). A $1 \times 1\ m^2$ x-y-slice of 128×128 pixels could be focused in less than 2 ms.

Concurrent operation of all Tx and Rx modules could not be achieved here, due to errors in the I/Q modulators employed in the current design. This will become available in a redesigned version of the system. The system has been tested with objects close to the spatial resolution limit of the system with a fixed position of the rotational triangular mirror.

Figure 9.9 illustrates a few of the test objects at 8 m distance from the scanner, and measured with 200 averages, leading to a measurement time of about 200 ms. The objects are an aluminum block with 7.5×7.5 cm size (left side of Fig. 9.9), and a metal block with a 2.5 cm-wide illuminated front side and a cylindrical aluminum post with a radius of 2.5 cm, respectively (right side of Fig. 9.9).

Results of measurements are shown in Fig. 9.10a–d for the objects of Fig. 9.9. The distance reading (around 7 m) is referenced to the mounting rack and not to the emitter/detector units (total distance with respect to these: about 10 m). The top panel shows the signature of the large aluminum block. A close-up look at the result at 0 dB level reveals a sharp line at 7.01 m with 7.5 cm width. This is in very

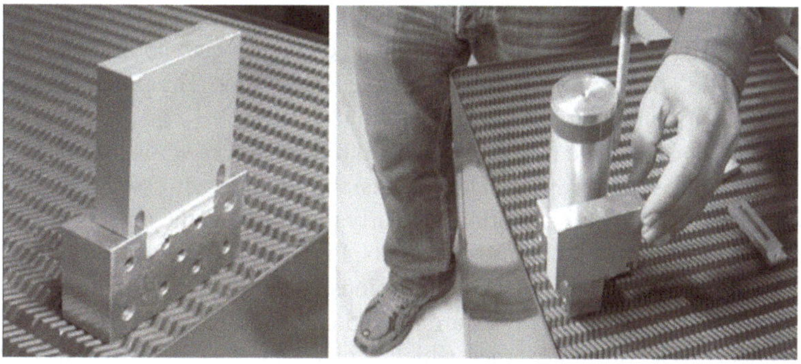

Fig. 9.9 Measured targets employed for the verification of the synthetic-aperture imaging system

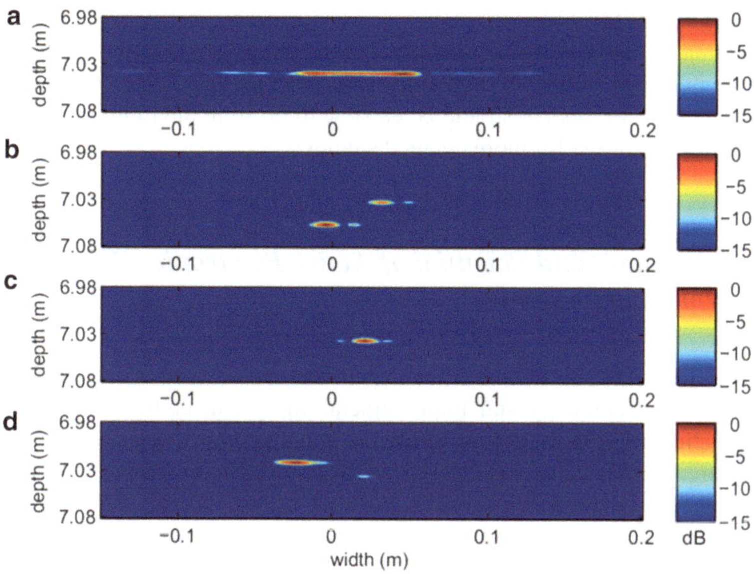

Fig. 9.10 Synthetically reconstructed images of objects presented in Fig. 9.9: (**a**) a 7.5 cm wide metal block, (**b**) two 2.5 cm wide metal blocks separated by 7.5 cm, (**c**) a cylindrical metal post with a radius of 2.5 cm, and (**d**) the cylindrical metal post together with a 2.5 cm wide metal block. Numbers on the axes are in units of m

good agreement with the object. However, there is a significant spill-over towards negative cross-range values. The source of this spill-over has not been identified unambiguously yet, possible causes being (i) a slight misalignment of the optics, to which the reconstruction algorithm is quite sensitive, or (ii) thermal fluctuations in one of the eight Rx blocks of the array. In addition, one can observe side-lobes in the figure. The side-lobes in all results are below 3 dB but can not easily be explained with parasitic reflections. We have therefore performed point target simulations with different receiver numbers and separations of the Rx and Tx elements. The simulations suggest that a part of the receiving elements were not fully operational, and that the Rx spacing differs from the envisaged 8 mm. The predicted side-lobe levels for the nominal spacing should be below −20 dB.

Figure 9.10b shows the results for two aluminum blocks whose width of 2.5 cm is close to the resolution limit. The objects should then act essentially as point-like scatterers. The object separation is clearly resolved and the objects indeed have signatures close to those of point-like targets. Figure 9.10c shows the reconstructed image of the cylindrical pillar. The dimensions of the pillar and its range position are well reconstructed, with values around 2.5 cm and 7 m, respectively. The pillar should appear as a point-like target, which is nicely the case in the reconstruction.

We have finally positioned the cylindrical pillar in direct contact with the narrow block, as indicated in Fig. 9.9. The corresponding data in Fig. 9.10d reveal that the pillar provides the strongest signal and the block contributes to the signal on the right hand side of the pillar.

In summary, a first proof-of-principle demonstration for synthetic-aperture imaging in the 300-GHz frequency regime has now been provided. The system does not yet operate in real-time, but should be capable to do so in the future; the system is being improved further for more detailed studies.

9.2.3 Summary and Potential of Active Electronic THz Imaging Systems

All-electronic real-time imaging has progressed rapidly due to advancement in semiconductor technology and improvements in system architecture and image reconstruction algorithms. As promising as it is, there are still major constraints to be overcome. The available output power from a single source is very limited beyond microwave frequencies, which limits the overall dynamic range. Low-noise amplification has seen tremendous progress, while output power improvement has advanced on a much slower pace. Comparing with the power level of kilowatts readily available in the microwave regime, it is amazing what has been achieved with the mWs of power and less, available to us at acceptable costs in the THz regime. This situation limits the current dynamic range of multi-element imaging radar systems to 40–70 dB, in contrast to ≥100 dB dynamic range at lower frequencies. Multi-static concurrently-measuring imaging radars are important for real-time operation, but their architecture is still a subject of intensive research, especially with regards to calibration issues.

Dramatic improvements have been observed in the area of image reconstruction algorithms. These have profited from the introduction of novel SAR (synthetic-aperture radar) systems at lower frequencies and from the work performed on 77-GHz systems for automotive industries. Several real-time algorithms have been proposed and implemented on GPU computer work-stations. The bottle-neck here is the data acquisition hardware, which has to sustain the very high data flow of gigabytes per second. With appropriate data reduction algorithms it is believed to be possible to reduce the data stream and to implement the algorithms on more convenient machines.

9.3 Optoelectronic Terahertz Imaging Systems

Here we describe the development of THz imaging systems employing multi-pixel EO detection, where the THz image information is transferred to the visible/near-infrared spectral range via EO mixing and subsequently measured using a commercial optical camera (e.g. CCD/CMOS). This allows one to project the complexity of multi-pixel detection onto a highly developed technology, albeit at the price of a limited EO conversion efficiency and hence a reduction in the dynamic range. EO detection provides a measure of the field amplitude and phase (similar to heterodyne electronic detection) but can also be extended to THz frequencies well above 1 THz.

In this context, the emphasis of our work has been placed on the realization of both a homodyne and a heterodyne THz imaging system. The homodyne approach takes advantage of a nanosecond pulsed quasi-CW THz optical parametric oscillator (OPO) in combination with a CCD camera for homodyne EO read-out, while the heterodyne approach is based on a hybrid system with a quartz-stabilized CW micro-electronic THz emitter in conjunction with an EO demodulating camera detector.

Here we describe each system in turn, including THz beam image measurements, a comparison of their relative current performance, and directions for further improvement of each system.

9.3.1 The Homodyne Concept

The homodyne approach has been guided by the elegant way of multi-pixel THz imaging introduced by the group of Zhang in conjunction with THz pulses generated by a femtosecond amplifier laser system [35]. Recently, this approach has allowed real-time imaging using a 1-kHz repetition rate laser and a state-of-the-art CMOS camera [36]. However, for practical applications, amplifier laser systems are typically too expensive and are difficult to deploy and maintain outside a laboratory environment. Multi-pixel EO THz imaging was also pursued with an optoelectronic CW source [37], although it was found that the THz power was too low for effective imaging; averaging over 40 min was required to measure the spatial profile of the focused THz beam.

Hence as a compromise between femtosecond- and CW-systems, we decided to explore the use of a THz optical parametric oscillator (OPO) pumped by a 10-kHz Q-switched Nd:YVO_4 laser at 1.06 μm. Such THz-OPOs emit quasi-CW THz radiation, with the emission in bursts with typical pulse duration of ~10 ns, at a repetition rate from anywhere between 10 Hz and 10 kHz [38, 39]. For instance, at a low repetition rate of 15 Hz, very high peak powers of >1 W can be generated [39], which translates into a peak field of >150 V/cm (assuming moderate focusing to a 1-mm beam radius). For the higher repetition rate of 10 kHz, such as the THz-OPO used here, peak powers of several 10 mW can still be readily achieved, corresponding to peak fields of several 10 V/cm (again assuming a beam radius of 1 mm).

Such THz-OPOs can reach considerably higher THz field amplitudes than typical truly-CW emitters, such as a multiplier-chain-based electronic source – with our 0.65-THz emitter, we obtain a power of ~1 mW, which translates into a field amplitude of 5 V/cm. To put this value further into perspective, we note that a high-repetition-rate fs Ti:sapphire laser can generate THz pulses with an average power of 40 μW [40, 41], corresponding to a peak power of about 400 mW which, when focused, provides a field amplitude of the order of 100 V/cm. This is much less, though, than what can be achieved with amplifier-based Ti:sapphire sources, which achieve tens to one hundred kV/cm with semiconductor-based THz emitters [42], and even reach higher with laser-induced plasmas [43, 44].

If one estimates with the peak fields of our THz-OPO and the known parameters of EO detection, over how large an area of an object one may be able to take diffraction-limited images with a camera, one finds that 100-pixel parallel detection should be possible in real-time mode for reflective imaging[1] [45]. While a limited parallelism alone will not yield detailed pictures per se, one can imagine a parallel imaging modality involving an additional scanning of the object, e.g. via a line scanner.

In the following, we experimentally explore multi-pixel detection with our THz-OPO and a Si CCD camera. The experimental set-up is displayed in Fig. 9.11 [46]. Quasi-CW THz radiation bursts are generated at a repetition rate of 10 kHz directly in a periodically poled $LiNbO_3$ crystal in the OPO cavity together with the signal wave [47]. Seeding at the signal wavelength enhances the conversion efficiency. The difference frequency of the pump and signal waves is equal to the frequency of the THz radiation. A key advantage of this OPO is that the two near-infrared beams, whose superposition is required for the EO detection, exit the OPO collinearly. After cleaning of the beam modes in an optical fiber, the dual-color beam is available for EO detection of the THz radiation, which is coupled out from the $LiNbO_3$ crystal

[1]To reach such a result, we consider a system with an input average THz power of $<P> = 4\ \mu W$, corresponding to a peak power of $P_0 = 40$ mW. Assuming a power collection efficiency of 10^{-4} from the scattering object, we have a THz peak power of $P = 4\ \mu W$ at the EO detector, which translates to an electric field amplitude of $E = 6$ V/m in the EO detection (assuming 100 effective pixels, each an area of $(3\lambda)^2$ at 1 THz). Based on the typical sensitivity of crossed-polarizer EO detection [35] (modulation depth coefficient $k_{EO} = 2.5 \cdot 10^{-6}\ (V/m)^{-1}$), the achievable modulation depth is $1.5 \cdot 10^{-5}$. This is equal to the relative shot noise of a CCD/CMOS camera system (such as the CCD employed here, described below) after binning from 10^6 CCD pixels down to a 100 THz-image pixels with a 50 ms integration time.

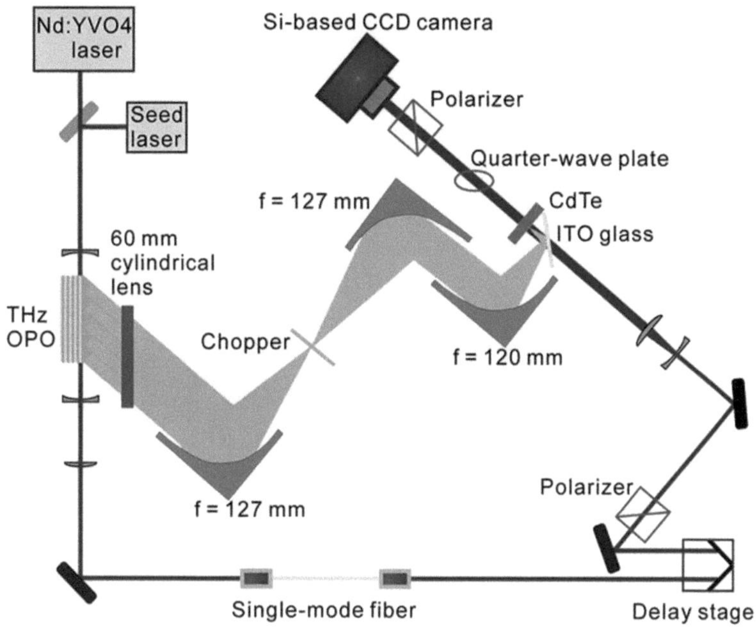

Fig. 9.11 Schematic of the THz imaging system based on a ns quasi-CW THz-OPO with multi-pixel cross-polarizer EO detection

via a set of silicon prisms, and overlaid with the read-out beam with the help of an indium-tin-oxide dichroic beam combiner [48]. The frequency of the THz radiation of 1.5 THz is fixed by the poling of the crystal. We have chosen a 1-mm-thick CdTe (110) crystal as a suitable EO detection crystal for this optical wavelength and THz frequency [49]. The peak power of the THz radiation was found to be 13 mW directly behind the Si prism coupler [46]. For a focal spot with a radius of 1 mm, the corresponding field strength is ~18 V/cm.

We first performed single-pixel raster-scan imaging with EO detection in a crossed-polarizer geometry and Ge photodetectors. The THz beam profile was measured without a sample in the beam. The near-infrared dual-color read-out beam was tightly focused and raster-scanned across the EO crystal.

These measurements yielded a dynamic range of 28.3 dB/Hz and a NEP of 1.87 nW/Hz at the peak of the focal spot. Note that the values in Meng et al. [46] are revised here to correctly account for the effective detection bandwidth $B_{eff} = 1/(6\tau)$ (filter slope of 12 dB/oct) where $\tau = 50$ ms is the lock-in time constant, and we now use the correct units for coherent detection, i.e. dB/Hz (instead of $dB/Hz^{1/2}$). The dynamic range is calculated from the ratio of the maximum EO lock-in signal (S) to the rms noise level in the absence of the THz beam (σ) via $DR = 20\log_{10}\left(\sqrt{B_{eff}}S/\sigma\right)$, while the NEP is obtained from $NEP = B_{eff}(\sigma/S)^2P$ (where $P = 1.3\ \mu W$ is the average THz power). While these single-pixel measurements employed differential detection with a reference Ge detector, the limiting

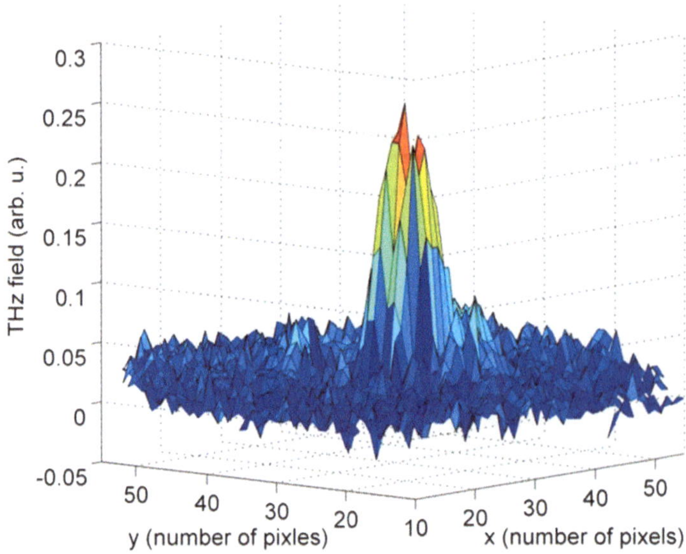

Fig. 9.12 CCD-recorded EO field image of the focused THz beam from a THz-OPO. The x-y-scaling represents binned camera pixels resulting in an effective pixel size in the focus of 192×192 m^2

noise source is still due to residual laser fluctuations (i.e. not shot-noise limited) due to a small loss in the common mode fluctuations induced by the optics in the EO detection arm.

The diameter of a diffraction-limited Airy disk on the EO detector is about 0.7 mm under the given experimental conditions. With a diameter of the THz beam of 2 mm, the number of effective optical pixels covered is about 10. Hence, for a pixel number of 100 (i.e. with a three-times larger spot diameter) the achievable dynamic range would then be reduced to 18.3 dB/Hz.

We now come to parallel camera-based imaging. The Ge detector was replaced by a Si CCD camera (Dalsa 1M60CL) having $1{,}024 \times 1{,}024$ pixels, a large full-well capacity (FWC, $1.5{\cdot}10^5$ electrons), and a 12-bit low-noise read-out. The optical beam was expanded to cover an area of about 1 cm^2. The camera acquires 50 fps (with fast CPU processing of the 150 MB/s frame data), and the THz beam is mechanically chopped at 25 Hz allowing one to extract a differential THz field image by subtracting bright and dark frames.

Figure 9.12 shows a measured field-amplitude profile of the THz beam for a fixed position of the delay stage (fixed relative phase between THz wave and amplitude modulation of optical beam), using an integration time of 120 s (i.e. 3,000 light/dark frames), where we apply 16×16 digital binning to yield a 64×64-pixel image. The relative modulation depth of the peak signal was $3.5{\cdot}10^{-4}$, which is close to that expected from the field strength in the EO crystal predicted from the power measurements, taking into account THz losses, the crossed-polarizer geometry and the imperfect temporal overlap of the dual-color pulses [46]. Based on the noise floor (from a reference measurement without the THz beam) the dynamic range relative to the peak signal is 17 dB. Taking into account the effective measurement time of 120 s, the integration-time-corrected value is then −3 dB/Hz. However, very recent measurements with a higher THz field have achieved a relative modulation depth of $1.2 \cdot 10^{-3}$ and a dynamic range of +9 dB/Hz [45].

Nevertheless, this performance is considerably lower than that of single-pixel measurements. This is attributed to signal fluctuations both from the laser (which in this case, are not normalized at all by a reference detector) and the asynchronous frame acquisition/chopping, both of which lead to a relative noise of $>10^{-4}$ for a 1-s integration time.

Hence two clear directions for improvement of the current system would be to add a reference detector to normalize each acquired frame (although here care must be taken to correctly account for the detection of the two wavelength components of the dual-color optical beam, which are not correlated due to the fact that the signal beam depletes the pump beam in the OPO), and to trigger the camera synchronously to a sub-harmonic of the pulse train.

If these noise sources can be sufficiently well suppressed, one can approach the significantly lower detection limit due to shot noise (which exceeds the readout noise here). Given the current pixel FWC and assuming binning to a 64×64-pixel image, this sets a relative noise level of $4.6 \cdot 10^{-5}$ for the differential THz image with a 1-s integration time (i.e. over an order of magnitude smaller than the modulation depth) which would result in a dynamic range of ~28 dB/Hz.

In principle, it is possible to improve the performance of this ns-THz-OPO system even further. Firstly, the EO modulation depth could be improved by either increasing the THz power, or by improving the EO mixing by using a crystal with a larger nonlinear coefficient or coherence length. Moreover, as the THz radiation here is quasi-CW, narrow-band quasi-phase matching schemes are also applicable (in contrast to broadband pulsed THz systems). Secondly, the use of emerging state-of-the-art CCD/CMOS with a larger photo-electron throughput (FWC pixel number frame-rate) would allow the shot noise level to be further reduced. Without such improvements in the throughput, it is also not possible to exploit the far higher detection quantum efficiency of InGaAs-based cameras at this ~1 μm wavelength, as one simply reaches the pixel saturation at a lower incident optical power.

9.3.2 The Heterodyne Concept

Our novel heterodyne EO imaging concept is based on a hybrid set-up, in which the incident radiation of a quartz-stabilized CW THz source is electro-optically mixed [50] with either a femtosecond-laser [8, 51, 52] or a synchronized CW diode-laser pair [53], whose EO modulation is captured with a photonic-mixing device camera (PMDtec PMD[vision] 3k-S [54]) [55, 56]. Such PMD cameras were originally developed for 3D time-of-flight imaging with near-infrared radiation [57, 58], and

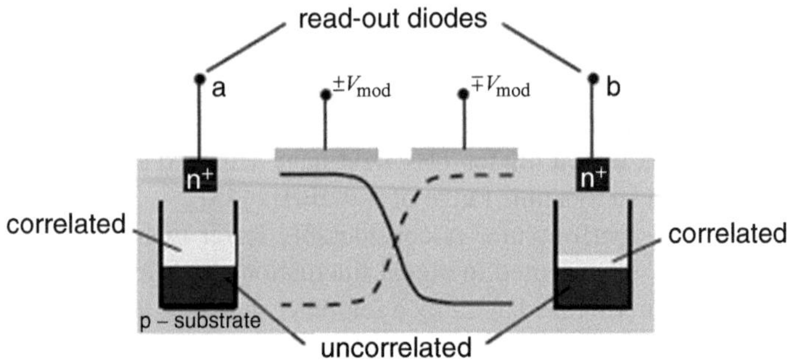

Fig. 9.13 Schematic of a single PMD pixel

allow phase sensitive detection of a modulated optical signal. As the repetition rate of the optical pulses/intensity modulation is incommensurate with the CW THz frequency, the EO signal is modulated at an intermediate frequency (in our case, 10 MHz), with a modulation phase dependent on the phase of the THz wave. This EO signal is then demodulated by the PMD camera, allowing one to recover both the THz amplitude and phase. The camera used here has a 64 × 48-pixel sensor with 0.1-mm pitch.

As shown in Fig. 9.13, each pixel consists of two transparent photo-gates between two read-out diodes, which are connected to the read-out circuitry. A push-pull modulation of the photo-gates controls the charge carrier transport to the read-out diodes and hence acts like a charge swing. As the small EO signal modulation is synchronized with the pixel modulation frequency, one measures a charge difference between the two output channels proportional to the EO signal (and hence, THz field amplitude) depending sinusoidally on the phase shift between these two modulations (whereas the constant back-ground light creates equal output signal components which cancel). Consequently, the phase information of the detected signal can be retrieved by several suitable measurements with equidistant phase shifts of the pixel modulation. To extend the dynamic range of acceptable unmodulated background, a circuit for the suppression of background illumination (SBI) is integrated in every pixel. This circuit drains integration capacitors equally from charge carriers resulting from uncorrelated background [57].

The scheme in Fig. 9.14 illustrates the implementation of the PMD-camera into the hybrid set-up using a femtosecond-laser system for the EO imaging of a THz focal spot. Both a reference and an imaging branch are provided with an equal portion of the radiation from a quartz-stabilized microelectronic 0.65-THz emitter (Radiometer Physics GmbH) with 1.1 mW output power. Each branch contains an EO detection unit, which uses the pulses from a high-repetition-rate femtosecond Ti:Sapphire laser system (Spectra Physics, Tsunami) as the optical probe light. Indium-tin-oxide dichroic beamsplitters [48] are used to combine the THz and optical beams, which then propagate collinearly to the EO crystal (<110>-oriented ZnTe).

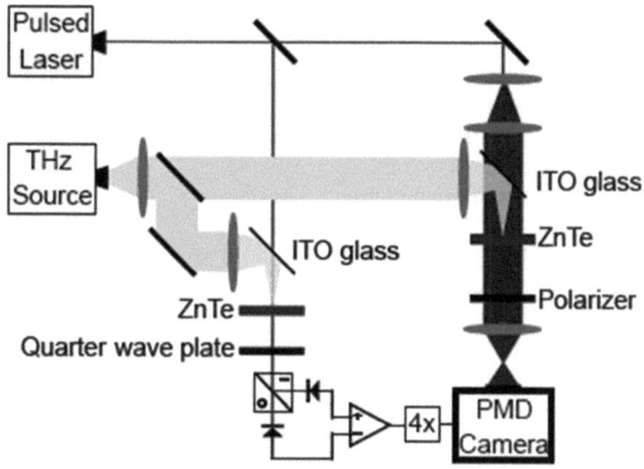

Fig. 9.14 Scheme of the heterodyne hybrid EO THz imaging system employing an electronic THz source, fs optical pulse train for detection, and a PMD camera for demodulation of the EO signal

The THz field is electro-optically mixed with the optical pulse train, which acts as the LO signal in this heterodyne set-up, producing an optical modulation whose lowest frequency component lies at an intermediate frequency of ~10 MHz (due to mixing of the 0.65-THz wave with the ~8,100th harmonic of the repetition rate of the optical pulse train). This IF can be tuned by adjusting the length of the laser cavity, and hence the pulse repetition rate. Due to the inherent stability of the femtosecond laser, no active synchronization scheme is required.

The reference branch is based on a (single-pixel) balanced EO scheme with a focused optical beam [59], and is used to provide a strong reference signal at the IF, which is required for the external pixel modulation input of the PMD camera. Due to the design of the PMD hardware, we must generate the fourth harmonic of the reference signal for this input, as this is used internally to generate four 90°-shifted reference signals at the IF, as is needed to retrieve the EO signal amplitude and phase.

The read-out unit of the imaging branch comprises a large-area 1-mm-thick ZnTe crystal in a crossed-polarizer set-up [60] with an expanded optical beam (average intensity 200 mW/cm^2). For these measurements, the incident THz radiation is focused into the EO crystal, where we estimate a field amplitude of ~1.5 V/cm based on the measured focal beam radius of ~1.4 mm. The EO crystal is imaged with a reduction factor of 2 onto the sensor of the PMD camera.

The image of the THz focal spot in Fig. 9.15 was effectively recorded within 2.5 s and shows a dynamic range of 23 dB/Hz. The measurements were taken close to the saturation limit of the PMD camera, which resulted in the best possible SNR. However, the measurements suffer from slight inhomogeneities of the pixels, obstructing the detection of the weak EO signal by a considerable background signal, which varies from pixel to pixel. This necessitated the additional measurement of dark frames (i.e. with the THz beam blocked), subsequent to THz image recording, to allow subtraction of this spatially noisy background. Hence the 2.5-s effective integration time of the THz images comprises both 2,500 bright and 2,500 dark frames, each with a frame integration time of 0.5 ms. Due to the

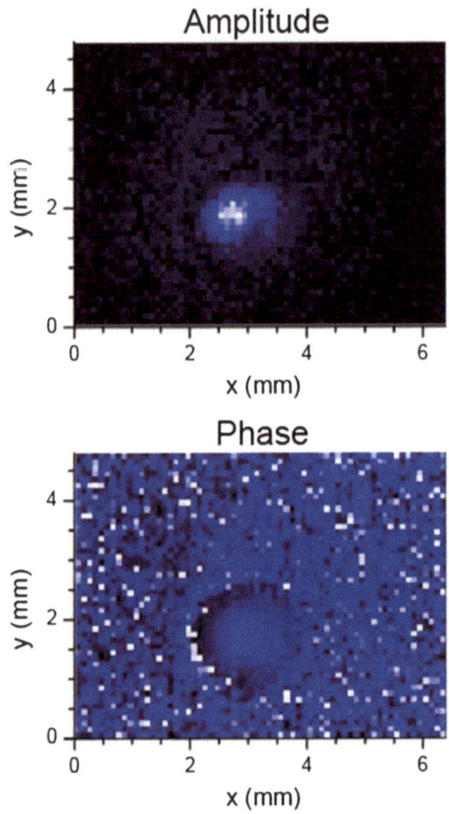

Fig. 9.15 EO image of the THz field amplitude and phase for a focused beam with the heterodyne hybrid system of Fig. 9.14, using an image reduction factor of 2 (Data taken from Friederich et al. [56]). The size of the image is 6.4 mm × 4.8 mm. The radius of the spot of well-defined phase is 0.8 mm; the corresponding radius on the crystal is 1.6 mm

restricted pixel capacitance, longer integration times were not possible. The limited speed of the camera's embedded computer for data pre-processing and IEEE-1394-Firewire communication extends the total data acquisition time of the image to several minutes.

Although the present system possesses these drawbacks, it is important to note that the current commercial PMD cameras are designed for time-of-flight measurements of directly modulated near-infrared radiation, and thereby are not optimized for the small EO modulation depth. The development of PMD cameras with larger pixel capacitance and more homogeneous read-out channels in the detector array could significantly improve the performance of the heterodyne EO THz imaging system presented here.

In order to take advantage of the higher electric fields of THz pulses, we also implemented a THz-pulse-based imaging system in which the EO signal is demodulated by means of a PMD camera [55]. The set-up, which is displayed in Fig. 9.16, is again a THz-illumination/optical-probing arrangement, but now the optical pulses are used for both the generation of the (pulsed) THz radiation and for the EO read-out.

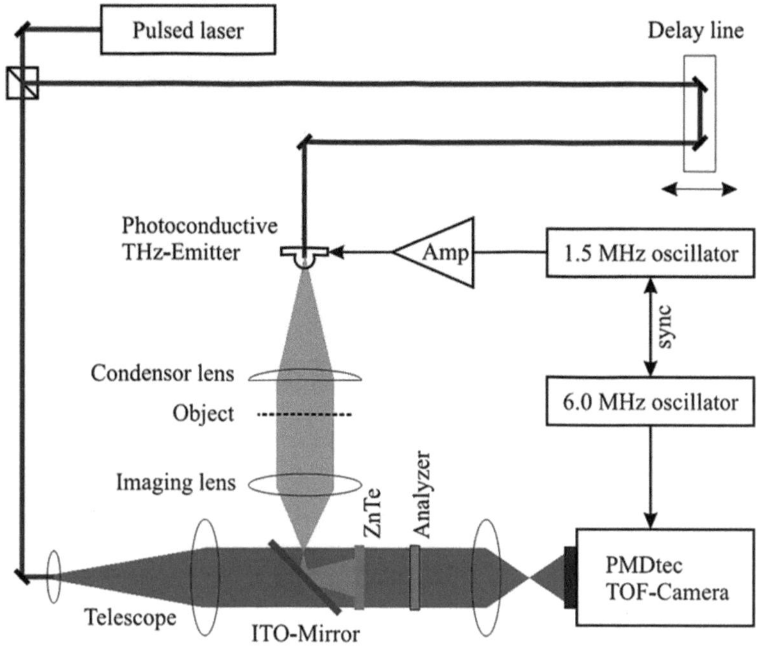

Fig. 9.16 Scheme of the THz-pulse EO imaging system comprising a modulated photoconductive THz-pulse emitter and a PMD camera for the demodulation of the EO signal. The NIR-converter of Fig. 9.14 is utilized; it is depicted here in more detail. In this system, the laser beam is split, one portion driving the photoconductive emitter, the other serving for EO sampling. The relative timing of the THz pulses and the optical pulses can be adjusted via the delay line. This allows imaging with high temporal resolution for the monitoring of the THz-pulse evolution e.g. for the extraction of spectral information (not exploited here)

Femtosecond laser pulses from a high-repetition-rate Ti:sapphire laser (Coherent Mira 900) generate THz pulses in a photoconductive antenna whose bias voltage – and with it the amplitude of the THz pulse train – is modulated. The average power of the THz radiation was determined to be ~2 W. With the THz beam, we perform transmission measurements. The THz signal is then read out in an EO crystal, where the information of the image and the modulation of the THz beam are transferred onto the optical read-out beam. There is no need for a reference detector, because the modulation signal of the emitter is fed into the PMD camera synchronously with the reference signal which allows use the full available THz power for the imaging path.

In contrast to the heterodyne set-up using the hybrid system described above, here, the THz beam is collimated and not focused, allowing imaging of small objects.

Figure 9.17 shows examples of THz images. Frame (a) depicts the beam profile whose amplitude distribution was then used to normalize the other THz images in order to enhance weakly illuminated areas at the boundary regions of the field of view.

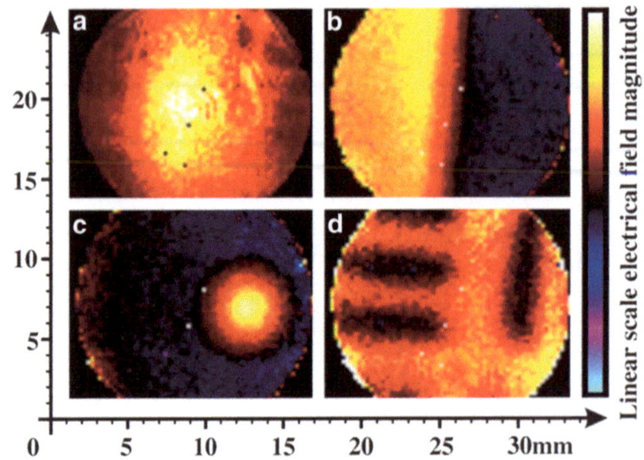

Fig. 9.17 THz images recorded with the THz-pulse imager comprising the PMD camera: (**a**) Beam profile; (**b**) edge of a steel sheet; (**c**) 4-mm diameter aperture; (**d**) part of element −2.2 of an USAF resolution target. Images (**b**)–(**d**) have been normalized to beam profile to enhance weakly illuminated regions

The following frames show: (b) an edge of a steel sheet, (c) an aperture with a diameter of 4 mm, and (d) element −2.2 of an USAF resolution target. The images altogether indicate that the lateral resolution is in the order of 2–3 mm. The image data were recorded with an effective integration time of 8 s, 2.5 s, 0.6 s, and 2.0 s, respectively. Owing to the comparatively high peak field of the femtosecond pulses from the Ti:sapphire, this imaging setup yields a dynamic range of 44 dB/Hz. These results demonstrate that the PMD-based EO-imaging system performance improves rapidly with higher THz fields (i.e. EO modulation depths), and is already an attractive option for a range of fs-optoelectronic-based THz imaging applications.

9.3.3 Summary and Potential of Optoelectronic THz Imaging Systems

In comparing the two (quasi)-CW EO THz imaging systems presented in the previous sections (Sects. 9.3.1 and 9.3.2), despite certain issues for each system, the current performance of the heterodyne system (dynamic range 23 dB/Hz) is significantly higher than that of the homodyne system (9 dB/Hz). Indeed, this could be expected to be the case, as the CW-hybrid heterodyne system with a PMD camera can exploit the far better noise rejection at the 10 MHz detection frequency (which is above the noise bandwidth of many laser systems).

We note, however, that for both systems the multi-pixel performance is degraded significantly compared to that of the corresponding single-pixel detection – i.e.

with single PIN photodiodes and a lock-in amplifier (with the same detection frequency) – even though our imaging measurements here were made with focused THz beams. As described above for each system, this indicates that the key direction for improvement at present is on the implementation of the optical camera detection (CCD, PMD), in order to approach the shot-noise detection limit where a far higher dynamic range can be expected. For applications where a reduced pixel number would be acceptable, an interesting alternative is to use PIN photodiode arrays (e.g. a line array with ~50 detectors) [14], although for the heterodyne system this would require large-scale multi-channel lock-in amplifier electronics, which are already integrated elegantly and inexpensively in the PMD camera design.

Based on our existing results, the dynamic range of the PMD-based EO-imaging system is significantly improved when using a fs-laser-pumped THz emitter, due to the high peak fields of the THz pulses. Moreover, the all-optoelectronic system allows one to gain broadband spectral information by measuring the time-domain THz waveform at each pixel by scanning the detection-pulse delay. However, for such imaging applications where spectral information is not desired, the fs-approach comes at the price of having to scan the delay to find the peak field signal for each pixel. Hence for many applications, e.g. long-range stand-off imaging of objects with depth variation on the scale of millimeters to centimeters, this can significantly reduce the acquisition rate.

9.4 THz Focal Plane Arrays

Coming back to electronic techniques of THz detection, we now address monolithically integrated receiver arrays suitable for room temperature operation.

It was often expected and hoped that high-sensitivity detector arrays would be developed on the basis of Schottky diodes, especially considering that they cannot only be used as power detectors but are also suitable for heterodyne detection. Until now, however, fabrication of such arrays suffers from an insufficient yield and significant performance fluctuations, although this situation may be changing gradually [61].

In 2005 and 2006, two unrelated developments gave the search for cost-effective and high-performance arrays a push into new and somewhat unexpected directions. The first development relates to detectors of infrared radiation. The group of Q. Hu at the MIT found out that microbolometer arrays, made for the detection of infrared radiation, are surprisingly good detectors of radiation with a frequency of several THz, emitted from quantum-cascade lasers and molecular gas lasers [62, 63]. These findings have initiated numerous activities, e.g. at LETI, Grenoble, aimed at the optimization of microbolometers for the THz frequency regime [64]. Also other types of IR sensor, such as pyroelectric detectors, are now explored more vigorously [65].

The second development, which has triggered our work outlined below, relates to Si CMOS field-effect transistors (FETs). The group of W. Knap, Montpellier, investigated un-optimized commercial MOSFETs as THz detectors. They had performed similar studies with III/V-based transistors before, but the investigation of the MOSFETs was the first where the responsively and the NEP were determined tentatively [66]. The numbers obtained indicated much promise for further development.

9.4.1 Si FET-Based THz FPAs

The physical mechanism underlying this novel path towards real-time-capable FPAs is rectification of the THz signal in the FET. To our knowledge, it was the work of Diakonov and Shur [67], which, for the first time, drew attention to rectification of high-frequency signals whose frequency exceeds the FET's transit-time-limited cut-off frequency. They considered the possibility of plasma resonances in the gated two-dimensional electron gas of the sub-micron-long transistor channels. Although the excitation of collective plasma oscillations had been an object of intensive studies for quite a long time [68], a novelty of the Diakonov-Shur approach was the prediction that the existence of plasmonic resonances should also manifest itself in the device's DC properties. This prediction ignited considerable interest to search for plasmon-based rectification phenomena in various implementations of FETs by simply measuring a THz-field-induced potential difference between the source and drain terminals.

In the meantime, the theory of detection had been refined for various modalities of electron transport with one of them being the so-called case of strongly-damped plasmon excitation, also termed the non-resonant limit of plasmonic mixing [69]. This operation regime was then found by Knap's group to be relevant for Si MOSFETs with a gate longer than 100 nm and operated at room temperature. For this regime, a good sensitivity to THz radiation was reported. NEP values of the order of 100 pW/Hz$^{1/2}$ [66] were estimated, which are quite comparable to those of many other practical power detectors operating in the THz frequency range [70].

Lisauskas et al. [70] and Ojefors et al. [71] later pointed out that plasma-wave-based rectification in the non-resonant limit can be understood to be an extension of classical resistive mixing, well-known in micro-electronics. Resistive mixing is based on the quasi-stationary behavior of FETs; if extended to the non-quasi-stationary case, plasma-wave excitation has to be included. The relationship to resistive mixing is reflected now in the newly-coined term of *distributed resistive (self-)mixing*.

The experimental results led to the first demonstration of imaging with transistors above their cut-off frequency, in this case employing GaAs HEMTs [72]. It followed the first dedicated attempt to fabricate monolithically integrated FPAs, where each detector pixel contained an on-chip integrated antenna, a differential pair of FETs and an amplifier, with a commercial foundry process [70, 71, 73, 74]. Figure 9.18 shows a photograph and some layout information of the best detectors of these first designs which were fabricated with a 250-nm (Bi)CMOS process. For these detectors of 0.65-THz radiation, a responsively of 80 kV/W and a NEP of 300 pW/Hz$^{1/2}$ were obtained.

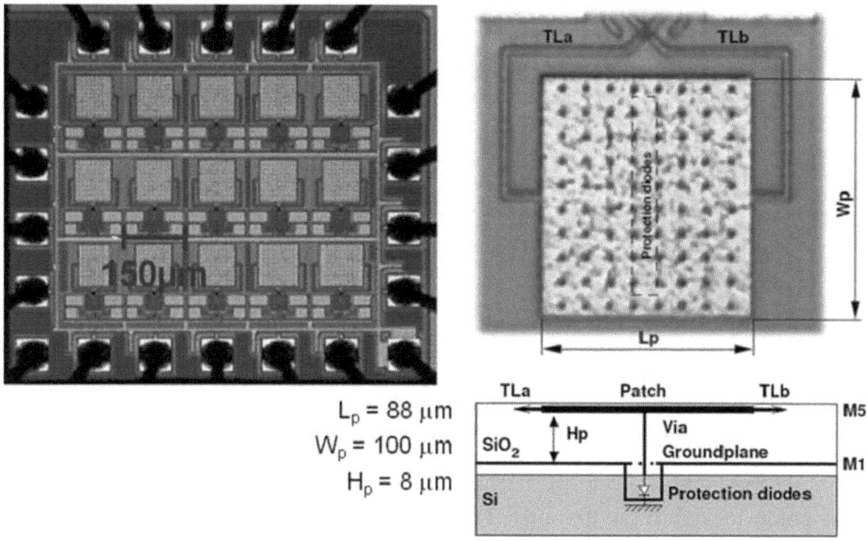

Fig. 9.18 *Left side*: 3 × 5-pixel FPA for the detection of 0.65-THz radiation. *Top right side*: one of the patches of the array. The transmission lines (TLa and TLb) guiding the THz signals to the FETs are visible. *Bottom right*: Cross-section through a patch antenna (Image data from Ojefors et al. [71])

In the meantime, various commercial Si process technologies and different circuit designs have been explored. Two papers report significantly improved performance. First, an implementation of this detection principle in a high-end 65-nm CMOS-on-SOI technology allowed to reach NEP values as low as 50 $pW/\sqrt{Hz}$ [75]. The applied design has room for improvement. A second paper reports slightly better NEP values of 43 $pW/Hz^{1/2}$; in this case, a more relaxed and significantly less costly 150-nm CMOS technology was used [76]. It is worth remarking, that – in the lowest-order approximation of the transport models – the detector's responsivity does not depend on channel length, which partially explains why a very similar NEP of about 50 $pW/Hz^{1/2}$ can be achieved using different channel lengths.

It is important to note, that the detection principle is entirely based on the field effect, and not on real carrier transport, thus the high-frequency limit is determined by dielectric relaxation time which typically lies in the range of several THz and strongly exceeds the transistor's cut-off frequency. The latter represents an ultimate limiting factor for signal modulation frequencies [77] or the response time. Therefore, in contrast to thermal-time-limited detectors with similar NEP, FET-based detectors allow modulation speeds reaching up to 1 MHz and above (limited by capacitive loading because of the high impedance of the transistor's channel [70, 78]). In their high speed, they are similar to Schottky diode detectors.

This property allows increase the sensitivity of the FETs by application of the heterodyne detection principle. We demonstrated heterodyne imaging with a 0.65-THz emitter [79, 80]. As LO, we employed a second radiation source which

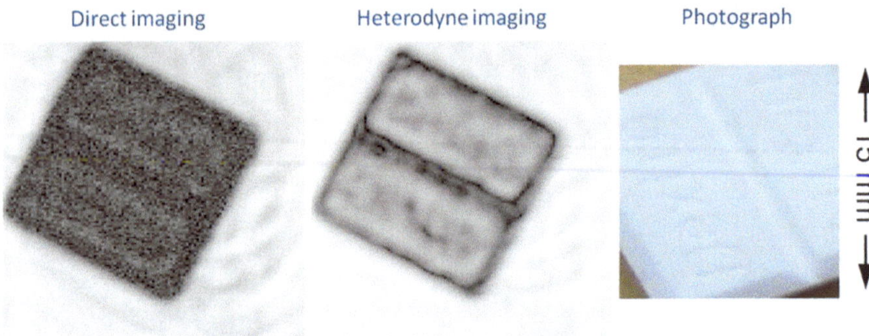

Fig. 9.19 THz image taken through a dextrose tablet (see photograph on the *right side*) hidden in a letter envelope. *Left side*: Detection of transmitted power alone; *middle*: heterodyne measurement. The measurements were taken with two FET detectors in parallel, the object was raster-scanned

was phase-locked to the first one. The radiation of the first source was focused to illuminate the object which was raster scanned through the focus. The transmitted radiation was then overlaid with that from the LO at a beam combiner, guided to the detectors, and distributed over the entire FPA. We estimate that each FET detector received only $\sim 2\ \mu W$ of LO power, far less than needed for optimal performance [80]. This was sufficient, however, to improve the dynamic range by 29 dB [79]. The NEP was estimated to be 8 fW/Hz (-112 dBm/Hz).

The improvement of the contrast can be seen from the imaging data shown in Fig. 9.19. It presents data of transmission measurements through a paper envelope containing a dextrose tablet. In power-detection mode (radiation from LO source blocked), the received signal is weak, and details such as the central notch of the tablet are nearly invisible. In heterodyne mode, the considerably higher contrast permits to clearly identify the notch. The writing on the tablet can be noticed, although the image shown here does not allow discern individual characters. In addition to the improvement in sensitivity, a heterodyne approach gives access to the information on the THz phase which can be used for reconstruction of depth information, hence offers some 3D imaging capabilities. An illustration of this aspect is given in Fig. 9.20 which presents the power and the phase information obtained by a heterodyne scan (single scan, 10-ms integration time per pixel) of a leaf of a ficus tree. The processed phase information renders a pseudo-3D representation of the leaf.

9.4.2 Summary and Potential of Si FET-Based THz FPAs

We point out that, until now, all images have been taken in raster scan mode, albeit with fairly short integration time in the ms to 10-ms range. Currently, we and others are working to-wards real-time implementations of MOSFET-based imaging.

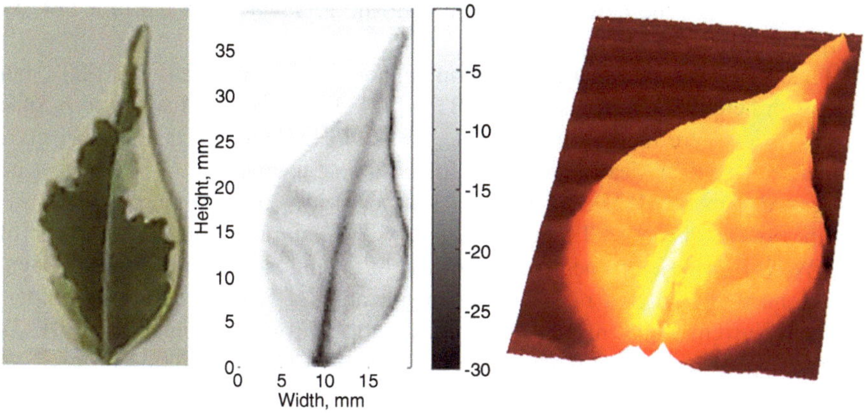

Fig. 9.20 Transmission image of a freshly cut leaf of a ficus benjamina, at 592 GHz. *Left panel*: photograph of the leaf; *middle panel*: transmitted-power image; *right panel*: pseudo-3D image obtained after unwrapping of the phase information (Image data from Boppel et al. [81]). The measurements were taken with a single FET detector, the object was propagated in x-y-direction for raster scanning

This requires both a large number of pixel, distributed over an area of the FPA of several to many cm^2, and implementation of a proper read-out electronics.

Based on the data given above, we can estimate the expected performance of an FPA for 600 GHz in heterodyne mode. Assuming a power of 0.5 mW of each the LO and the THz imaging beam directly in front of the FPA, and further assuming equal distribution of this power across it (containing 256×256 detectors over an area of $50 \times 50\ mm^2$), then a dynamic range of 30 dB would be achieved at 30-Hz frame-rate.

In many real applications, the power of the THz imaging beam at the FPA is likely to be lower, as 1 mW is about the maximum available source power at 600 GHz today (considering oscillator-multiplier sources; backward-wave oscillators may provide somewhat more). Coupling and imaging losses have to be taken into account. Furthermore, one of the challenges for the success of an imaging technology with monolithically integrated FPAs is the large wavelength of THz radiation (1 THz corresponding to $\lambda_{vac} = 300\ \mu m$). Diffraction aspects suggest that a FPA with a large number of monolithically integrated pixels is more practical at high frequencies (0.5 THz and above) and in conjunction with large numerical-aperture optics, while its usefulness is questionable at frequencies approaching 100 GHz. FPAs based on CMOS FETs will be used at first for short-distance, high-resolution imaging, rather than compete with all-electronic imaging approaches of stand-off security scanners.

The FET FPAs investigated until now have been designed for narrow-band operation. The underlying detection principle is not of resonant nature, though. The bandwidth limitation results from the choice of the integrated antenna. The future will certainly see the development of broadband FET detectors, e.g., for FMCW purposes or spectroscopic applications.

Finally, we mention that rectification cannot only be achieved with FETs, but also with heterobipolar transistors (HBTs) [82]. This was tested with *n-p-n* SiGe HBTs which were fabricated with a $f_{max} = 220$ GHz SiGe:C process. Each pixel contained a monolithically integrated folded-dipole antenna for 645 GHz, but no integrated amplifier. While the measured responsivity (850 V/W) at 645 GHz was comparable to that of FETs, the NEP was, with minimum values in the range of 30 pW/Hz$^{1/2}$, inferior by nearly three orders of magnitude. The main reason being that – unlike the case of FETs – detector operation of HBTs requires an emitter-collector current which introduces significant noise. From this point of view, HBTs do not appear to be an attractive alternative to FETs and HEMTs. However, an interesting sub-harmonic mixing approach, detecting 650-GHz radiation by mixing with the fourth subharmonic at about 162 GHz, has recently been demonstrated with them [83]. Subharmonic mixing may help to render heterodyne detection of THz radiation more practical and less cost-intensive because it alleviates the need for additional power sources as LOs at the fundamental frequency.

9.5 Conclusions and Outlook

This paper has introduced respectively reviewed five novel THz imaging modalities. The main thrust of the work has been to identify viable approaches of parallel multi-pixel detection with the goal to achieve real-time imaging capabilities.

Of the THz imaging systems described in this paper, the all-electronic 812-GHz imager with mechanical scanning has already demonstrated operation close to real-time mode, albeit with limited dynamic range. The second all-electronic system – combining synthetic-aperture techniques and mechanical scanning with range determination by FMCW sweeping in the range of 220–320 GHz – is designed to achieve several frames per second, but this could not yet be reached because of the limited output power and sensitivity of the components at 220–320 GHz. Therefore a similar system will be realized at 75–110 GHz within the new LiveDetect3D project. We have verified already that components show sufficient dynamic range so that real-time operation should be possible even with a 10-m working distance.

The two laser-based approaches, which employ EO THz-NIR conversion to permit multi-pixel detection with NIR cameras (i.e. either homodyne detection with a CCD or heterodyne detection with a PMD), still have a way to go before they reach a frame-rate of 1 fps and more. They are attractive because they can principally cover a significant portion of the THz frequency regime (hundreds of GHz to at least several THz) and thus may well be able to combine imaging with spectroscopic distinction of objects. However, camera-based multi-pixel detection for the (quasi-) CW systems still requires improvement. THz-OPO-based imaging would benefit enormously from an OPO with higher peak power than was available here, while the approach using a time-of-flight camera needs a camera with faster read-out. For both systems, noise suppression leaves room for optimization.

The fifth system, again an electronic one, employs a monolithically integrated Si MOSFET FPA for detection around 600 GHz (but the approach is also principally capable of broad-band operation). The focus of the work has been on optimization of the performance of these novel detectors. The results are favorable and strongly suggest that the near future will see implementation of larger FPAs with suitable camera read-out for imaging at high frame rates.

In summary, the capability to capture THz images of actively illuminated scenes in or close to real time has progressed significantly. We have identified promising approaches which exploit various ways of multi-pixel parallel detection. With the exception of the silicon CMOS FPA, which also allows direct power detection, all approaches employ coherent detection for better noise suppression, either in homodyne mode, or via heterodyne mixing for additional enhancement of the sensitivity. It has become clear that the limited available THz power for scene illumination necessitates to push sensitivity enhancement and noise suppression in the detectors to their limits in order to achieve real-time operation of imaging with good dynamic range. We have identified ways for specific improvements for each of the various approaches which now have to be implemented.

Acknowledgments The work presented here was funded by numerous sources, including the Federal Ministry of Education and Research Germany (BMBF) through projects LYNKEUS, TERACAM, TEKZAS, and LiveDetect3D; the German Research Foundation (DFG) through PAK-73 "Dynamisches 3D Sehen mit PMD"; the European Space Agency ESA/ESTEC (contract no. 21155/07/NL/ST); WI Bank Hessen; Oerlikon AG. We are grateful for contributions by TOPTICA GmbH (Munich); Xiton GmbH (Kaiserslautern); D. Molter, J. Jonuscheit, and R. Beigang of the Fraunhofer-Institut IPM (Kaiserslautern); J. Dall, A. Kusk, V. Zhurbenko and T. Jensen of Denmark Technical University.

References

1. Liu H-B, Zhong H, Karpowicz N, Chen Y, Zhang X-C (2007) Terahertz spectroscopy and imaging for defense and security applications. Proc IEEE 95:1514–1527
2. Fischer BM, Demarty Y, Schneider M, Löffler T, Keil A, Quast H (2010) THz all-electronic 3D imaging for safety and security applications. Proc SPIE 7671, 767111–767111-7
3. Davies AG, Burnett AD, Fan WH, Linfield EH, Cunningham JE (2008) Terahertz spectroscopy of explosives and drugs. Mater Today 11:18–26
4. Krumbholz N, Hochrein T, Vieweg N, Hasek T, Kretschmer K, Bastian M, Mikulic M, Koch M (2009) Monitoring polymeric compounding processes inline with THz time-domain spectroscopy. Polym Test 28:30–35
5. Hasegawa N, Löffler T, Thomson M, Roskos HG (2003) Remote identification of protrusions and dents on surfaces by terahertz reflectometry with spatial beam filtering and out-of-focus detection. Appl Phys Lett 83:3996–3998
6. Banerjee D, von Spiegel W, Thomson MD, Schabel S, Roskos G (2008) Diagnosing water content in paper by terahertz radiation. Opt Express 16(12):9060–9066

7. Kawase K, Shibuya T, Hayashi S'i, Suizu K (2010) THz imaging techniques for nondestructive inspections. Comptes Rendus Physique 11:510–518
8. Hils B, Thomson MD, Löffler T, von Spiegel W, am Weg C, Roskos H, de Maagt P, Doyle D, Geckeler RD (2008) Terahertz profilometry at 600 GHz with 0.5 m depth resolution. Opt Express 16:11289–11293
9. Roggenbuck A, Schmitz H, Deninger A, Cámara Mayorga I, Hemberger J, Gsten R, Grüninger M (2010) Coherent broadband continuous-wave terahertz spectroscopy on solid-state samples. New J Phys 12:043017
10. Strachan CJ, Taday PF, Newnham DA, Gordon KC, Zeitler JA, Pepper M, Rades T (2005) Using terahertz pulsed spectroscopy to quantify pharmaceutical polymorphism and crystallinity. J Pharm Sci 94:837–846
11. Nagel M, Richter F, Haring Bolívar P, Kurz H (2003) A functionalized THz sensor for marker-free DNA analysis. Phys Med Biol 48:3625–3636
12. Cooper KB, Dengler RJ, Llombart N, Talukder A, Panangadan AV, Peay CS, Mehdi I, Siegel PH (2010) Fast, high-resolution terahertz radar imaging at 25 meters. Proc SPIE 7671:76710Y
13. Cooper KB, Dengler RJ, Chattopadhyay G, Schlecht E, Skalare A, Mehdi I, Siegel PH (2008) Penetrating 3-D imaging at 4- and 25-m range using a submillimeter-wave radar. IEEE Microw Theory Tech 56(12):2771–2778
14. Pradarutti B, Müller R, Matthäus G, Brückner C, Riehemann S, Notni G, Nolte S, Tunnermann A (2007) Multichannel balanced electro-optic detection for terahertz imaging. Opt Express 15:17652–17660
15. Pradarutti B, Müller R, Freese W, Matthäus G, Riehemann S, Notni G, Nolte S, Tunnermann A (2008) Terahertz line detection by a microlens array coupled photoconductive antenna array. Opt Express 16:18443–18450
16. Dai JM, Liu J, Zhang X-C (2011) Terahertz wave air photonics: terahertz wave generation and detection with laser-induced gas plasma. IEEE J Sel Top Quantum Electron 17:183–190
17. Malcolm G (2011) Terahertz laser sources based on optical parametric oscillators. In: Perenzoni M, Paul D (eds) Proceedings of the 6th Optoelectronics and Photonics Winter School, Fai della Paganella, 20–26 Feb 2011
18. Wohnsiedler S, Theuer M, Herrmann M, Islam S, Jonuscheit J, Beigang R, Hase F (2009) Simulation and experiment of terahertz stand-off detection. Proc SPIE 7215:72150H
19. von Spiegel W, am Weg C, Henneberger R, Zimmermann R, Löffler T, Roskos HG (2009) Active THz imaging system with improved frame rate. Proc SPIE 7311:73110O
20. von Spiegel W, am Weg C, Henneberger R, Zimmermann R, Roskos HG (2010) Illumination aspects in active terahertz imaging. IEEE Trans Microw Theory Tech 58:2008–2013
21. am Weg C, von Spiegel W, Henneberger R, Zimmermann R, Löffler T, Roskos HG (2009) Quasioptical system design. Proc SPIE 7215:72150R
22. am Weg C, von Spiegel W, Henneberger R, Zimmermann R, Löffler T, Roskos HG (2009) Fast active THz cameras with ranging capabilities. Infrared Millim THz Waves 30:1281–1296
23. am Weg C, von Spiegel W, Henneberger R, Zimmermann R, Löffler T, Roskos HG (2009) Fast active THz camera with range detection by frequency modulation. Proc SPIE 7215:72150F
24. Kemp MC (2006) Millimetre wave and terahertz technology for the detection of concealed threats a review. Proc SPIE 6402:64020D
25. Krozer V, Löffler T, Dall J, Kusk A, Eichhorn F, Olsson RK, Buron J, Jepsen PU, Zhurbenko V, Jensen T (2010) THz imaging systems with aperture synthesis techniques. IEEE Trans Microw Theory Tech 58:2027–2039
26. Miyashiro K, Schellenberg J, Loveberg J, Kolinko V, McCoy J (2007) An E-band electronically scanned imaging radar system. In: Proceedings of the IMS, IEEE/MTT-S International Microwave Symposium, Honolulu, Hawaii
27. Manasson V, Sadovnik L, Mino R, Rodionov S (2000) Novel passive millimeter-wave imaging system: prototype fabrication and testing. Proc SPIE 4032:2–13
28. Natarajan A, Komijani A, Guan X, Babakhani A, Wang Y, Hjimiri A (2006) A 77GHz phased-array transmitter with local LOPath phase-shifting in silicon. IEEE J Solid State Circ 41: 2795–2806

29. Schulwitz L, Mortazawi A (2005) A compact dual-polarized multibeam phased-array architecture for millimeter-wave radar. IEEE Trans Microw Theory Tech 53(11):3588–3594
30. Skou N, Le Vine D (2006) Microwave radiometer systems; design and analysis, 2nd edn. Artech House, Boston
31. Ruf CS (1993) Numerical annealing of low-redundancy linear arrays. IEEE Trans Antenna Propag 41:85–90
32. Yegulalp AF (1999) Fast backprojection algorithm for synthetic aperture radar. In: Radar conference, 1999. The record of the 1999 IEEE, Waltham, MA, USA, pp 60–65
33. Basu S, Bresler Y (2002) O(N3logN) backprojection algorithm for Radon transform. IEEE Trans Med Imaging 21:76–88
34. Ulander LMH, Hellsten H, Stenström G (2003) Synthetic-aperture radar processing using fast factorized back-projection. IEEE Trans Aerosp Electron Syst 39:760–776
35. Wu Q, Hewitt TD, Zhang X-C (1996) Two-dimensional electro-optic imaging of THz beams. Appl Phys Lett 69:1026–1028
36. Yasuda T, Kawada Y, Toyoda H, Takahashi H (2007) Terahertz movies of internal transmission images. Opt Express 15:15583–15588
37. Nahata A, Yardley JT, Heinz TF (2002) Two-dimensional imaging of continuous-wave terahertz radiation using electro-optic detection. Appl Phys Lett 81:963–965
38. Kawase K, Ogawa Y, Minamide H, Ito H (2005) Terahertz parametric sources and imaging applications. Semicond Sci Technol 20:S258–S265
39. Edwards TJ, Walsh D, Spurr MB, Rae CF, Dunn MH, Browne PG (2006) Compact source of continuously and widely-tunable terahertz radiation. Opt Express 14:1582–1589
40. Zhao G, Schouten RN, van der Valk N, Wenckebach WT, Planken PC (2002) Design and performance of a THz emission and detection setup based on a semi-insulating GaAs emitter. Rev Sci Instrum 73:1715–1719
41. Dreyhaupt A, Winnerl S, Dekorsy T, Helm M (2005) High-intensity terahertz radiation from a microstructured large-area photoconductor. Appl Phys Lett 86:121114
42. Loffler T, Kreß M, Thomson M, Hahn T, Hasegawa N, Roskos HG (2005) Comparative performance of terahertz emitters in amplifier-laser-based systems. Semicond Sci Technol 20:S134–S141
43. Bartel T, Gaal P, Reimann K, Woerner M, Elsaesser T (2005) Generation of single-cycle THz transients with high electric-field amplitudes. Opt Lett 30:28052807
44. Thomson MD, Blank V, Roskos HG (2010) Terahertz white-light pulses from an air plasma photo-induced by incommensurate two-color optical fields. Opt Express 18:23173–23182
45. Meng F, Molter D et al THz imaging with a THz-OPO and a CMOS camera (unpublished)
46. Meng FZ, Thomson D, Molter D, Löffler T, Bartschke J, Bauer T, Nittmann M, Roskos HG (2010) Coherent electro-optical detection of THz radiation from an optical parametric oscillator. Opt Express 18:11316–11326
47. Molter D, Theuer M, Beigang R (2009) Nanosecond terahertz optical parametric oscillator with a novel quasi phase matching scheme in lithium niobate. Opt Express 17:6623–6628
48. Bauer T, Kolb JS, Löffler T, Mohler E, Roskos HG, Pernisz UC (2002) Indium-tin-oxide-coated glass as dichroic mirror for far-infrared electromagnetic radiation. J Appl Phys 92:2210–2212
49. Meng F, Thomson MD, Blank V, von Spiegel W, Loffler T, Roskos G (2009) Characterizing large-area electro-optic crystals toward two-dimensional real-time terahertz imaging. Appl Opt 48:51975204
50. Wu Q, Zhang X-C (1995) Free-space electro-optic sampling of terahertz beams. Appl Phys Lett 67:3523–3525
51. Loffler T, May T, am Weg C, Alcin A, Hils B, Roskos HG (2007) Continuous-wave terahertz imaging with a hybrid system. Appl Phys Lett 90:091111
52. May T, am Weg C, Alcin A, Hils B, Loffler T, Roskos HG (2007) Towards an active real-time THz camera: First realization of a hybrid system. Proc SPIE 6549:654907
53. Friederich F, Schuricht G, Deninger A, Lison F, Spickermann G, Haring Bolívar P, Roskos HG (2010) Phase-locking of the beat signal of two distributed-feedback lasers to oscillators working in the MHz to THz range. Opt Express 18:8621–8629

54. Spickermann G (2012) Terahertz-Bildgebung mit demodulierendem Detektorarray, dissertation, Universität Siegen, Germany
55. Spickermann G, Friederich F, Roskos HG, Haring Bolívar P (2009) A high signal-to-noise ratio electrooptical THz imaging system based on an optical demodulating detector array. Opt Lett 34:3424–3426
56. Friederich F, Spickermann G, Roggenbuck A, Deninger A, am Weg C, von Spiegel W, Lison F, Haring Bolívar P, Roskos HG (2010) Hybrid continuous-wave demodulating multipixel terahertz imaging systems. IEEE Trans Microw Theory Tech 58:2022–2026
57. Ringbeck T, Möller T, Hagebeuker B (2007) Multidimensional measurement by using 3-D PMD sensors. Adv Radio Sci 5:135–146
58. Ringbeck T (2007) A 3D time of flight camera for object detection. Presented at the 8th conference on optical 3-D measurement techniques, ETH Zurich, Zurich, Switzerland, July 2007
59. Gallot G, Grischkowsky D (1999) Electro-optic detection of terahertz radiation. J Opt Soc Am B 16:1204–1212
60. Jiang Z, Sun FG, Chen Q (1999) Electro-optic sampling near zero optical transmission point. Appl Phys Lett 74:1191–1193
61. Ortolani M, Di Gaspare A, Casini R (2011) Progress in producing terahertz detector arrays. SPIE Newsroom. Feb. 14, 2011, doi:10.1117/2.1201101.003449
62. Lee AWM, Hu Q (2005) Real-time, continuous-wave terahertz imaging by use of a microbolometer focal-plane array. Appl Phys Lett 30:2563–2565
63. Lee AWM, Qin Q, Kumar S, Williams BS, Hu Q (2006) Real-time terahertz imaging over a standoff distance (> 25 meters). Appl Phys Lett 89:141125
64. Simoens F, Durand T, Meilhan J, Gellie P, Maineult W, Sirtori C, Barbieri S, Beere H, Ritchie D (2009) Terahertz imaging with a quantum cascade laser and amorphous-silicon microbolometer array. Proc SPIE 7485:74850M
65. Li Q, Ding S-H, Yao R, Wang Q (2010) Real-time terahertz scanning imaging by use of a pyroelectric array camera and image denoising. J Opt Soc Am A 27:2381–2386
66. Tauk R, Teppe F, Boubanga S, Coquillat D, Knap W, Meziani YM, Gallon C, Boeuf F, Skotnicki T, Fenouillet-Beranger C, Maude K, Rumyantsev S, Shur MS (2006) Plasma wave detection of terahertz radiation by silicon field effects transistors: responsivity and noise equivalent power. Appl Phys Lett 89:253511
67. Dyakonov M, Shur M (1996) Detection, mixing, and frequency multiplication of terahertz radiation by two-dimensional electronic fluid. IEEE Trans Electron Devices 43:380–387
68. Allen SJ, Tsui DC, Logan RA (1977) Observation of the two-dimensional plasmon in silicon inversion layers. Phys Rev Lett 38:980983
69. Knap W, Kachorovskii V, Deng Y, Rumyantsev S, Lü J-Q, Gaska R, Shur MS, Simin G, Hu X, Asif Khan M, Saylor CA, Brunel LC (2002) Nonresonant detection of terahertz radiation in field effect transistors. J Appl Phys 91:9346–9353
70. Lisauskas A, Pfeiffer U, Ojefors E, Haring Bolívar P, Glaab D, Roskos HG (2009) Rational design of high-responsivity detectors of terahertz radiation based on distributed self-mixing in silicon field-effect transistors. J Appl Phys 105:114511
71. Ojefors E, Pfeiffer U, Lisauskas A, Roskos HG (2009) A 0.65 THz focal-plane array in a quarter-micron CMOS process technology. IEEE J Solid State Circ 44:1968–1976
72. Lisauskas A, von Spiegel W, Boubanga-Tombet S, El Fatimy A, Coquillat D, Teppe F, Dyakonova N, Knap W, Roskos HG (2008) Terahertz imaging with GaAs field-effect transistors. Electron Lett 44:408–409
73. Lisauskas A, Glaab D, Roskos HG, Ojefors E, Pfeiffer U (2009) Terahertz imaging with Si MOSFET focal-plane arrays. Proc SPIE 7215:72150
74. Ojefors E, Lisauskas A, Glaab D, Roskos HG, Pfeiffer UR (2009) Terahertz imaging detectors in CMOS technology. J Infrared Millim THz Waves 30:1269–1280
75. Ojefors E, Baktash N, Zhao Y, Al Hadi R, Sherry H, Pfeiffer UR (2010) Terahertz imaging detectors in a 65-nm CMOS SOI technology. In: Proceedings of the ESSCIRC, Seville, Spain, pp 486–489, September 2010

76. Boppel S, Lisauskas A, Krozer V, Roskos HG (2011) Performance and performance variations of sub-1-THz detectors fabricated with a 0.15-m CMOS foundry process. Electron Lett 47(11):661–662 (appears on May 26, 2011)
77. Stillman WJ, Shur MS (2007) Closing the gap: plasma wave electronic terahertz detectors. J Nanoelectron Optoelectron 2:209–221
78. Stillman W, Shur MS, Veksler D, Rumyantsev S, Guarin F (2007) Device loading effects on nonresonant detection of terahertz radiation by silicon MOSFETs. Electron Lett 43:422–423
79. Glaab D, Boppel S, Lisauskas A, Pfeiffer U, Ojefors E, Roskos HG (2010) Terahertz heterodyne detection with silicon field-effect transistors. Appl Phys Lett 96:042106
80. Pfeiffer U, Ojefors E, Lisauskas A, Glaab D, Roskos HG (2009) A CMOS focal-plane array for heterodyne terahertz imaging. In: IEEE RFIC symposium, Boston, MA, USA, pp 433–436
81. Boppel S, Lisauskas A, Pfeiffer U, Ojefors E, Roskos HG (2010) Field effect transistors for power and heterodyne detection of terahertz radiation fabricated in CMOS technology. In: Proceedings of the of NATO SET Panel Meeting SET-159 Terahertz and other electromagnetic wave techniques for defence and security, Vilnius, 3–4 May 2010
82. Pfeiffer UR, Ojefors E, Lisauskas A, Roskos HG (2008) Opportunities for silicon at mmwave and terahertz frequencies. In: IEEE BCTM proceedings (Proceedings of the 2008 bipolar/BICMOS circuits and technology meeting), Monterey, CA, USA, pp 149–156
83. Ojefors E, Pfeiffer UR (2010) A 650 GHz receiver front-end for terahertz imaging arrays. In: IEEE International Solid-State Circuits Conference ISSCC, San Francisco, CA, USA, pp 430–432

Chapter 10
Recent Results on Broadband Nanotransistor Based THz Detectors

Wojciech Knap, Dimitry B. But, N. Dyakonova, D. Coquillat, A. Gutin, O. Klimenko, S. Blin, F. Teppe, M.S. Shur, T. Nagatsuma, S.D. Ganichev, and T. Otsuji

Abstract Nanometer size field effect transistors can operate as efficient detectors of terahertz radiation that means far beyond their fundamental cut-of frequency. This work is an overview of some recent results concerning the low temperatures operation, linearity, circular polarization studies and double grating gate structures of nanometer scale field effect transistors working as terahertz detectors.

W. Knap (✉)
Laboratoire Charles Coulomb UMR 5221 Universitè Montpellier 2 & CNRS, 24950 Montpellier, France

Institute of High Pressure Physics UNIPRESS PAN, 02-845 Warsaw, Poland
e-mail: knap.wojciech@gmail.com

D.B. But • N. Dyakonova • D. Coquillat • F. Teppe
Laboratoire Charles Coulomb UMR 5221 Universitè Montpellier 2 & CNRS, 24950 Montpellier, France

A. Gutin • M.S. Shur
Rensselaer Polytechnic Institute, Troy, 12180 New York, USA

O. Klimenko • S. Blin
Institute Electronique du Sud Universite Montpellier 2 & CNRS, 34950 Montpellier, France

T. Nagatsuma
Graduate School of Engineering Science, Osaka University, Toyonaka, Osaka, Japan

S.D. Ganichev
Terahertz Center, University of Regensburg, Regensburg 93040, Germany

T. Otsuji
RIEC, Tohoku University, 2-1-1 Katahira, Aoba-ku, Sendai 981-8577, Japan

C. Corsi and F. Sizov (eds.), *THz and Security Applications*, NATO Science for Peace and Security Series B: Physics and Biophysics, DOI 10.1007/978-94-017-8828-1_10,

10.1 Introduction

Terahertz (THz) imaging, wireless communication and other well-known applications suffer, from the lack of low-cost detector arrays operating at room temperature.

In the early 1990s the first theoretical works [10] predicted that the channel of a FET could act as a resonator for plasma waves. These waves have typical velocities $s \sim 10^6$ m/s. The fundamental frequency f of this resonator depends on its dimensions and, for nanometer gate lengths ($L_0 \leq 10^{-6}$ m), can reach the terahertz (1 THz $= 10^{-12}$ s^{-1}) range, since $f \sim s/L_0$. They have shown also that the nonlinear properties of the 2D plasma in the transistor channel could be used for detection and mixing of THz radiation [11]. The first works treated mainly the resonant case – when the channel of the transistor works as a resonator for plasma waves. However rectification and detection of THz radiation is also possible in the non-resonant case, when plasma waves decay at the distance smaller than the channel length. Typical length of this region, which effectively rectifies THz radiation, ranges from 20 to 300 nm. Both THz emission [12, 13, 15, 32], resonant detection [16, 28, 38, 39] and non-resonant [17, 30, 31, 38, 46, 56] detection, were observed at cryogenic and room temperatures, clearly demonstrating effects related to the plasma excitations. Currently, the most promising application appears to be the room temperature broadband THz detection in the non-resonant regime for imaging and communication applications.

The real large scale interest in using FETs as THz detectors started around 2004 after first experimental demonstration of sub-THz and THz detection in silicon CMOS FETs [34]. Soon later it was shown that Si-CMOS FETs can reach a noise-equivalent power competitive with the best conventional room temperature THz detectors [55]. Both pioneering works have clearly stated importance of Si-CMOS FETs, which present the advantages of room temperature operation, very fast response times, easy on-chip integration with read-out electronics and high reproducibility leading to straightforward array fabrication. Recent studies demonstrate the main detector characteristics, responsivity and noise equivalent power, within the same range as Schottky barrier diodes [23, 48, 49]. Recently focal plane arrays in silicon technology have been designed and used for imaging at frequencies reaching the 1 THz range [2, 3, 36, 41, 51]. Heterodyne detection using Si MOSFETs was also shown [20] and seems to be very promising approach for THz imaging applications. Improvement of plasmonic THz detection by using double-grating-gate field-effect transistor structures was also demonstrated [23, 43, 44, 58].

This paper is a short review of some recent results on terahertz detection using nanometer size FETs. The subjects were selected in a way to stress some new physical aspects and developments. More general and full review can be found in recently published a book chapter and a journal paper – see [29, 33].

Section 11.2 presents a summary of the plasmonic detection theory. Sections 11.3, 11.4, and 11.5 are devoted to the power dependence of the photoresponse, temperature dependence of the response, and helicity sensitive detection, respectively. Until now, most work on nanometer FETs detectors considered mainly THz imaging applications. In Sect. 11.6, we show the progress in overcoming the loading problems and demonstrate first results on the application of nanometer FETs as detectors in wireless communication with signal modulated in the GHz frequency range. In Sect. 11.7 we report on THz detectors based on asymmetric dual grating gate FETs that thanks to their original device structure get better coupling with THz radiation, and reach very high responsivities.

10.2 Hydrodynamic Theory of Plasmonic FETs Detection

Plasma waves propagating in a FET channel have a linear dispersion law [7, 14, 40]

$$\omega = sk. \tag{10.1}$$

Here ω is frequency, $s = (qU_0/m)^{1/2}$ is the plasma wave velocity, q is the electronic charge, m is the electron effective mass, V_g is the gate-to channel voltage swing, $U_0 = V_g - V_{th}$ is the gate to channel voltage, V_{th} is the threshold voltage, and k is the wave vector. THz radiation impinging on the device induces THz voltages between gate and source and/or gate and drain of the FET, depending on the contact configuration and THz field polarization. The device nonlinearities lead to the photoresponse in the form of a dc voltage between source and drain or DC current flowing in the device channel and external circuit.

The basic equations describing the plasma waves in a FET [10] are the equation of motion (the Euler equation), continuity equation, and the equation describing the gradual channel approximation [52], which is valid when the characteristic scale of the potential variation in the device channel is much larger than the gate-to-channel barrier thickness.

$$\frac{\partial v}{\partial t} + v\frac{\partial U}{\partial x} + \frac{q}{m}\frac{\partial U}{\partial x} + \frac{v}{\tau} = 0, \tag{10.2}$$

$$\frac{\partial n}{\partial t} + \frac{\partial (nv)}{\partial x} = 0, \tag{10.3}$$

$$qn_s = CU_0. \tag{10.4}$$

Here n_s is the surface carrier concentration, C is the gate capacitance per unit area, $\partial U/\partial x$ is the longitudinal electric field in the channel, $v(x, t)$ is the local electron velocity, τ is the momentum relaxation time. These equations do not account for the viscosity of the electronic fluid.

For an ideal intrinsic FET (with no parasitic elements), for the THz radiation inducing only the gate-to-source voltage and open boundary conditions at the drain,

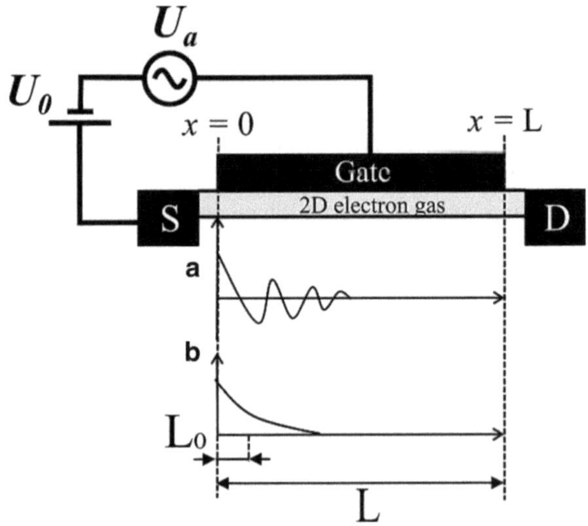

Fig. 10.1 Qualitative space dependence of electron oscillations excited by THz signal for (**a**) high frequency, high mobility detector where $\omega\tau \gg 1$ and (**b**) lower frequency, lower mobility detector where $\omega\tau \ll 1$. L_0 is the characteristic length of the decay of the plasma wave excited at source side of the channel

the detector response ΔU, which is the constant source-to-drain voltage induced by the incoming THz signal (see Fig. 10.1) [9]:

$$\frac{\Delta U}{U_o} = \frac{1}{4}\left(\frac{U_a}{U_o}\right)^2 f(\omega), \tag{10.5}$$

where

$$f(\omega) = 1 + \beta - \frac{1 + \beta\cos\left(2k_o' L\right)}{\sinh^2\left(k_o'' L\right) + \cos^2\left(k_o' L\right)}. \tag{10.6}$$

Here

$$\beta = \frac{2\omega\tau}{\sqrt{1 + (\omega\tau)^2}}, \tag{10.7}$$

and

$$k_o' = \frac{\omega}{s}\left(\frac{\left(1 + \omega^{-2}\tau^{-2}\right)^{1/2} + 1}{2}\right)^{1/2}. \tag{10.8}$$

This theory was generalized to account for the subthreshold response in [31], where Eq. 10.4 was replaced by the generalized equation of the unified charge control model [52]

$$n = \frac{C\eta k_B T}{q^2}\ln\left[1 + \exp\left(\frac{qU_0}{\eta k_B T}\right)\right] \tag{10.9}$$

Here η is the ideality factor of FET, T is temperature, and k_B is the Boltzmann constant. For large positive values of the gate voltage swing ($U_0 > \eta k_B T/q$), Eq. 10.1 coincides with Eq. 10.4

In the opposite case of large negative gate voltage swings, $U_0 < 0$, $|U_0| > \eta k_B T/q$, the electron concentration predicted by Eq. 10.9 is exponentially small

$$n = \frac{C\eta k_B T}{q^2}\exp\left(\frac{qU_0}{\eta k_B T}\right) \tag{10.10}$$

Another generalization was adding the gate leakage current density, j_0, by replacing the continuity Eq. 10.3 with

$$\frac{\partial n}{\partial t} + \frac{\partial(nv)}{\partial x} = \frac{j_o}{q}, \tag{10.11}$$

$$\Delta U = \frac{qU_a{}^2}{4ms^2}\left\{\frac{1}{1+\kappa\exp\left(-\frac{eU_0}{\eta k_B T}\right)} - \frac{1}{\left(1+\kappa\exp\left(-\frac{eU_0}{\eta k_B T}\right)\right)^2\left[\sinh^2 Q + \cos^2 Q\right]}\right\}, \tag{10.12}$$

where, $\kappa = \frac{j_0 L^2 q^2}{2C\mu\eta^2 k_B^2 T^2}$ and $\kappa = \frac{j_0 L^2 q^2}{2C\mu\eta^2 k_B^2 T^2}$ is the low filed mobility,

$$s^2 = s_0{}^2\left(1+\exp\left(-\frac{qU_0}{\eta k_B T}\right)\right)\ln\left(1+\exp\left(\frac{qU_0}{\eta k_B T}\right)\right) \tag{10.13}$$

The physical meaning of parameter Q could be understood by presenting it in the following form

$$Q = \sqrt{\frac{\omega}{2\tau}\frac{L}{s}} = \frac{L}{\sqrt{2}L_o}, \tag{10.14}$$

here L_0 is the characteristic length of plasma wave penetration into the channel

$$L_0 = \sqrt{\frac{\mu n}{\omega\,(dn/dU)\,|_{U=V_g}}}, \tag{10.15}$$

which simplifies above the threshold to

$$L_0 = \sqrt{\frac{\mu U_o}{\omega}}, \tag{10.16}$$

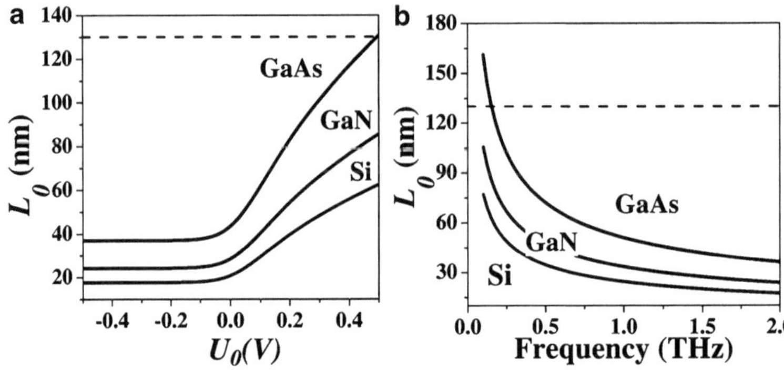

Fig. 10.2 (**a**) L_0 as a function of U_0 at 1.63 THz; (**b**) L_0 as a function of frequency below threshold. Values of mobility used: for InGaAs (3,500 cm^2/V · s), GaN (1,500 cm^2/V · s), and Si (800 cm^2/V · s)

Figure 10.1a shows the space profile of the plasma waves for $\omega\tau > 1$, but $\omega t_p < 1$ where t_p is the plasma wave transit time across the channel. Figure 10.1b shows the decaying plasma wave for the case when $\omega\tau > 1$ but $\omega t_p < 1$.

Figure 10.2 shows typical values of L_0 as a function of gate voltage and incident frequency for different materials.

For "long samples", such that ($L \gg L_0$) and a negligible gate current, Eq. 10.11 simplifies to

$$\Delta U = \frac{q{U_a}^2}{(4\eta k_B T)} \frac{1}{\left(1 + \exp\left(-\frac{qU_0}{\eta k_B T}\right)\right) \ln\left(1 + \exp\left(\frac{eU_0}{\eta k_B T}\right)\right)}, \qquad (10.17)$$

Hence, the maximum value of response

$$\Delta U_{\max} \approx \frac{q{U_a}^2}{4\eta k_B T}, \qquad (10.18)$$

and well above threshold (for $qU_0 \gg \eta k_B T$),

$$\Delta U = \frac{{U_a}^2}{4U_0}, \qquad (10.19)$$

A more detailed description of the physical mechanism of THz detection by FETs can be found in Knap et al. [29] and Knap and Dyakonov [30]. In the case of room temperature broadband detection (over-damped plasma) the detection process can be alternatively explained by the model of distributed resistive self-mixing. Although not treating all plasma related physics rigorously, the resistive mixing model allows a rational detector design [37, 42].

10.3 Detection of THz Radiation by FET in Low and High Power Limits

When THz radiation is coupled between gate and source of the FET, the THz AC voltage modulates simultaneously carrier density and the carrier drift velocity. This leads to nonlinearity and as a result, the photoresponse appears in the form of a DC voltage between source and drain. For high carrier mobility devices (e.g. III-V devices at cryogenic temperatures) the THz field can induce plasma waves that propagate in the channel, and resonant plasma modes can be excited leading to a resonant narrowband and gate bias tunable detection [11, 16, 27, 28, 38]. At room temperature, plasma waves are usually over-damped and the THz radiation only causes a carrier density perturbation that decays exponentially with the distance of the order of a few tens of nanometers.

A typical schematic of the photoresponse measurement circuit and a test element layout together with measured signal and transfer characteristics are shown in Fig. 10.3. Asymmetry between the source and drain is needed to induce photoresponse. There exist various ways to reach such asymmetry. One is the difference in the source and drain boundary conditions due to external (parasitic) or internal capacitances. Another is the asymmetry in feeding the incoming radiation, which

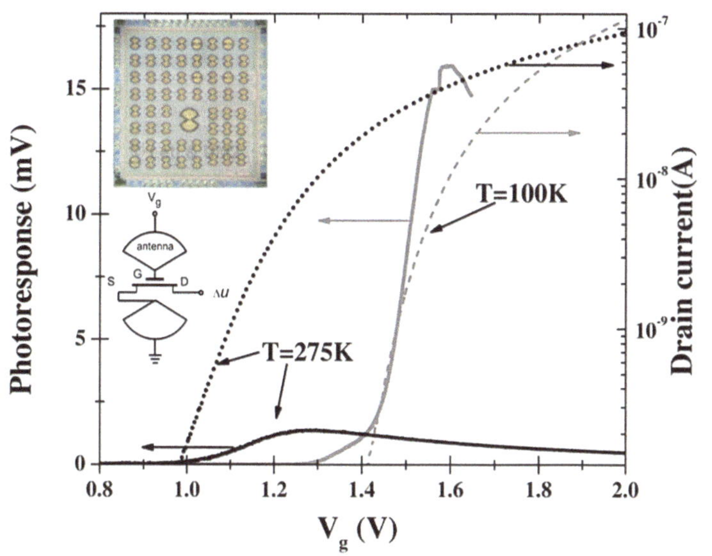

Fig. 10.3 Transfer current–voltage characteristics measured for drain voltage of 1 mV (*dash-lines*, right hand scale) and response of 0.13 mm CMOS FET to 0.3 THz radiation at 100 and 275 K (*solid lines*). *Upper inset* shows the layout of the test structure containing transistors with different antenna design. An array of 3 by 4 transistors in the *low left* corner of the *inset* represents a prototype of the imaging matrix. *Lower inset* on the *right* shows the schematic of the photoresponse measurement circuit [48]

can be achieved by special antenna connections as shown in the lower inset in Fig. 10.3. In this circuit the radiation predominantly creates an ac voltage between the source and the gate [48].

In case of nonresonant at detection by long transistors the THz photoresponse may be simply calculated using formula that relates the expected detector signal ΔU with the channel conductivity σ [46].

$$\Delta U = \frac{U_a^2}{4}\left[\frac{1}{\sigma}\frac{d\sigma}{dU}\right]_{U=V_0}. \tag{10.20}$$

Equation 10.20 is important because it allows calculating the expected photoresponse by a simple differentiation of the experimental transfer current–voltage characteristic.

In the sub-threshold range the channel conductivity decays as $\sigma(U_0) \sim \exp(-qV_0/_{I}\eta k_B T)$. Using Eq. 10.20 one obtains the maximum value of the response signal similar to Eq. 10.18. The same result was obtained in [31].

Equation 10.18 allows estimating the maximum responsivity. Assuming that all incident radiation power P is absorbed by the detector, one can approximate $U_a{}^2$ by the product of the input channel impedance $\sim L_0/(\sigma W)$ and power $U_a{}^2 \sim P \cdot L_0/(\sigma W)$. Here W is the gate width, σ is the channel conductivity and L_0 the characteristic length as defined above. In this case the voltage responsivity R_V can be approximated as

$$R_V \sim \frac{qL_0}{4sW\eta k_B T}. \tag{10.21}$$

More rigorous derivation of FET detector responsivity can be found in Kachorovskii et al. [24].

Since the conductivity entering the denominator decays $\sigma(U_0) \sim \exp(-qU_0/_{I}\eta k_B T)$ it becomes exponentially small below threshold. Therefore R_V may become exponentially large below threshold. For example using Eq. 10.5 with $\eta = 2$ and input channel resistance of 500 Ω one gets $R_V \sim 10$ kV/W. In the real experimental conditions responsivity is limited by antenna coupling and loading effects discussed in Sect. 11.6. The responsivity up to 5 kV/W and NEP in the range of a few $pW/\sqrt{Hz}$ was reported by many authors [41, 48]. These values are comparable with the best results reported for Schottky diodes based detectors placing FETs as important competitors. It should be mentioned however that FETs have the additional advantage of easy integration into arrays.

For low intensities of THz radiation the photoresponse is proportional to the radiation power (photovoltaic effect) and is described by Eq. 10.12. At higher intensities the photoresponse is no longer proportional to the incoming power. Sublinear dependence was observed in experiments (see Fig. 10.5). The theoretical model [21] provides analytical expressions ($U_0 > 0$):

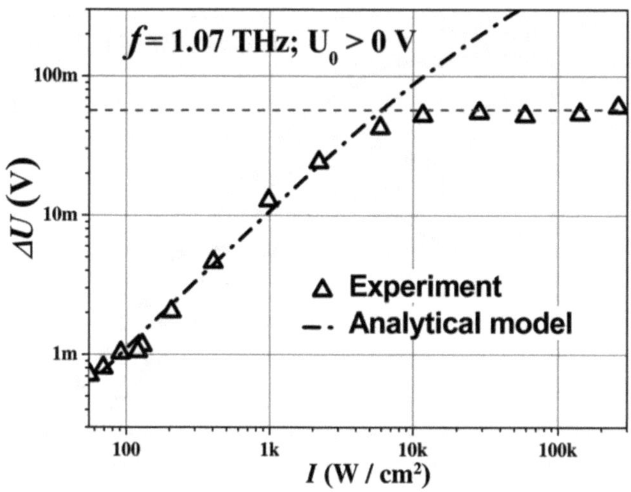

Fig. 10.4 Comparison between the analytical calculations using Eq. 10.22 (*dashed-dot lines*) and measured results (*rectangles*) for $V_g = -100$ mV. *Dashed lines* indicate the level of signal saturation

$$\Delta U = \frac{U_a^2}{2\left(\sqrt{U_0^2 + U_a^2/2} + U_0\right)} \qquad (10.22)$$

Below threshold ($U_0 < 0$), the response is given by

$$\Delta U = \frac{\eta k T}{q} \ln I_0 \left(\frac{q U_a}{\eta k_B T} \right) \qquad (10.23)$$

where I_0 is the Bessel function.

Figure 10.4 compares experimental results with theory for high THz power (intensity $I = P/S_b$, where S_b area of radiation beam) non-linear regime of operation [6]. InGaAs/GaAs HEMTs with a 0.13 μm gate length were used as a detector of 1.07 THz radiation produced by optically pumped gas laser with the maximum power of about 10 kW [18].

As can be seen in Fig. 10.4 after the linear region the response increases sub-linearly with the increase of the incoming THz intensity and then saturates.

Photoresponses of InGaAs/GaAs HEMT were studied at several frequencies (Fig. 10.5). The saturation is observed for all frequencies but the linear region extends to higher power (radiation intensity) with increasing frequency. I other words the saturation appears at higher radiation intensities with the frequency increase.

The change of the linear photoresponse range with frequency originates from combined effects of the frequency dependent properties of both the device and the antenna [35].

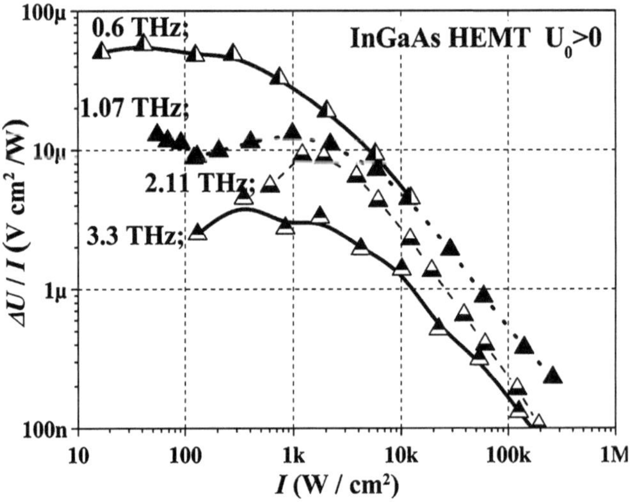

Fig. 10.5 Photoresponse to intensity ratio as a function of radiation intensity at different frequencies for InGaAs HEMTs (After But et al. [6])

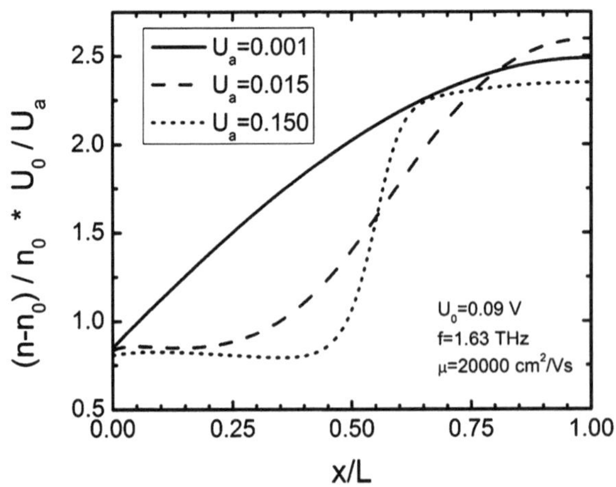

Fig. 10.6 Electron density profile in the channel of InGaAs HEMT at time t = 2.5 ps. In the strong intensity regime the plasma oscillations propagate in the channel as the shock waves (After Rudin et al. [45])

For the moment there is no theoretical explanation of neither saturation behavior. One of the hypothesis is that the photoresponse saturation could be due to non linear properties of 2D electron gas in high electric field namely to the drift velocity saturation.

Another hypothesis relates the saturation with propagation of shock plasma waves. Recently numerical calculations [45] provided the evidence that at high THz intensities shock plasma waves can propagate along the transistor channel (see Fig. 10.6).

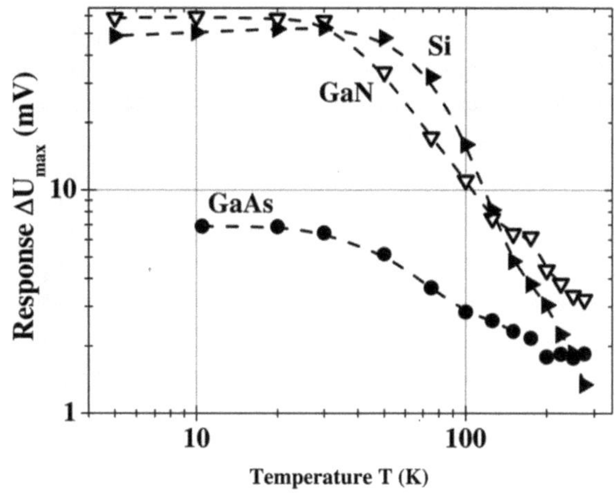

Fig. 10.7 Maximum response (at $V_g \sim V_{th}$) of GaAs, GaN HEMT and Si MOSFET to 300 GHz radiation as a function of temperature (After Klimenko et al. [26])

10.4 Temperature Dependence of the Nonresonant Detection

Temperature dependence of non-resonant THz detection by FETs was investigated in Klimenko et al. [26]. The experimental results in the temperature range $30 < T < 300$ K agree well with Eq. 10.18 that predicts that the maximal response is inversely proportional to temperature. At temperatures from 300 to 30 K the photoresponse of GaAs-, GaN-, and Si-based FETs was increased with temperature decrease. Below 30 K the photoresponse was temperature independent (see Fig. 10.7). (The difference in the response amplitude for different transistors is explained by the different antenna structures leading to different THz coupling to the devices.) As shown in [49] the low temperature saturation is correlated with the saturation of the sub-threshold slope observed in the static transfer characteristics of the transistors.

Therefore, the physical mechanism of the low temperature response saturation is attributed to the transport regime change from the collision/diffusion dominated one to the ballistic or trap dominated regime. These results clearly show that THz detectors based on FETs may significantly improve their responsivity with lowering temperature. Since noise also decreases with the temperature decrease, even higher improvement in noise equivalent power can be achieved. However, at low temperatures (below 30 K) the change of transport mechanism limits further improvements.

10.5 Helicity Dependent Terahertz Detection with FETs

The sensitivity of the THz plasmonic broadband detectors to linear polarization orientation has been relatively well established [47]. However, the photoresponse proportional to the degree of circular polarization has been only recently

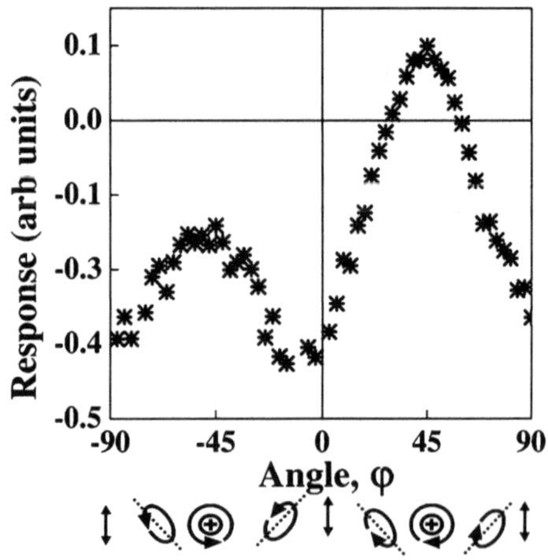

Fig. 10.8 Polarization dependence of the response, $f = 0.8$ THz. Ellipses on the top illustrate the polarization states (After Klimenko et al. [26])

observed. Drexler et al. [8] reported on the THz helicity sensitive photoresponse in GaAs/AlGaAs HEMTs and Si MOSFETs. The experiments were performed applying a CW and pulsed THz radiation in the frequency range of 0.6–2.5 THz. The important new experimental and theoretical discovery by [8] was that the photon helicity can be sensed through the interference of two ac currents generated on opposite sides of the transistor channel. The coupling of the radiation to the transistor channel can be modeled by two effective antennas producing an AC voltage between the source and gate, and between the drain and gate. In the case of long transistors, there is no interference between the currents induced at opposite sides of the channel, and the corresponding contributions to the total photoresponse are independent. For sufficiently short transistor channels, the source and drain AC currents interfere in the mid-section of the device leading to DC current component that depends on their phase difference. Such a phase difference appears when the source and the drain sides are excited by mutually orthogonal components of circular (or elliptically) polarized radiation. The interference term in this case is sensitive to the radiation helicity.

Figure 10.8 shows a polarization behavior of the signal obtained by varying the radiation ellipticity.

It is worth to stress that the observed helicity-dependent response in FETs is related to their particular THz rectification physical mechanism. It provides the basis for a very sensitive (more than two orders of magnitude compared to other known methods), fast and all-electric characterization of THz radiation polarization state, and, therefore can be used for development of the new methods of THz ellipsometry.

10.6 Loading Effects and Plasma-Wave Detectors for Terahertz Communication

The channel resistance exponentially increases below threshold and may become comparable or even higher than the input resistance of the measurements system. At this point the response signal decreases due to a simple voltage-dividing effect (loading effect). Due to this loading effect, the signal may be also modulation frequency dependent because even small capacitance can lead to big RC constants.

The loading effects related to resistances were first analyzed in [54] (see also [53] and references therein). The most complete approach to the loading effects is presented in Sakowicz et al. [46]. It was shown that to reproduce the experimental results one should divide the photovoltaic signal ΔU by a factor $(1 + R_{CH}/Z)$, where R_{CH} is the channel resistance and Z is the complex load impedance of the read-out setup. Since Z contains not only the load resistance of the preamplifiers but also all the parasitic capacitances, the amplitude of the response depends on the modulation frequency f_m.

Figure 10.9 shows the influence of the modulation frequency, f_m and load resistance on the gate voltage dependence of the response.

In general, both: (i) increase of the modulation frequency f_m and (ii) decrease of the input impedance of the read-out circuit lead to decrease of the signal in the sub-threshold range.

As seen from Fig. 10.9a, increase of the modulation frequency up to 10 kHz leads to a decrease of the signal in the high resistance range (close to threshold). However, far from the threshold when the transistor is in the low resistance state, the signal is frequency independent in the whole investigated frequency range.

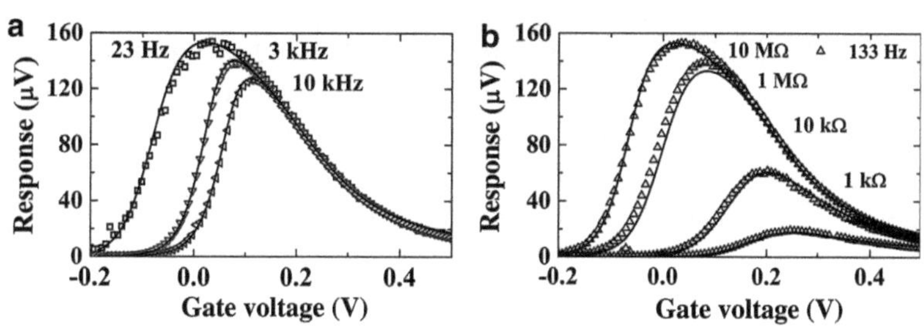

Fig. 10.9 (**a**) Gate voltage dependence of the 300 GHz photoresponse for constant load resistance 10 MΩ and different modulation frequencies. The modulation frequency was gradually increased from 23 Hz, 3 kHz to 10 kHz. (**b**) Gate voltage dependence of the response for different load resistances and two modulation frequencies. Results obtained with constant modulation frequency 133 Hz (*triangles*) and load resistance changing from 10 MΩ to 1 MΩ, 10 kΩ and 1 kΩ. Continuous *black lines* show results of calculations (After Sakowicz et al. [46])

In Fig. 10.9b we show results obtained with the modulation frequencies 133 Hz for load resistances decreasing from 10 MΩ to 1 kΩ. Similar to results from Fig. 10.9a, one can see gradual decrease of the signal in the high resistance state (close to threshold). Results in Fig. 10.9 show the trade-off between bandwidth and response amplitude. The experimental results could be very well reproduced (solid lines) without any fitting parameters using a phenomenological approach (Eq. 10.20) and normalization factor $(1 + R_{CH}/Z)$ with Z measured by a standard *LCR* bridge.

As can be seen in Fig.10.9, the bandwidth of detectors operated in their high impedance mode (close to the threshold) is limited to a few kHz. The limitation of bandwidth is a result of three main factors: the output resistance of the detector, the input resistance of the amplifier and the parasitic capacitances. In the case presented in Fig. 10.9 the parasitic capacitances of the cables between transistor and amplifier were relatively high (of the order of 150 pF), the output resistance close to the threshold was around 1 MΩ, and the input resistance of the preamplifier was 10 MΩ. There are two ways to increase the bandwidth:

(i) Operate transistor far from the threshold where the channel resistance is lower (close to 50 Ω). In Fig. 10.9 it is clear that the bandwidth for higher gate voltages is higher.
(ii) Integrate transistor, load and amplifier to decrease parasitic capacitances (see, for example, [22]).

As shown below the bandwidth up to the GHz range can be achieved by combining both methods.

An important application of THz systems is wireless communication. In theory, the modulation bandwidth could be in the sub THz range, exceeding 100 GHz [25]. For example, for the heterodyne detection, the intermediate frequency can be of the order of 50–100 GHz for a 200 nm gate transistor operating at room temperatures in the above-threshold regime, and of the order of 5–10 GHz in the below threshold regime [19]. As was demonstrated in Blin et al. [1], the modulation up to 8 GHz can be reached in wireless communication systems using a 250 nm gate-length plasmonic GaAs/AlGaAs field-effect transistor as a detector.

Figure 10.10 presents the amplitude of the detected modulation signal measured using a spectrum analyzer. The detector was mounted directly on 50 Ω microstrip lines. The gate voltage was applied to keep the detector in the open state (far from the threshold) providing its output resistance close to 50 Ω. The signal was amplified using a 50 Ω and 30 dB amplifier. The modulation signal was observed for frequencies between 0.3 MHz and 8 GHz.

There is usually a trade-off between bandwidth and response amplitude. Far from the threshold this kind of detector has the 3 dB bandwidth up to the GHz range but a relatively low sensitivity (1.3 V/W) and a relatively high noise-equivalent power (13.3 nW/$\sqrt{\text{Hz}}$). As mentioned above, an improvement can be obtained by integrating the transistor with a fast amplifier.

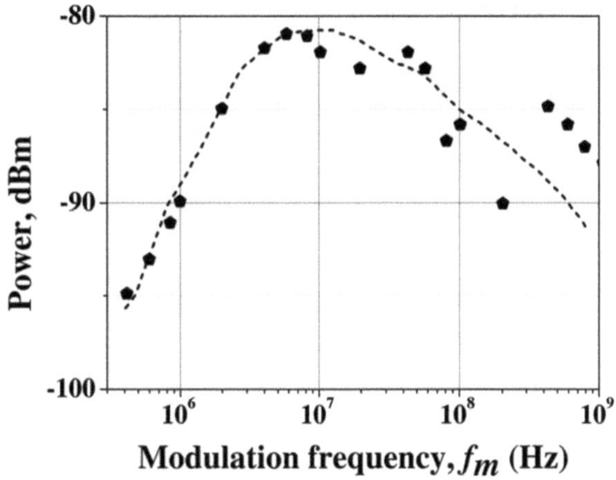

Fig. 10.10 Detection power as function of the modulation frequency, f = 0.305 THz (After Blin et al. [1])

10.7 Double Grating Gate Structures for THz Detection

The asymmetric dual grating gate (A-DGG) structure is the special structure with improved THz coupling and rectification efficiency obtained thanks to implementation of two grating gates with asymmetric inter-finger spaces. These structures show giant enhancement of the responsivity (by four orders of magnitude) with respect to the symmetric double grating structures and can provide ultra-high sensitive THz detection [44]. This is because the unit cell of the A-DGG structure can create strong build-in asymmetric field. The THz photoresponse dramatically increases when the parts of 2D channel under the fingers of one of the two sub-gratings are depleted. InAlAs/InGaAs/InP A-DGG HEMTs were processed and investigated as room temperature THz detectors [58]. Monochromatic THz pulsed waves with frequencies from 1 to 3 THz from a ring-cavity THz parametric oscillator source were focused on the detector [4, 57, 58]. The photovoltaic signal was observed as a V_d variation with a lock-in technique. The responsivity was estimated as $R_V = \Delta U \cdot S_t / P_t \cdot S_d$ where P_t is the total power of the source on the detector plane, S_t is the radiation beam spot area, and S_d is the active area of the detector.

Figure 10.11 shows the measured responsivity of the detector at 1 THz under zero-V_d condition as a function of gate voltage swing ($V_{g1,2}$–V_{th}); dc voltage of Gate 1: V_{g1} (Gate 2: V_{g2}) is swept while V_{g2} (V_{g1}) is floated (biased at 0 V) [57, 58]. The best result with R_V = 2.2 kV/W was obtained when sweeping V_{g1} to the threshold V_{th}. The detector sample # 1–1 was characterized at 1.5 THz under V_d-biased conditions. With increasing V_d from 0 to 0.4 V, the maximum responsivity increased up to 6.4 kV/W. All these values were, to the best of our knowledge, the best ever reported over for any fast-response detector at 300 K [4]. In terms of the noise performance, these detectors exhibit extremely low NEP with the minimal value 15 pW/$\sqrt{}$Hz at 1 THz under V_d-unbiased conditions [58]. These values are lower than those of any commercial room temperature THz detectors such as Golay cells or Schottky barrier diodes [57].

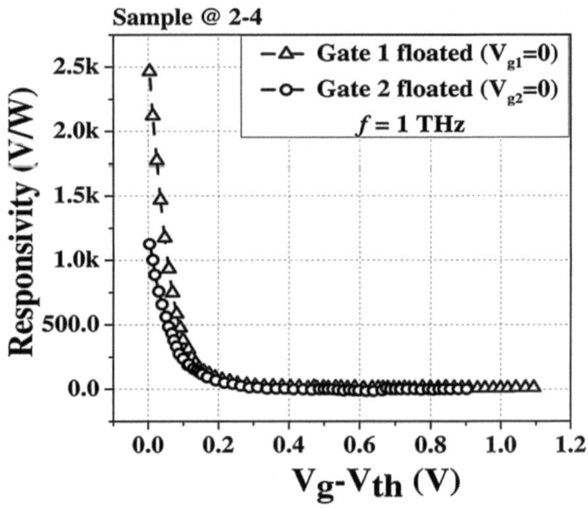

Fig. 10.11 Responsivity of detector #2–4 as a function of the gate swing voltage ($V_{g1}-V_{th1}$ or $V_{g2}-V_{th2}$) at 1 THz under drain-unbiased condition (After Watanabe et al. [58])

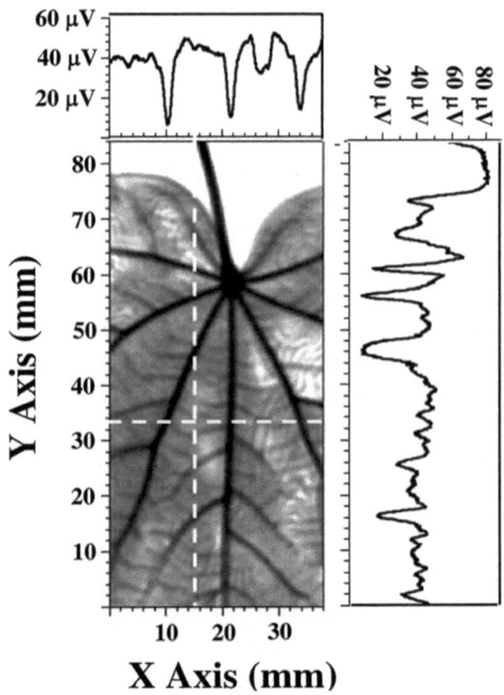

Fig. 10.12 Fast large-area room temperature terahertz imaging of a fresh tree leaf using A-DGG structures as detector in the 650-GHz transparency window. *Top* and *right* hand graphs show intensity profile along the horizontal and vertical *dotted lines* correspondingly

The A-DGG structures are already exploitable, for example for large-area fast imaging at 655 GHz. As test object we used a tree leaf. Figure 10.12 shows the THz image, consisting of 320 × 1,120 scanned points, collected by raster-scanning the object in the beam focus, with an integration time of 10 ms per point. The main veins of about equal size, all of which extend from a common point at the base of the blade can be visualized with a reasonably good spatial resolution, as well as the pattern of the secondary and tertiary veins. The vein pattern can be useful for plant identification. This example shows that our A-DGG devices can already be used in a realistic setting, enabling large-area, fast imaging of macroscopic samples.

10.8 Discussion and Conclusions

Nanometer size FETs operating as broadband THz detectors, mixers, phase shifters and frequency multipliers at THz frequencies compete with commercially available Schottky diodes. Recent studies show that FETs reach the responsivity, NEP and speed of the same order of magnitude as Schottky diodes while having advantages in CMOS VLSI compatibility.

Further improvement of the FET THz detectors should relate to (i) improving coupling with external radiation and (ii) improving the transistor design as well as its integration with impedance matching amplifiers.

The typical wavelength of sub-THz or THz radiation is in the range from one mm to tens of microns. With transistor dimensions in nanometer scale, direct efficient coupling through antennas is difficult due to (i) unknown transistor input impedance in the THz frequency range and (ii) radiation coupling to the substrate instead of the transistor itself.

The most spectacular progress can be achieved by improving the transistor and reaching the resonant detection regime. Resonant detection can be more sensitive, spectrally resolved and gate voltage tunable. There are two main ways to reach this goal: (i) increase carrier mobility and (ii) improve device geometry. The high mobility can be attained by using InSb or graphene based channels with the carrier mobility exceeding 8,000–10,000 $cm^2/V \cdot s$.

The improvement of device geometry is necessary to eliminate the existence of oblique plasma modes. Most transistors are devices with the gate width much greater than the gate length. In such devices, plasma waves propagating in the oblique directions travel in different directions and for different distances. This leads to the broadening of the plasma wave spectrum [16, 29]. One may decrease the role of oblique modes using narrow channels transistors or transistor based on nanowires [5, 50]. However, there is a tradeoff between the channel width and carriers mobility because for very narrow channels additional scattering from the channel borders limits the mobility. Therefore, the development of the narrow channel nanowires with high carrier mobility for resonant plasma THz detection is still a future technological challenge.

In conclusion: we have presented an overview of recent experimental results on the detection of terahertz radiation by nanometer-scale FETs. Currently, the most promising application appears to be the broadband THz detection in the non-resonant regime. Different kinds of FETs have exhibited such detection including Si, GaAs, GaN, nanowire, and graphene devices.

New applications like THz ellipsometry and polarization sensitive imaging have been discussed. We have shown also that operation of THz FET detectors at cryogenic temperatures can improve their performances. An important achievement is the demonstration of broadband terahertz communications with up to 8 GHz modulation frequency.

Further progress with THz detection by FETs should be related to the improvement of the THz coupling and improvement of the transistor's design. Particularly, suspended graphene transistors with the carrier mobility exceeding 10,000–20,000 $cm^2/V \cdot s$ have a promise for the resonant detection. Other promising directions are the development of grating gate structures and nanowire transistors. They may lead to efficient, resonant and voltage tunable THz detectors.

Acknowledgments We thank Prof. M. Dyakonov for many helpful discussions. This work was supported by ANR project "WITH" and by CNRS and GDR-I project "Semiconductor sources and detectors of THz frequencies" and by the US – French initiative "PUF". The Montpellier team affiliated with the physics and electronics departments was supported by the "Scientific Interest Groupement" GIS –TERALAB. The work at RPI was supported by the US NSF under the auspices of the NSF EAGER program and by the ARL Cooperative research Agreement.

References

1. Blin S, Teppe F, Tohme L, Hisatake S, Arakawa K, Nouvel P et al (2012) Plasma-wave detectors for terahertz wireless communication. IEEE Electron Device Lett 33:1354–1356. doi:10.1109/led.2012.2210022
2. Boppel S, Lisauskas A, Krozer V, Roskos HG (2011) Performance and performance variations of sub-1 THz detectors fabricated with 0.15 μm CMOS foundry process. Electron Lett 47:661
3. Boppel S, Lisauskas A, Max A, Krozer V, Roskos HG (2012) CMOS detector arrays in a virtual 10-kilopixel camera for coherent terahertz real-time imaging. Opt Lett 37:536–538. doi:10.1364/OL.37.000536
4. Boubanga-Tombet S, Tanimoto Y, Watanabe T, Suemitsu T, Wang Y, Minamide H et al (2012) Asymmetric dual-grating gate InGaAs/InAlAs/InP HEMTs for ultrafast and ultrahigh sensitive terahertz detection. In: 2012 37th international conference on Infrared, Millimeter, and Terahertz Waves (IRMMW-THz), pp 1–2. doi:10.1109/IRMMW-THz.2012.6380401
5. Boubanga-Tombet S, Teppe F, Coquillat D, Nadar S, Dyakonova N, Videlier H et al (2008) Current driven resonant plasma wave detection of terahertz radiation: toward the Dyakonov–Shur instability. Appl Phys Lett 92:212101–212103. doi:10.1063/1.2936077
6. But DB, Dyakonova N, Drexler C, Drachenko O, Romanov K, Golenkov OG, Sizov FF, Gutin A, Shur MS, Ganichev SD, Knap W (2013) The dynamic range of THz broadband FET detectors. In: Razeghi M (ed) Proceedings of SPIE 8846, Terahertz emitters, receivers, and applications IV, San Diego, USA, pp 884612–884617. doi:10.1117/12.2024226
7. Chaplik AV (1972) Possible crystallization of charge carriers in the inversion layer of low density. Sov Phys JETP 35:395
8. Drexler C, Dyakonova N, Olbrich P, Karch J, Schafberger M, Karpierz K et al (2012) Helicity sensitive terahertz radiation detection by field effect transistors. J Appl Phys 111:124504–124506. doi:10.1063/1.4729043
9. Dyakonov M, Shur M (1996) Detection, mixing, and frequency multiplication of terahertz radiation by two-dimensional electronic fluid. Electron Devices IEEE Trans 43:380–387

10. Dyakonov M, Shur M (1993) Shallow water analogy for a ballistic field effect transistor: new mechanism of plasma wave generation by dc current. Phys Rev Lett 71:2465. doi:10.1103/PhysRevLett.71.2465
11. Dyakonov MI, Shur MS (1996) Plasma wave electronics: novel terahertz devices using two dimensional electron fluid. Electron Devices IEEE Trans 43:1640–1645. doi:10.1109/16.536809
12. Dyakonova N, El Fatimy A, Lusakowski J, Knap W, Dyakonov MI, Poisson MA et al (2006) Room-temperature terahertz emission from nanometer field-effect transistors. Appl Phys Lett 88:141906–3. doi:10.1063/1.2191421
13. Dyakonova N, Teppe F, Lusakowski J, Knap W, Levinshtein M, Dmitriev AP et al (2005) Magnetic field effect on the terahertz emission from nanometer InGaAs/AlInAs high electron mobility transistors. J Appl Phys 97:114313–114315. doi:10.1063/1.1921339
14. Eguiluz A, Lee TK, Quinn JJ, Chiu KW (1975) Interface excitations in metal-insulator-semiconductor structures. Phys Rev B 11:4989–4993
15. El Fatimy A, Dyakonova N, Meziani Y, Otsuji T, Knap W, Vandenbrouk S et al (2010) AlGaN/GaN high electron mobility transistors as a voltage-tunable room temperature terahertz sources. J Appl Phys 107:024504–4. doi:10.1063/1.3291101
16. El Fatimy A, Teppe F, Dyakonova N, Knap W, Seliuta D, Valušis G et al (2006) Resonant and voltage-tunable terahertz detection in InGaAs/InP nanometer transistors. Appl Phys Lett 89:131926–3. doi:10.1063/1.2358816
17. Elkhatib TA, Kachorovskii VY, Stillman WJ, Rumyantsev S, Zhang XC, Shur MS (2011) Terahertz response of field-effect transistors in saturation regime. Appl Phys Lett 98:243505–3. doi:10.1063/1.3584137
18. Ganichev SD, Prettl W (2006) Intense terahertz excitation of semiconductors, vol 14. Oxford University Press, Oxford
19. Gershgorin B, Kachorovskii VY, Lvov YV, Shur MS (2008) Field effect transistor as heterodyne terahertz detector. Electron Lett 44:1036–1037. doi:10.1049/el:20080737
20. Glaab D, Boppel S, Lisauskas A, Pfeiffer U, Ojefors E, Roskos HG (2010) Terahertz heterodyne detection with silicon field-effect transistors. Appl Phys Lett 96:042106–3. doi:10.1063/1.3292016
21. Gutin A, Kachorovskii V, Muraviev A, Shur M (2012) Plasmonic terahertz detector response at high intensities. J Appl Phys 112:014508–5. doi:10.1063/1.4732138
22. Gutin A, Ytterdal T, Kachorovskii V, Muraviev A, Shur M (2013) THz spice for modeling detectors and nonquadratic response at large input signal. Sensors J IEEE 13:55–62. doi:10.1109/jsen.2012.2224105
23. Han R, Zhang Y, Coquillat D, Videlier H, Knap W, Brown E et al (2011) A 280-GHz Schottky diode detector in 130-nm digital CMOS. Solid State Circ IEEE J 46:2602–2612. doi:10.1109/jssc.2011.2165234
24. Kachorovskii VY, Rumyantsev SL, Knap W, Shur M (2013) Performance limits for field effect transistors as terahertz detectors. Appl Phys Lett 102:223505–4. doi:10.1063/1.4809672
25. Kachorovskii VY, Shur MS (2008) Field effect transistor as ultrafast detector of modulated terahertz radiation. Solid State Electron 52:182–185. doi:10.1016/j.sse.2007.08.002
26. Klimenko OA, Knap W, Iniguez B, Coquillat D, Mityagin YA, Teppe F et al (2012) Temperature enhancement of terahertz responsivity of plasma field effect transistors. J Appl Phys 112:014506–5. doi:10.1063/1.4733465
27. Knap W, Deng Y, Rumyantsev S, Lu JQ, Shur MS, Saylor CA et al (2002) Resonant detection of subterahertz radiation by plasma waves in a submicron field-effect transistor. Appl Phys Lett 80:3433–3435. doi:10.1063/1.1473685
28. Knap W, Deng Y, Rumyantsev S, Shur MS (2002) Resonant detection of subterahertz and terahertz radiation by plasma waves in submicron field-effect transistors. Appl Phys Lett 81:4637–4639. doi:10.1063/1.1525851
29. Knap W, Dyakonov M (2013) Field effect transistors for terahertz applications. In: Saeedkia D (ed) Handbook of terahertz technology. Woodhead Publishing, Waterloo, pp 121–155

30. Knap W, Dyakonov M, Coquillat D, Teppe F, Dyakonova N, Łusakowski J et al (2009) Field effect transistors for terahertz detection: physics and first imaging applications. J Infrared Millim Terahertz Waves 30:1319–1337. doi:10.1007/s10762-009-9564-9
31. Knap W, Kachorovskii V, Deng Y, Rumyantsev S, Lu JQ, Gaska R et al (2002) Nonresonant detection of terahertz radiation in field effect transistors. J Appl Phys 91:9346–9353. doi:10.1063/1.1468257
32. Knap W, Lusakowski J, Parenty T, Bollaert S, Cappy A, Popov VV et al (2004) Terahertz emission by plasma waves in 60 nm gate high electron mobility transistors. Appl Phys Lett 84:2331–2333. doi:10.1063/1.1689401
33. Knap W, Rumyantsev S, Vitiello M, Coquillat D, Blin S, Dyakonova N et al (2013) Nanometer size field effect transistors for terahertz detectors. Nanotechnology 24:214002
34. Knap W, Teppe F, Meziani Y, Dyakonova N, Lusakowski J, Boeuf F et al (2004) Plasma wave detection of sub-terahertz and terahertz radiation by silicon field-effect transistors. Appl Phys Lett 85:675–677. doi:10.1063/1.1775034
35. Kreisler AJ (1986) Submillimeter wave applications of submicron Schottky diodes. In: Izatt JA (ed) Proceedings of SPIE 666 Quebec symposium. International Society for Optics and Photonics, Quebec, Canada, pp 51–63. doi:10.1117/12.938820
36. Lisauskas A, Glaab D, Roskos HG, Oejefors E, Pfeiffer UR (2009) Terahertz imaging with Si MOSFET focal-plane arrays. In: Linden KJ (ed) Proceedings of SPIE 7215: Terahertz technology and applications II. International Society for Optics and Photonics, San Jose, p 72150J. doi:10.1117/12.809552
37. Lisauskas A, Pfeiffer U, Ojefors E, Bolivar PH, Glaab D, Roskos HG (2009) Rational design of high-responsivity detectors of terahertz radiation based on distributed self-mixing in silicon field-effect transistors. J Appl Phys 105:114511–114517. doi:10.1063/1.3140611
38. Lu J-Q, Shur MS (2001) Terahertz detection by high-electron-mobility transistor: enhancement by drain bias. Appl Phys Lett 78:2587–2588. doi:10.1063/1.1367289
39. Lu J-Q, Shur MS, Hesler JL, Liangquan S, Weikle R (1998) Terahertz detector utilizing two-dimensional electronic fluid. Electron Device Lett IEEE 19:373–375. doi:10.1109/55.720190
40. Nakayama M (1974) Theory of surface waves coupled to surface carriers. J Phys Soc Jpn 36:393–398. doi:10.1143/jpsj.36.393
41. Ojefors E, Pfeiffer UR, Lisauskas A, Roskos HG (2009) A 0.65 THz focal-plane array in a quarter-micron CMOS process technology. Solid State Circ IEEE J 44:1968–1976. doi:10.1109/JSSC.2009.2021911
42. Perenzoni D, Perenzoni M, Gonzo L, Capobianco AD, Sacchetto F (2010) Analysis and design of a CMOS-based terahertz sensor and readout. In: Berghmans F (ed) SPIE 7726, Optical sensing and detection, Brussels, Belgium, pp 772618–772612. doi:10.1117/12.854442
43. Popov VV, Ermolaev DM, Maremyanin KV, Maleev NA, Zemlyakov VE, Gavrilenko VI et al (2011) High-responsivity terahertz detection by on-chip InGaAs/GaAs field-effect-transistor array. Appl Phys Lett 98:153504-3. doi:10.1063/1.3573825
44. Popov VV, Fateev DV, Otsuji T, Meziani YM, Coquillat D, Knap W (2011) Plasmonic terahertz detection by a double-grating-gate field-effect transistor structure with an asymmetric unit cell. Appl Phys Lett 99:243504–4. doi:10.1063/1.3670321
45. Rudin S, Rupper G, Gutin A, Shur M (2014) Theory and measurement of plasmonic terahertz detector response to large signals. J Appl Phys 115(6):064503–064511. doi:10.1063/1.4862808
46. Sakowicz M, Lifshits MB, Klimenko OA, Schuster F, Coquillat D, Teppe F et al (2011) Terahertz responsivity of field effect transistors versus their static channel conductivity and loading effects. J Appl Phys 110:054512–054516. doi:10.1063/1.3632058
47. Sakowicz M, Lusakowski J, Karpierz K, Grynberg M, Knap W, Gwarek W (2008) Polarization sensitive detection of 100 GHz radiation by high mobility field-effect transistors. J Appl Phys 104:024519-5. doi:10.1063/1.2957065
48. Schuster F, Coquillat D, Videlier H, Sakowicz M, Teppe F, Dussopt L et al (2011) Broadband terahertz imaging with highly sensitive silicon CMOS detectors. Opt Express 19:7827–7832. doi:10.1364/OE.19.007827

49. Schuster F, Knap W, Nguyen V (2011) Terahertz imaging achieved with low-cost CMOS detectors. Laser Focus World 47(7):37–41
50. Shchepetov A, Gardes C, Roelens Y, Cappy A, Bollaert S, Boubanga-Tombet S et al (2008) Oblique modes effect on terahertz plasma wave resonant detection in InGaAs/InAlAs multichannel transistors. Appl Phys Lett 92:242105–3. doi:10.1063/1.2945286
51. Sherry H, Grzyb J, Yan Z, Al Hadi R, Cathelin A, Kaiser A et al (2012) A 1kpixe CMOS camera chip for 25fps real-time terahertz imaging applications. In: Fujino LC (ed) Solid-state circuits conference digest of technical papers (ISSCC), vol 55, 2012 IEEE international, pp 252–254. doi:10.1109/isscc.2012.6176997
52. Shur M (1996) Introduction to electronic devices. Wiley, New York, p 608. ISBN 9780471103486
53. Stillman W, Donais C, Rumyantsev S, Shur MS, Veksler D, Hobbs C et al (2011) Silicon FinFETs as detectors of terahertz and sub-terahertz radiation. Int J High Speed Electron Syst 20:27–42. doi:10.1142/s0129156411006374
54. Stillman W, Shur MS, Veksler D, Rumyantsev S, Guarin F (2007) Device loading effects on nonresonant detection of terahertz radiation by silicon MOSFETs. Electron Lett 43:422–423. doi:10.1049/el:20073475
55. Tauk R, Teppe F, Boubanga S, Coquillat D, Knap W, Meziani YM et al (2006) Plasma wave detection of terahertz radiation by silicon field effects transistors: responsivity and noise equivalent power. Appl Phys Lett 89:253511–3. doi:10.1063/1.2410215
56. Veksler D, Teppe F, Dmitriev AP, Kachorovskii VY, Knap W, Shur MS (2006) Detection of terahertz radiation in gated two-dimensional structures governed by dc current. Phys Rev B 73:125328. doi:10.1103/PhysRevB.73.125328
57. Watanabe T, Boubanga-Tombet SA, Tanimoto Y, Fateev D, Popov V, Coquillat D et al (2013) InP- and GaAs-based plasmonic high-electron-mobility transistors for room-temperature ultrahigh-sensitive terahertz sensing and imaging. Sensors J IEEE 13:89–99. doi:10.1109/jsen.2012.2225831
58. Watanabe T, Tombet SB, Tanimoto Y, Wang Y, Minamide H, Ito H et al (2012) Ultrahigh sensitive plasmonic terahertz detector based on an asymmetric dual-grating gate HEMT structure. Solid State Electron 78:109–114. doi:10.1016/j.sse.2012.05.047

Chapter 11
THz Screening for Civil and Military Security

Norbert Palka, Marcin Kowalski, Radosław Ryniec, Mieczysław Szustakowski, and Elżbieta Czerwińska

Abstract We report on the developed system for the detection of the objects hidden under people's clothes. The system is based on a commercially available THz camera equipped with specially designed software for image processing and fusion. Additionally, a camera for visible light registers the image that is later processed and combined with the THz image by means of the designed fusion algorithm. Such final fused image significantly increases the visibility of the detected elements. A very important aspect of the usefulness of various image processing techniques is the proper evaluation of the image quality. We propose a combination of two image quality assessment methods as a methodology used for the quality estimation of fused images and for comparing image fusion algorithms. Moreover, we designed and constructed a moving, wirelessly-controlled thermal phantom for testing of the functioning of the cameras in the terahertz range. The terahertz camera operating at 0.25 THz and the standard thermal camera were used for tests. We compared the images taken for the phantom and a man and we obtained satisfactory similarities both for naked and dressed targets with hidden items (guns, knives, bombs). The temperature stability of the phantom is sufficient for the evaluation of those cameras.

11.1 Introduction

The law enforcement agencies are currently faced with the problem of countering the terrorism-related threats. Especially explosive devices and weapon hidden beneath the clothing are the things of particular interest. Therefore, many research and development institutions are looking for safe and accurate stand-off

N. Palka • M. Kowalski (✉) • R. Ryniec • M. Szustakowski • E. Czerwińska
Institute of Optoelectronics, Military University of Technology, 2 S. Kaliski Street, Warsaw, Poland
e-mail: mkowalski@wat.edu.pl

C. Corsi and F. Sizov (eds.), *THz and Security Applications*, NATO Science for Peace and Security Series B: Physics and Biophysics, DOI 10.1007/978-94-017-8828-1_11,

technologies for non-invading people inspection. Electromagnetic waves at terahertz frequencies (0.1–3.0 THz) seem to be a promising candidate for this purpose. Many explosives, e.g., Hexogen, Penthrite, and Octogen, have characteristic transmission and reflection features in the THz range [1, 2] that could help to distinguish them from other common materials. Moreover, THz radiation can be transmitted through clothes with small attenuation [3] and it is strongly reflected by metallic objects like knives or guns [4]. THz waves pose minimal health risk to human beings and they do not affect the operation of various systems because photon energy is very small (4.4 meV @ 1 THz) [4]. The main factor that limits its propagation through the air is strong molecular absorption by water vapor [1, 4]. The above mentioned features cause that THz radiation can be applied in two security-orientated areas: explosives detection [2, 5] and people screening [6–8].

Taking into consideration the second application there are two possibilities: portals [9, 10] and cameras [11, 12] for people screening, both can be either passive or active and both are working mainly in the range between 0.1 and 0.3 THz. Passive systems working in this wave region can detect hidden objects on people due to the difference in naturally emitted radiation between the human body and those objects. Human skin emits more radiation than hidden items because its radiometric temperature is usually higher than the radiometric temperature of those items [13]. Moreover, emissivity of the human skin, which consists of 70 % of water, is usually different than emissivity of hidden dangerous metallic or dielectric materials, which makes them easier to distinguish.

In this paper we describe the system developed for the detection of items hidden under clothes on people (Sect. 11.2). The system is based on the commercially available camera which is additionally equipped with our software for image processing and fusion (Sect. 11.3). Such numerical algorithm increases the visibility of the detected elements. The evaluation of the processed image quality is presented in Sect. 11.4.

To avoid the time-consuming, tedious and expensive testing on human subjects we built the moving platform with the human phantom, which mimics a moving person. In Sect. 11.5 we focused on the construction of the thermal phantom and on the results of tests carried out on it. This phantom was developed for testing THz cameras and our image processing software.

11.2 System for the Detection of Concealed Objects

Terahertz vision systems are becoming more and more popular due to the possibility to utilize very interesting properties of the terahertz radiation. Nowadays, terahertz cameras are in the early stage of development, nevertheless this development is very fast. Manufacturers offer both passive and active cameras (the latter use an external source of radiation to register an image).

Terahertz imaging devices can be used to build systems for early detection of threats [14–16]. Terahertz imaging makes it possible to acquire images, for

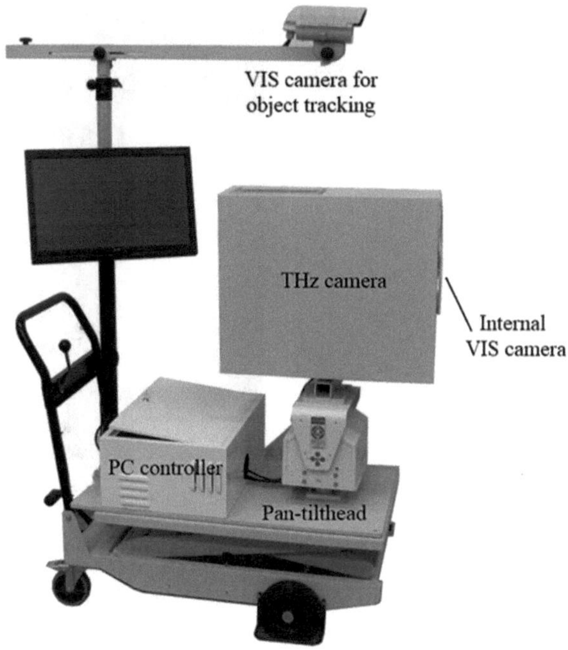

Fig. 11.1 A system for the detection of concealed objects

example of objects concealed underneath clothing, by measuring the radiometric temperatures of these objects that are clearly visible on a human body background. Finding dangerous objects hidden under clothes is one of the key issues of the security in public places. It can be realized by a multispectral, terahertz – based screening system with implemented image fusion algorithm. The final image after fusion algorithm gives more information and is more convenient for the human perception than typical image from THz camera or camera for the visible light.

The terahertz camera is the core element of the developed system for the detection of concealed objects. Its role is to provide basic data necessary to detect hidden objects. The terahertz camera is placed on a pan-tilt head, so that the direction of the observation can be easily selected (Fig. 11.1).

The system is also equipped with two cameras for the visible light – the first one is tracking the objects and has the resolution of 10 MPix and wide field of view and the second one is placed inside a housing of the terahertz camera. The internal camera for the visible light is supposed to provide the image background, which will be fused with the terahertz image after software processing. The entire system is controlled by a central processing unit and the data visualization is done with the touch screen display. The system is placed on the movable platform to ensure satisfactory mobility.

In our study, the ThruVision TS4 [11] terahertz camera was used. It is a passive camera operating at the frequency of 250 GHz. This camera is the core of the system and its main task is to provide the image, that is later processed to detect hidden objects in the registered scene.

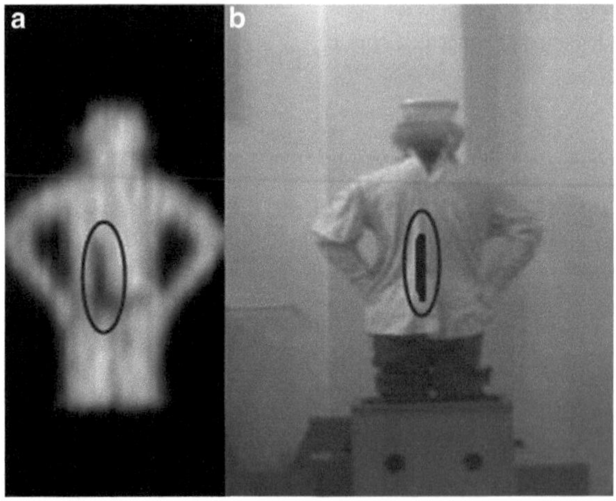

Fig. 11.2 Complementary images from two ranges of spectrum – (**a**) THz, (**b**) VIS

The terahertz camera records an image with the resolution of 124 × 271 pixels, which has a high level of noise and large heterogeneity. Thus, it is not possible to utilize the camera to identify the detected object. The low image resolution, the high level of noise and the limited range of imaging are the fundamental issues that must be dealt with in the terahertz imaging. All of these problems are directly related to the current level of technology. Therefore, to improve the quality of THz images simultaneously taking the advantage of the unique properties of the THz radiation, we decided to apply advanced image processing methods and combine selected elements of the THz image with images registered in other spectral bands [17]. In order to increase the potential of THz images, which are the base for the detection of concealed objects, an additional camera for visible light (VIS) was applied. The role of the VIS camera is to provide the image that will be used as the background for the illustration of the hidden objects detected by THz camera. Thanks to the use of VIS camera, the final image is easy to interpret for the system operator and it provides significant support in the decision making process (Fig. 11.2).

Unfortunately, none of the cameras provide the functionality of the image processing. In order to find hidden objects and to develop the system for object detection it is needed to integrate both cameras. There is a two-level integration – in the hardware and in the software (programming) domain. In terms of the former, the time synchronization of the acquisition process of cameras was carried out. Taking in the consideration the latter, the software for image processing (pre-processing and image fusion) was prepared.

11.3 Image Processing

The image processing methodology for our system was developed on the basis of the results of investigations carried out during last years. The methodology consists of three steps – acquisition, processing and fusion. In the first step, images from VIS

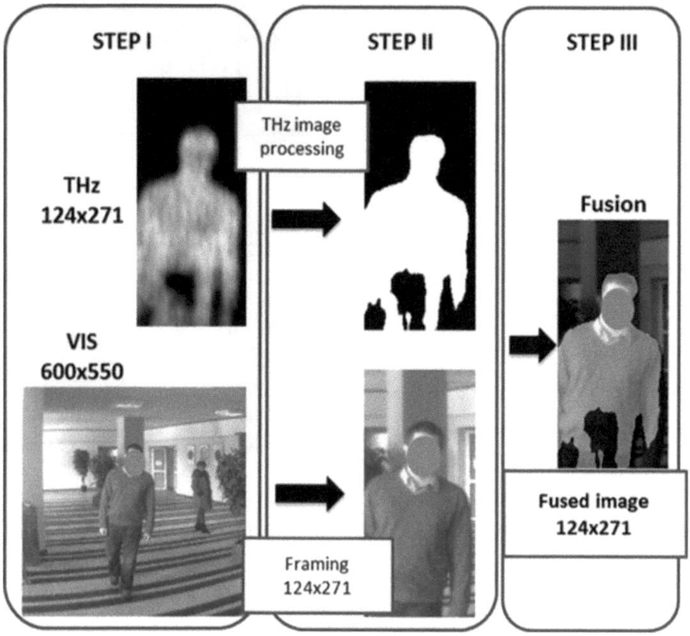

Fig. 11.3 The methodology of processing images (fusion algorithm)

and THz cameras are registered. After the acquisition, the recorded images are being processed. At first, both the visible and the terahertz images are being pre-processed. The VIS image is cropped to match the geometry of the THz image. At the same time, the THz image is subjected to various image processing operations in order to locate anomalies which, in practice, can be the objects hidden under clothing. In the third step of the image processing, the fusion of images from these two various spectral bands is performed. The algorithm of the image fusion is shown in Fig. 11.3.

After the acquisition process (first step), the first essential task of the image processing software is to geometrically match the images for further fusion process. A VIS camera has the resolution of 600×550 pixels and what is more it has different field of view than the THz camera. Due to this fact the VIS image has to be cropped to fit the THz image.

The second step of the algorithm contains the processing of the THz image. Since the image data from the terahertz camera is characterized by high level of noise, image filtering is necessary to use the THz image in further steps of algorithm (fusion). The main goal of image filtering is to recover the original pixel data from noisy measurements without introducing additional distortion to the original data.

Image filtering can be represented by the following expression:

$$v(i) = u(i) + n(i), \tag{11.1}$$

where: $v(i)$ is the measured value, $u(i)$ is the original value, $n(i)$ is the noise and i is the pixel number.

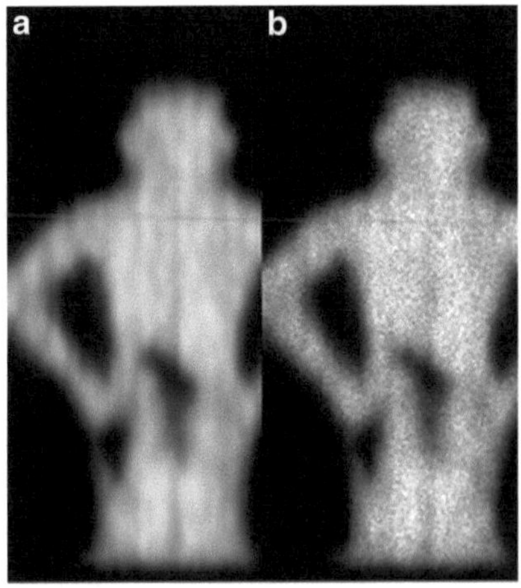

Fig. 11.4 THz image filtering – (**a**) RAW THz image, (**b**) filtered THz image

For image v, pixel values after filtering (NL) are computed using the formula:

$$NL\,[v]\,(i) = \sum_{j\in I} w\,(i,j)v(j), \tag{11.2}$$

where $w(i, j)$ factors are described as follows:

$$w\,(i,j) = \frac{1}{Z(i)}\exp\left(-\frac{\left\|v\,(\mathrm{N}_i) - v\left(\mathrm{N}_j\right)\right\|_{2,a}^2}{h^2}\right). \tag{11.3}$$

In the Eq. 11.3, $Z(i)$ is the normalized constant described by the following equation:

$$Z(i) = \sum_{j}\exp\left(-\frac{\left\|v\,(\mathrm{N}_i) - v\left(\mathrm{N}_j\right)\right\|_{2,a}^2}{h^2}\right). \tag{11.4}$$

where h is the degree of filtering and $v(\mathrm{N}_i)$ is the neighborhood of the i-th pixel.

The exemplary THz image after filtering and the corresponding blurred image are shown in Fig. 11.4. After filtering, the image is then segmented using thresholding of the THz image in order to detect hidden objects (Fig. 11.5).

Then, the final image is created after combining the THz and VIS images. During the research 12 various image fusion methods were investigated. Different versions of the image fusion methodology were considered. In the first scenario, the VIS image was superimposed on the binary image. The incorporating of the binary image into the VIS image creates a shadow corresponding to the hidden item on the body of a human. A location of the hidden object is then clearly visible. However, the shadow on the human body in the resulting image can interfere with subjective perception of the image and can lower the visibility of objects' details

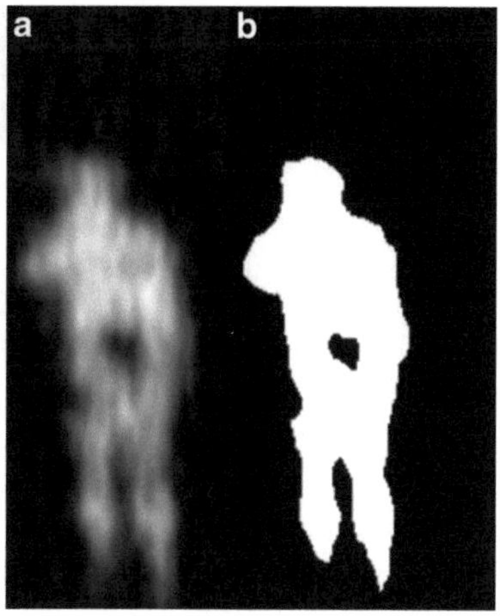

Fig. 11.5 THz image thresholding – (**a**) THz image, (**b**) binary THz image

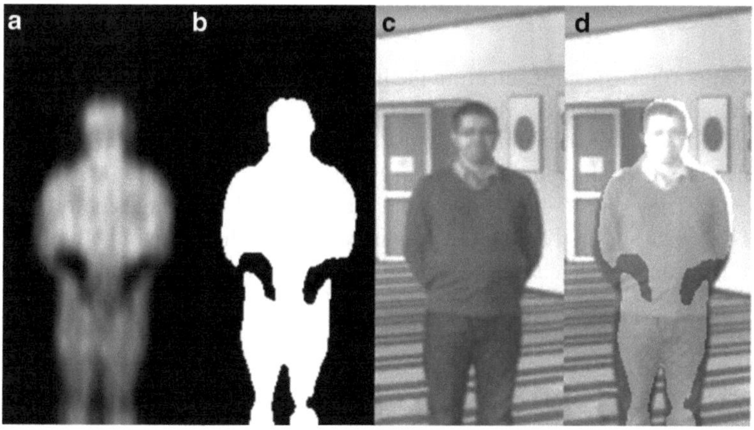

Fig. 11.6 Fusion of VIS and binary THz image – (**a**) THz image, (**b**) THz binary, (**c**) VIS, (**d**) fused image

in the VIS image. The exemplary images before and after the fusion using the first version of the method are presented in Fig. 11.6.

The second variant considered during our study was created to achieve a visualization which was composed only of the detected object that was superimposed on the VIS image. The shadow from the first variant is in this case eliminated. The second option requires significantly more complex methods of image segmentation. The exemplary image processed with the second version of fusion is presented in Fig. 11.7.

It should be noted that the visualization of the processed information is crucial to system operation. The goal of the visualization is to support the system operator in the decision making process.

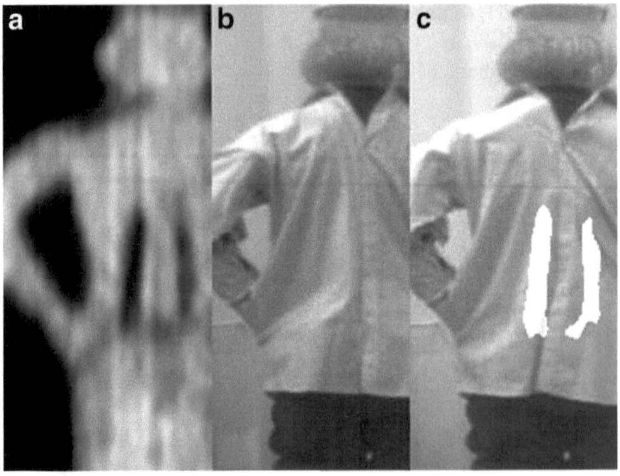

Fig. 11.7 Fusion of VIS image and detected object – (**a**) THz image, (**b**) VIS, (**c**) fused image

11.4 Image Quality Evaluation

A very important aspect of selecting the image processing method is to compare the performance of all different versions of the algorithm based on the required criteria. It is crucial that all the benchmarks and assessment criteria should be repeatable and objective [18]. For this purpose several methods of image quality evaluation were developed. They should enable assessing the quality of the images before and after processing. During the research, it was necessary to choose one method of image fusion. In order to be able to objectively compare the performance of the image fusion methods and to select only one method, an image quality assessment (IQA) methodology was developed. It should be noted that the assessment methodology of all image fusion methods allows to evaluate not only the result of the fusion algorithm but also all of the image processing operations applied to the image before the fusion.

All of the IQA methods can be divided into two groups – reference and non-reference. The former use a reference image to measure the quality of the final image, while the latter use only a resultant image to compute an image quality rate.

The main problem of assessing the quality of images is concerning fused images. They do not appear in the real world and thus they should not be assessed with methods invented for the evaluation of natural images. IQA in the case of fused images is a complex task because after a process of image fusion one obtains an image that is an effect of the transformation of two images (THz and VIS). It is not possible to find any reference image. A very popular method of evaluating the quality of fused images is the calculation of values of the Peak Signal to Noise Ratio, Mean Squared Error or Average Difference [17].

Our approach of evaluating fused image quality is to measure the perception of images by a human eye. The methodology of evaluating fused images is presented in Fig. 11.8.

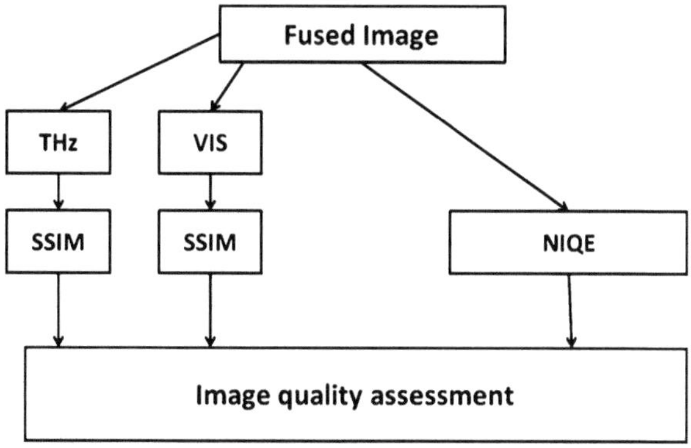

Fig. 11.8 The method of evaluating fused images

The structural similarity (SSIM) index was invented for measuring the similarity between the two images, thus it is a full reference method. The SSIM was designed to improve traditional methods which seemed to be inconsistent with human eye perception. This method takes under consideration the image degradation as a perceived change in structural information. The structural information describes how strong are inter-dependencies between the pixels, especially when they are spatially close to each other [8]. These dependencies carry important information about the structure of the object in the observed scene. SSIM index values are in the range of <0,1>, where 1 is the value corresponding to the two identical images. The SSIM index can be computed using the following equation [19]:

$$SSIM\,(x,y) = \frac{\left(2\mu_x\mu_y + c_1\right)\left(2\sigma_{xy} + c_2\right)}{\left(\mu_x^2 + \mu_y^2 + c_1\right)\left(\sigma_x^2 + \sigma_y^2 + c_2\right)} \tag{11.5}$$

where *x*, *y* are images, μ_x (μ_y) is the average of *x* (*y*), $\sigma^2{}_x$ ($\sigma^2{}_y$) is the variance of *x* (*y*), σ_{xy} is the covariance of *x* and *y*,$c_1 = (k_1L)^2$ and $c_2 = (k_2L)^2$ are two variables to stabilize the division with weak denominator, *L* is the dynamic range of the pixel-values (typically $2^{\#\text{bits per pixel}}-1$), $k_1 = 0.01$ and $k_2 = 0.03$.

The second IQA method proposed in this paper is Natural Image Quality Evaluator (NIQE). The measurable deviations from statistical regularities observed in natural images are the basis of this method. NIQE utilizes only the natural images so that the method do not need training on human-rated distorted images and do not need any exposure to distorted images.

NIQE is based on the construction of a quality aware collection of statistical features based on a simple and successful space domain natural scene statistic model. These features are derived from a corpus of natural, undistorted images [20].

The quality of the distorted image can be described as the distance between the natural scene statistic feature model and the multivariate Gaussian fit to the

Table 11.1 Criteria of selecting the image fusion method

Method	Range	Expected values
NIQE	<0, ∞)	0
SSIM (VIS-fused)	<0, 1>	1
SSIM (THz-fused)		<0,2

Table 11.2 The comparative results of three chosen image fusion methods

	Averaging	Shift invariant DWT	Contrast pyramid
	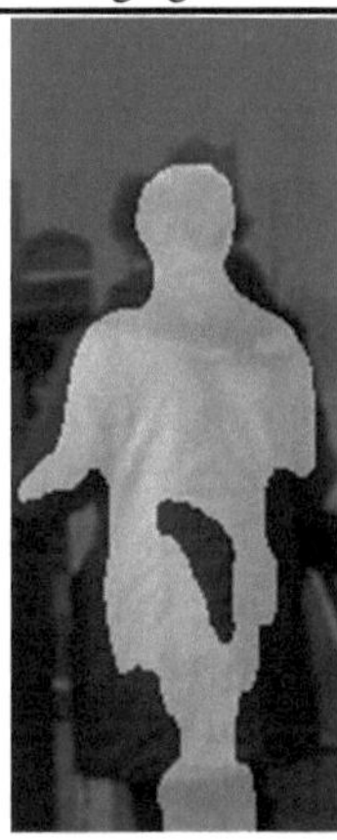		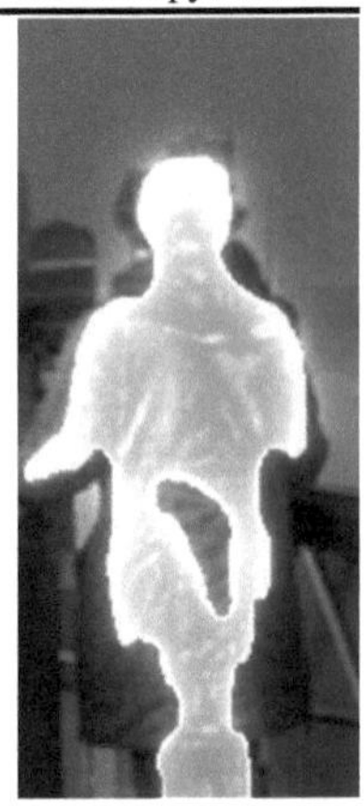
NIQE	8.1235	35.1217	12.0712
SSIM VIS-fused	0.6889	0.0336	0.2980
SSIM THz-fused	0.0457	0.3039	0.2665

features extracted from the distorted image. It can be computed using the following equation [8]:

$$D\left(v_1, v_2, \Sigma_1, \Sigma_2\right) = \sqrt{\left((v_1 - v_2)^T \left(\frac{\sum_1 + \sum_2}{2}\right)^{-1} (v_1 - v_2)\right)} \tag{11.6}$$

where ν_1, ν_2 are the mean vectors of the natural multivariate Gaussian model and the distorted image's multivariate Gaussian model, Σ_1, Σ_2 are the covariance matrices of the natural multivariate Gaussian model and the distorted image's multivariate Gaussian model. NIQE values are in the range of <0, ∞) (Table 11.1).

According to our initial assumptions the value of the SSIM index of the VIS-fused images should tend to 1 while the value of SSIM index of the THz-fused images and the value of NIQE should tend to 0. This means that the correlation index between the fused and the VIS images is bigger than the correlation between the fused and the THz ones. It seems to be evident because the VIS image is the background image for the fusion process, and only a few necessary and relevant elements of THz image are used. Comparative results of three chosen image fusion methods are presented in Table 11.2.

Despite the fact that the THz images have low signal to noise ratio and low contrast, we have successfully achieved our goal to automatically detect and segment concealed objects. We build the system for hidden objects detection based on the passive THz camera and the VIS camera. It should be noted that this system should operate in specific conditions. Very important aspect of the system operation is to ensure proper environmental conditions like temperature, humidity and the camera-to-object distance. All of these parameters should be controllable.

The most recent developments in the THz technology and new explorations of THz imaging have advanced THz applications in the defense and security fields. There are still some challenges and obstacles in implementing THz technologies for defense and security applications in real-world.

11.5 Thermal Phantom

The moving platform with the human phantom, which mimics a moving person was constructed to avoid the time-consuming, tedious and expensive testing on people. This section focuses on the construction and test results associated with the thermal phantom, which was developed for testing THz cameras and our image processing software.

11.5.1 Construction of the Thermal Phantom

It is commonly known that a naked human body does not have uniform temperature distribution and it varies in the range from about 29 to 36 °C. Therefore, the phantom was not designed to have homogeneous temperature on the whole surface but we assumed some range in which the temperature can vary. Moreover, in case of a dressed person, the temperature in some places can decrease down to about 25 °C and even lower depending on the number and kind of worn clothes.

During designing process of the phantom we patterned upon the human body, whose skin consists of many veins filled up with circulating and heated blood. The veins, through the arteries, are connected with heart, which pump the blood into all parts of the body. We built a substitute, which in our opinion fulfill all the requirements for testing mentioned THz devices and software (Fig. 11.9).

The core of the phantom is the 110-cm long polystyrene manikin. Then it was vertically covered with 150 plastic pipes with internal and external diameters of 4 and 5 mm, respectively. The distance between the pipes is not bigger than 3 mm to avoid in homogeneity in the heat transfer.

The pipes were glued to the manikin and next coated with a silicone resin. Therefore, the developed phantom is light and quite fragile. We decided to replace lower parts of the legs (from mid-thigh) with the box for the temperature controller. To increase its stiffness we inserted a long plastic rod into each leg of the manikin,

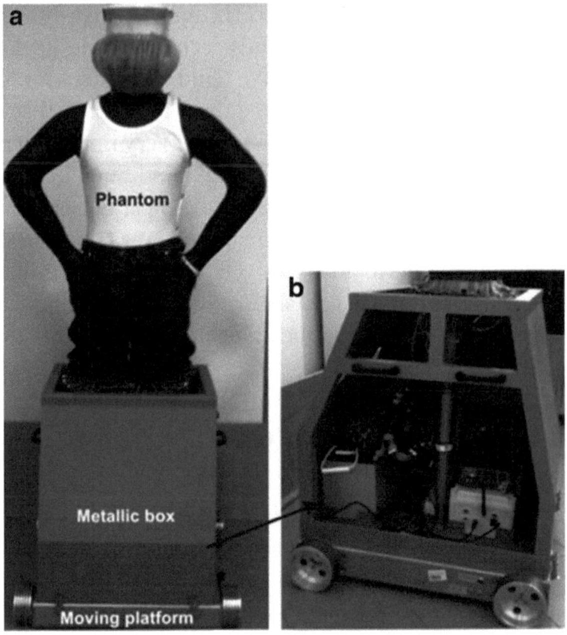

Fig. 11.9 The photograph of the human phantom (**a**) with the moving platform and the metallic box (**b**)

because especially during starting and stopping of the platform the relatively tall phantom is susceptible to deflection. The weight of the phantom with pipes and the silicone resin "skin" is about 5 k.

As expected, we observed temperature differences between the upward and the downward water streams due to the heat transfer mainly to the silicone skin. Therefore, to minimize this phenomenon we applied complicated arrangements of pipes (Fig. 11.10).

The upper ends of the pipes are connected to the water reservoir (a head of the phantom), while the lower ends of the pipes through a set of pipes with wider diameters (rings: I–V) reach inlet and outlet arteries (Fig. 11.10b). We determined experimentally that the water from the inlet artery should be directed upwards through the 80 % of pipes to the reservoir. Only 20 % of pipes is exploited to direct water downwards gravitationally from the reservoir. For each leg we applied four upward rings (Fig. 11.10b, solid line, no. I, II, IV, V) and one downward ring (dashed line, no. III). We grouped the smallest pipes which leave the rings into sections (Fig. 11.10b). Each section consists of the downward pipe (colder, dashed line), which is surrounded by a few upward pipes (hotter, solid lines).

If the temperature (T_{in}) rises above or falls down below the limit, the controller switches the heater off or on, respectively. T_{in} can be adjusted in the range up to 33 °C, which means that the maximal temperature of the naked body of the phantom is about 37 °C. We usually tested the phantom with $T_{in} = 31$ °C because it resulted in biggest similarity to a human.

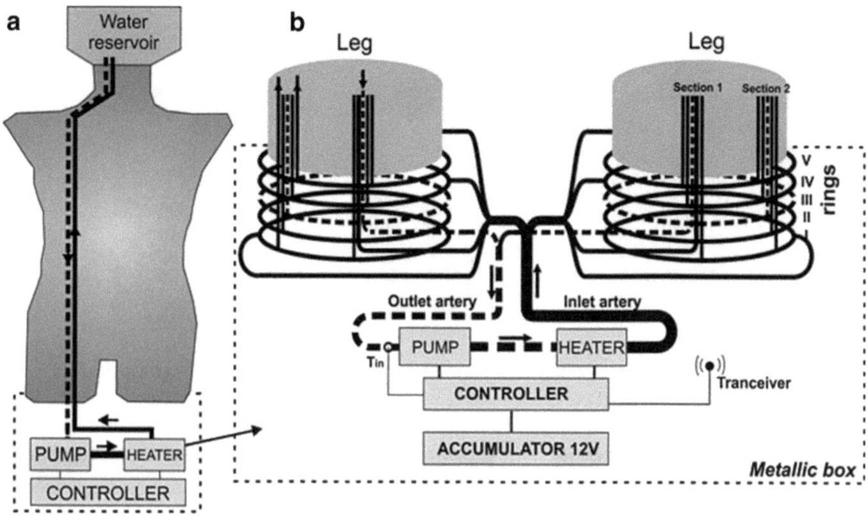

Fig. 11.10 The phantom of human body: a scheme (**a**) and the detailed layout of *upward* and *downward* pipes (**b**)

The flat, moving and wirelessly-controlled platform (Fig. 11.9) with four wheels provides the required maneuverability and load up to 100 kg. Thanks to four independent DC motors, this robot can operate on flat surfaces and can go forward and backward with the speed up to 400 mm/s as well as turn and rotate. The platform is equipped with the controller with the 868 MHz radio transceiver, and a set of four ultrasonic sensors to localize the platform and to avoid collision with obstacles. The LabView-based software can be used for the remote control of the movement. The phantom with the metallic box installed on the moving platform is 185 cm high and its weight is equal to about 70 kg including about 5 l of water.

The phantom is equipped with a set of ordinary clothes (made mainly of cotton and other popular materials): boxer shorts, vests, T-shirts, trousers, shirts, sweaters, jackets, dresses and belts (Fig. 11.11a). We purchased some fake weapons (small and regular pistols, a revolver, knives etc.) with necessary holsters (Fig. 11.11b). Moreover, we prepared five imitations of Body Borne Improvised Explosive Devices (BBIED) – metallic and non-metallic items that can be hidden under the clothes and that can mimics the original bombs (Fig. 11.11c).

11.5.2 The Measurement Arrangement and Results

The phantom should be visible in the THz range in the similar way to a man to prove the usefulness for testing. During tests both targets (phantom and human) were first

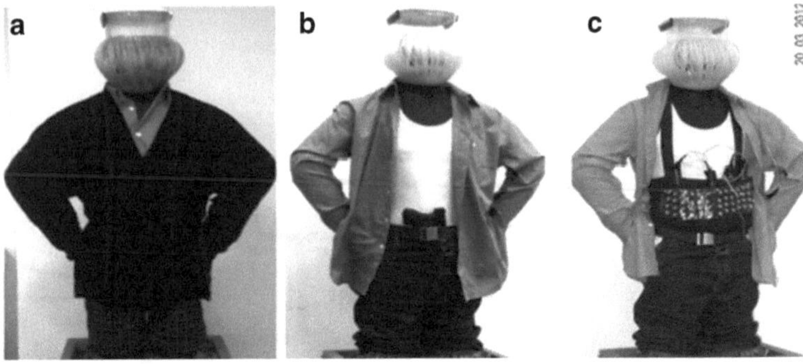

Fig. 11.11 The photographs of the ready-to-use moving platform. The phantom dressed with: clothes (**a**) gun (**b**) and BBIED (**c**)

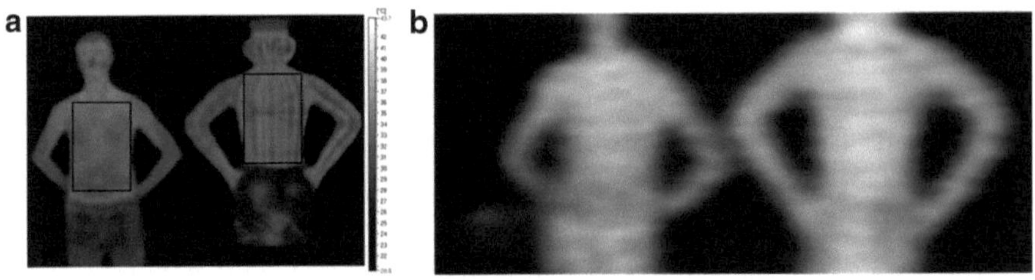

Fig. 11.12 The comparison of images of the naked phantom (*right*) and the person (*left*) acquired by: the thermal (**a**) and the THz (**b**) cameras

naked and then dressed with an ordinary cotton shirt and a sweater. Since the THz image is rather blurred, we also measured both targets by means of a thermal camera to see the temperature distributions and the temperature stability in time.

For the thermal tests we applied the VIGOcam v50 from VIGO Company. This thermal camera uses a microbolometer Focal Plane Array with 384×288 pixels working in the 8–14 μm range. The camera was connected to a PC by means of Ethernet and the thermal images could be acquired, visualized and recorded by THERM software. We determined the maximal and minimal temperature on the phantom surface during stabilization.

A distance between both cameras (THz and thermal) and the target was 6 m. The measurements were carried out in a laboratory with a stable temperature of about 22 °C and a relative humidity of about 40 %. The phantom was stabilized at the temperature $T_{in} = 31$ °C.

Figure 11.12a shows the thermal image of both naked targets. The most probable place to hide dangerous items is situated in the chest and stomach region, therefore we focused our attention on the rectangular areas marked in Fig. 11.12a and we determined the mean temperature inside these regions. We obtained 31.7 ± 0.2 °C for the person and 31.6 ± 0.2 °C for the phantom, which seems to be satisfactory result. Simultaneously, we acquired the THz image of both targets by means of TS4 camera (Fig. 11.12b). Due to the poorer resolution and smaller sensitivity of the THz

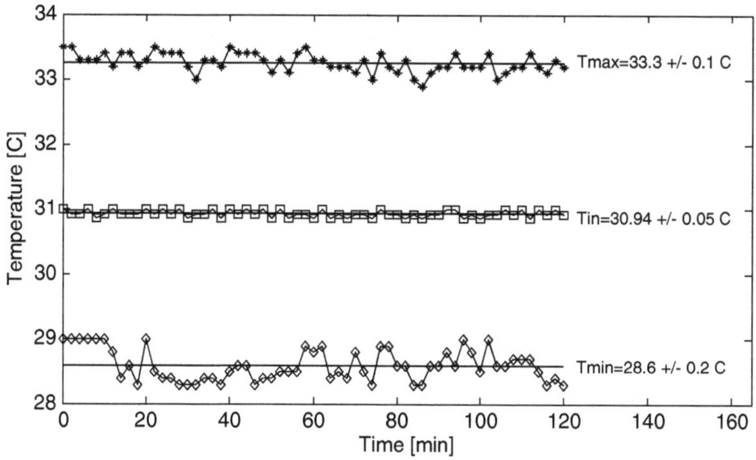

Fig. 11.13 Maximal (T_{max}) and minimal (T_{min}) temperatures measured on the phantom's surface inside the area marked in Fig. 11.12a during 2 h. The temperature of the heating liquid (T_{in}) during the same time

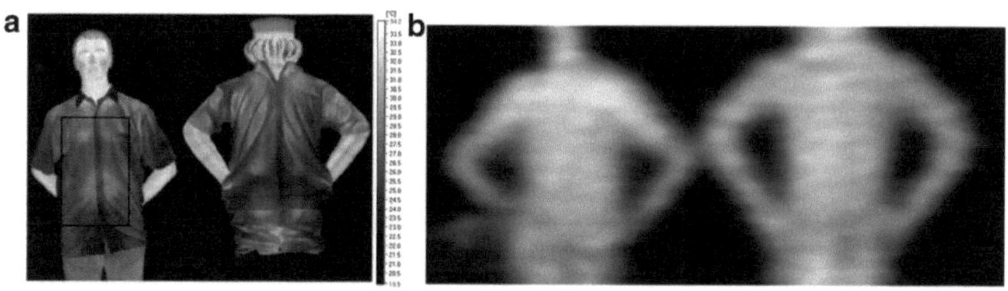

Fig. 11.14 The targets dressed in the shirt with short sleeves: thermal (**a**) and THz image (**b**)

camera, the obtained images are blurred but the contours of the targets are clearly visible. The intensity of both targets is similar and in homogeneity visible in the case of the thermal image is not observed.

Moreover, every 2 min we measured the maximal (T_{max}) and minimal (T_{min}) temperatures on the phantom's surface inside the rectangular area depicted in Fig. 11.12a during 2 h. The obtained temperature stabilities (Fig. 11.13) and their standard deviations in the range 0.1–0.2 °C seem to be sufficient for the expected applications. We also present the temperature stability of the heating liquid measured by the digital thermometer – $T_{in} = 30.94 \pm 0.05$ °C.

Next figures present the thermal and the THz images of the phantom and the man clothed in trousers and either a cotton shirt with short sleeves (Fig. 11.14) or a cotton shirt with long sleeves and a sweater (Fig. 11.15). We can notice nearly the same temperature of a person and the phantom both when targets are naked and when they are dressed. The brightness of both targets is nearly the same in the THz

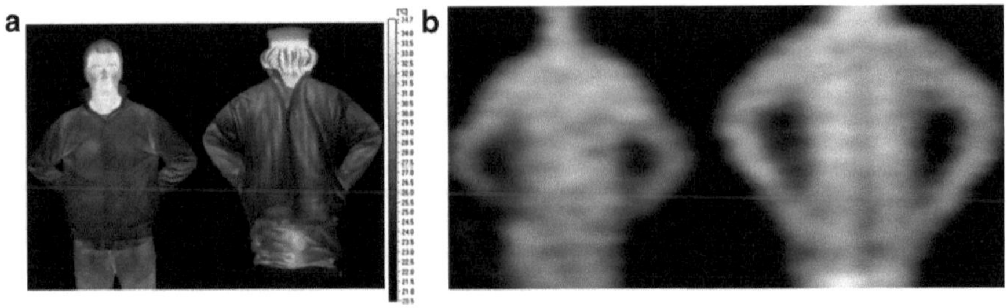

Fig. 11.15 The targets dressed in the shirt with long sleeves and the sweater: thermal (**a**) and THz image (**b**)

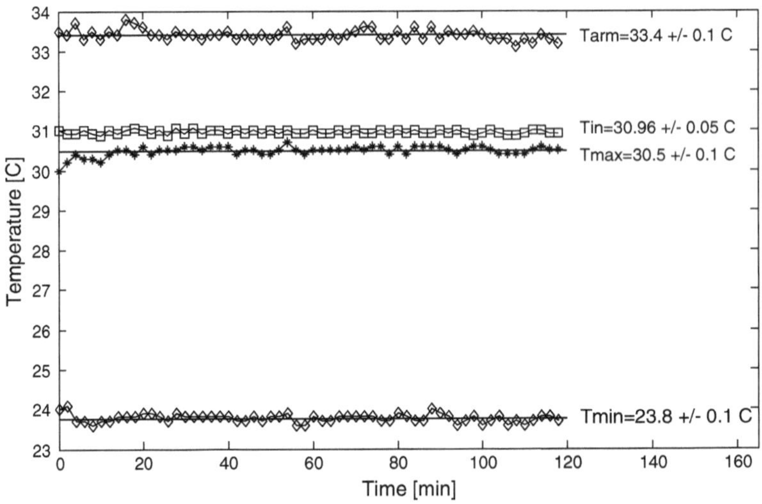

Fig. 11.16 The maximal (T_{max}) and the minimal (T_{min}) temperatures measured on the phantom's surface inside the area marked in Fig. 11.12a during 2 h. The temperature of the heating liquid (T_{in}) and the maximal temperature of the phantom's arm (T_{arm}) during the same time

image. The resolution of the THz camera is poorer in comparison to the thermal camera and therefore the THz images are less clear but shape of the body is clearly seen.

Moreover, we measured the temperature stability of the phantom dressed in the shirt with short sleeves during 2 h – the maximal (T_{max}) and the minimal (T_{min}) temperatures inside the rectangular area marked in Fig. 11.14. We also recorded the maximal temperature on the phantom's forearm, which was not covered with the shirt (T_{arm}) and the temperature stability of the heating liquid measured by the digital thermometer – $T_{in} = 30.96 \pm 0.05$ °C. Also in this case the stability depicted in Fig. 11.16 is satisfactory.

Finally, we compared THz images of the targets with a 25-cm long pistol hidden under clothes (Fig. 11.17). Both of them were dressed in the cotton shirts. The pistol is seen in the same way – as a dark region with the shape resembling the hidden item.

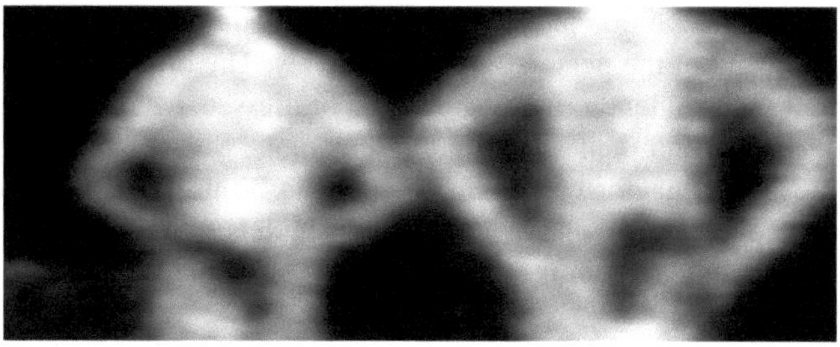

Fig. 11.17 The THz image of the targets (a person on the *left* and the phantom on the *right*) dressed in the cotton shirt with the hidden pistol

11.6 Summary

The results presented in this paper show that the computer processing of images from the THz camera is really promising and cost-effective way for demanding security and defense applications. The exploitation of exploratory data analysis tools and the preparation of predictive models is the next step in our future research.

During the investigation process many image processing methods and algorithms were tested taking into account effectiveness and efficiency. The effectiveness of developed algorithms and of software code mainly depends on the distance between the human body and the camera, on the size of the target and its type. It should be highlighted that the use of only one image processing method for all situations can be ineffective. To solve the problem of the image quality enhancement we need to combine various algorithms.

A very promising result of our research is the image fusion algorithm. Our recent investigations on combining THz and VIS images show that there is a great potential in combining two images from different spectral ranges, especially for the security applications. In that part of our research we focused on extracting the most valuable features from each image and combing them into one. The future research will lead to the further development of the image fusion algorithms as well as combination of VIS, THz and thermal images.

The methodology of evaluating the image quality presented in the paper shows that it is possible to assess the quality of the images and the human perception of those images by using a combination of various IQA methods. This methodology was very useful during investigations of the image fusion techniques and it allowed us to choose the most appropriate fusion algorithm for further implementation in the final system.

It is worth noticing that we obtained satisfactory similarities between the man and the phantom observed in the terahertz and the infrared ranges of radiation. The phantom was tested by means of the thermal and the terahertz cameras. The similarities between the images of a person and of the phantom, both dressed in various sets of clothes, are sufficient for the considered applications. The temperature stability of the phantom was also analyzed in time with satisfactory results.

Due to the poor resolution of the THz camera, we could compare the THz images of the phantom and of the person only qualitatively. The brightness of both targets is very similar independently if they are naked or dressed. Moreover, the intensities of the hidden items and their shapes are comparable. Such phantom construction, which mimics the behavior of human beings, can be useful for the performance evaluation of the terahertz cameras in the way which is free from all disadvantages connected with people screening.

References

1. Kemp MC (2006) Millimetre wave and terahertz technology for the detection of concealed threats – a review. Proc SPIE 6402:64020D
2. Palka N (2011) THz reflection spectroscopy of explosives measured by time domain spectroscopy. Acta Phys Polon A 120:713–715
3. Gatesman AJ et al (2006) Terahertz behaviour of optical components and common materials. Proc SPIE 6212:62120E
4. Yun-Shik L (2008) Principles of terahertz science and technology. Springer, New York
5. Palka N et al (2012) THz spectroscopy and imaging in security applications. 19th international conference on microwaves, radar and wireless communications, pp 265–270
6. Heinz E et al (2012) Development of passive submillimeter-wave video imaging systems for security applications. Proc SPIE 8544:854402
7. Cooper KB, Dengler RJ, Llombart N, Bryllert T, Chattopadhyay G, Mehdi I, Siegel PH (2009) An approach for sub-second imaging of concealed objects using terahertz (THz) radar. J Infrared Millim Terahz Waves 30:1297–1307
8. Appleby R, Wallace HB (2007) Standoff detection of weapons and contraband in the 100 GHz to 1 THz region. IEEE Trans Antennas Propag 55:2944–2956
9. Millivision, website: www.millivision.com
10. ProVision ATD, website: http://www.sds.l-3com.com
11. ThruVision System Ltd., website: www.truvision.com
12. Brijot imaging systems Inc., website: www.brijot.com
13. STANAG 4349 (ED. 1) Nato standardization agreement: measurement of the minimum resolvable temperature difference of thermal cameras (09-AUG-1995)
14. Trofimov VA, Trofimov VV, Chao Deng, Yuan-meng Zhao, Cun-lin Zhang, Xin Zhang (2011) Possible way for increasing the quality of imaging from THz passive device. Optics and photonics for counterterrorism and crime fighting VII, Optical materials in defence systems technology VIII and quantum-physics-based information security, 81890L, October 13
15. Murrill SR et al (2008) Terahertz imaging system performance model for concealed-weapon identification. Appl Opt 47(9):1286–1297
16. Jansen C (2010) Terahertz imaging: applications and perspectives. Appl Opt 49(19):E48–E57
17. Kowalski M, Piszczek M, Palka N, Szustakowski M (2012) Phot Lett Pol 4:3
18. Sumathi M, Barani R (2012) Qualitative evaluation of pixel level image fusion algorithms. In: Informatics and Medical Engineering (PRIME), international conference on pattern recognition, 21–23 March, pp 312–317
19. Brunet D, Vrscay ER, Wang Z (2004) On the mathematical properties of the structural similarity index. IEEE Trans Image Process 21(4):1488–1499
20. Mittal A, Soundarajan R, Bovik AC (2013) Making a 'completely blind' image quality analyzer. IEEE Signal Process Lett 20(3):209–212

Chapter 12
Clinotron Based Terahertz Imaging System

D.M. Vavriv, A.V. Somov, K. Schünemann, and V.A. Volkov

Abstract The clinotron is a vacuum tube that can effective operate in the THz frequency region. A series of clinotron based oscillators have been developed at the Institute of Radio Astronomy of the National Academy of Sciences of Ukraine. In this review, the design and characteristics of these devices are presented. Prospects for the further improvement of clinotron oscillator characteristics are considered. As an example of the application of clinotron oscillators, a 300 GHz beam steering imaging system is presented. Potentialities of the clinotron based imaging systems are considered for future implementations in 3D THz scanners for security and other applications.

12.1 Introduction

The THz frequencies technology is just beginning to come of age, and thus becoming increasingly attractive for various applications [1–5]. The main factor that still restrains a wide application of THz frequencies is related to a limited choice of suitable sources operating within this region. Most of the applications are pressing for compact, high-power, tunable, efficient, and room-temperature oscillators. At present time, available sources of the terahertz radiation are direct multiplier-based sources, frequency mixers, gyrotrons, backward wave oscillators (BWO), far infrared lasers, optically pumped lasers, free electron lasers, synchrotron type sources, and others. Each of these sources has its own advantages and drawbacks,

D.M. Vavriv (✉) • A.V. Somov • V.A. Volkov
Institute of Radio Astronomy of the National Academy of Sciences of Ukraine,
61002 Kharkov, Ukraine
e-mail: vavriv@rian.kharkov.ua

K. Schünemann (✉)
Technical University Hamburg-Harburg, Hochfrequenztechnik, D 21071 Hamburg, Germany
e-mail: schuenemann@tuhh.de

C. Corsi and F. Sizov (eds.), *THz and Security Applications*, NATO Science for Peace and Security Series B: Physics and Biophysics, DOI 10.1007/978-94-017-8828-1_12,

but none of them meets completely the formulated requirements for compact sources. As for the low-frequency part of the THz-region – up to about 3 THz, the BWO [6, 7] can be considered as the most suitable candidate for wide practical applications, but the main disadvantage of this type of oscillators is related to a low output power. In this review, we describe a vacuum tube, called the clinotron, in which the abovementioned shortcoming of the BWO is mitigated.

The clinotron was invented by Ukrainian scientists [8–10], and it has already proved its efficiency for the frequencies up to 500 GHz. Section 12.2 briefly introduces characteristics of clinotron oscillators. A high output power level and a wide frequency tuning range of the clinotron tubes make them attractive for the development of various imaging systems, for instance, designed for security and surveillance applications.

It has been already demonstrated that imaging at sub-millimeter wavelengths offer the definite advantage of being able to scan concealed objects through different materials opaque in other regions of frequency spectrum [1, 11, 12]. Various approaches have already been proposed for the development of such imaging systems [11–14]. For example, a beam steering technique is typically used to focus power from the single source on small area of investigated object and to collect reflected signal from the scanned area segments onto single or multiple receivers. In Sect. 12.3, an example of a recently developed by us beam steering imaging system is given. This system is based on a 0.3 THz clinotron source with 200 mW output power level. The system demonstrates advantages related to a long operation range in this atmospheric frequency window, where the attenuation of propagating electromagnetic waves is as low as about 10 dB/km. In Sect. 12.4, plans for future modifications of clinotron based radars, achievable parameters, and their potential applications are discussed.

12.2 Clinotron Design and Characteristics

The clinotron was proposed as a modification of the conventional backward-wave oscillator (BWO) in order to eliminate the main drawbacks of the latter, like its low efficiency and low level of output power. A schematic representation of the clinotron design is shown in Fig. 12.1. The tube is composed of a sheet electron beam, which is inclined to the surface of a grating at an angle α. The grating is placed on a wall of a rectangular cavity with an output coupling waveguide. The beam is directed by a longitudinal static magnetic field.

In the clinotron, like in the conventional BWO, the electron beam interacts with a spatial harmonic whose phase velocity is close to the electron velocity. However, due to the beam inclination, the characteristic interaction length is determined by the tilt angle and by the characteristic thickness of the field layer of the synchronous wave concentrated near the grating surface.

The clinotron is similar to the BWO in sense that it utilizes the interaction of an electron beam with spatial harmonic components of the electromagnetic field of a

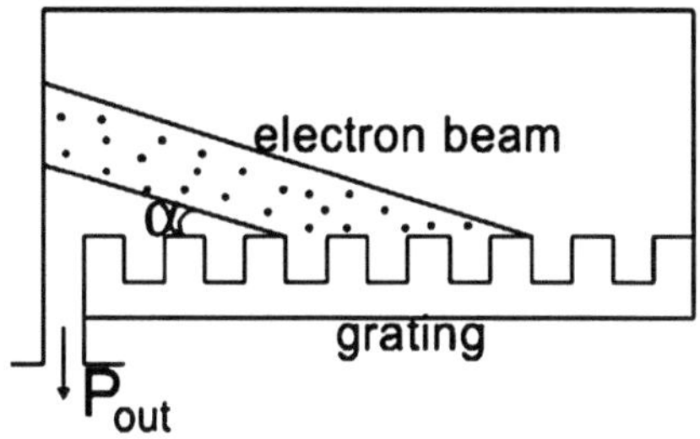

Fig. 12.1 Schematics of the clinotron

Table 12.1 Characteristics of clinotrons

Model	Freq. range [GHz]	Output power [W]	Anode voltage [kV]	Anode current [mA]	Weight [kg]
C-5M3	53–63	11.0	4.0	200	1.2
C-5M2	56–63	4.0	4.0	150	1.2
C-3M3	79–98	5.0	5.0	150	1.2
C-3M2	84–98	1.5	5.0	150	1.2
C-2.5M3	113–122	3.0	4.3	180	1.2
C-2.2M3	120–141	2.0	4.5	160	1.2
C-2.2M4	124–141	1.5	4.5	160	1.2
C-2.2M2	125–140	1.0	4.5	160	1.2
C-2.0M3	137–151	2.0	4.5	140	1.2
C-0.8M8	345–390	0.2	5.0	160	12
C-0.8	345–390	0.1	5.0	180	12
C-0.5	442–510	0.1	5.5	200	12

slow-wave structure. However, some essential modifications in the tube design were introduced. The distinguishing features of the clinotron are as follows:

1. The electron beam is inclined to the surface of a grating. By varying the tilt angle α, it is easy to optimize the length of the "effective" interaction space without changing the geometry of the tube,
2. The beam thickness is large in comparison to that in the conventional BWO,
3. The electrons are bunched in an exponentially growing field,
4. A wide electron beam is used, that enables increase of the beam current and the output power as well,
5. The clinotron is usually built as a resonance device. Nevertheless, the electronic tuning of the operating frequency is large that is achieved by successive exciting the resonator modes with beam voltage variation.

These approaches have resulted in the development of a series of clinotron oscillators through the millimeter and submillimeter wavelength bands [8–10]. Characteristics of the tubes produced at the Institute of Radio Astronomy of the National Academy of Sciences of Ukraine are given in Table 12.1.

The output power level of these tubes is at least an order of magnitude greater than that for conventional BWOs while small physical dimensions, low weight,

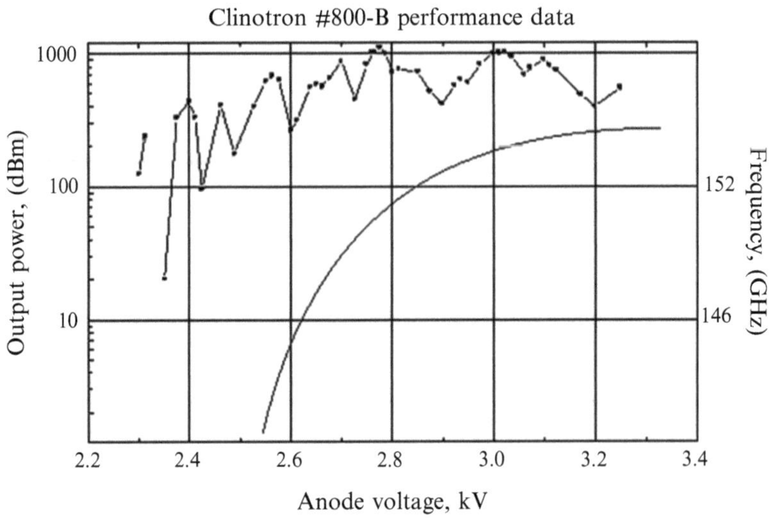

Fig. 12.2 Typical dependencies of the output power and the operation frequency versus the anode voltage

tunability and other advantages of BWO's have been retained. For example, 300 GHz and 500 GHz tubes provide the power level of about 500 mW and 100 mW, respectively. A wide range of the electronic frequency tunability is another attractive feature of the clinotron. The operating frequency range varies from 10 to 20 % depending on the central frequency value.

A typical frequency tuning characteristic is shown in Fig. 12.2 together with the dependence of the output power versus the accelerating voltage. In this case, about 20 resonator modes are successively excited with accelerating voltage variation. The tuning characteristic is rather smooth, and individual modes manifest themselves only at the low-voltage end of the tuning characteristic. Physical dimensions, weight, and the operating voltage of clinotrons are comparable with those of BWO's. Manufacturing technology of both subcomponents and of the complete device is simple, resulting in low-cost tubes.

The energy output in the clinotron is arranged by means of a waveguide, as schematically shown in Fig. 12.1. However, tubes with distributed energy output, where the energy is directly radiated via transparent window, as shown in Fig. 12.3, have been developed and produced as well [8–10]. The generated high-frequency energy is brought out here by means of the Smith-Purcell radiation. Such clinotron tubes are easily matched with quasioptical transition lines and, therefore, this modification is especially interesting for applications in THz-systems. Due to unique characteristics, the clinotron tubes have already been used in various electronic systems, like plasma diagnostic instruments, short-range radars, local oscillators, etc.

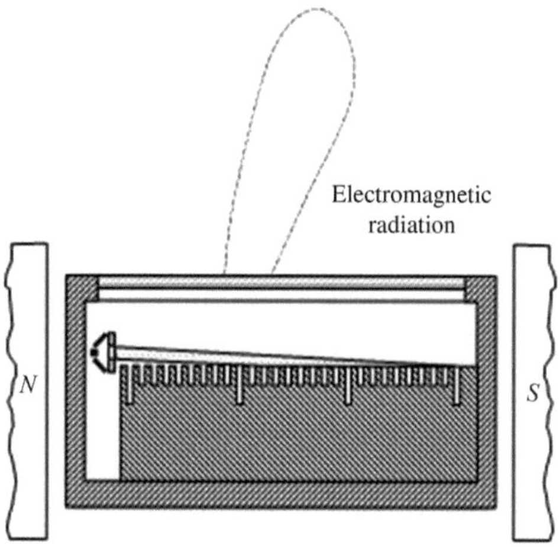

Fig. 12.3 Schematic view of the clinotron with distributed energy output

To meet such applications requirements, a compact, solid-state, high-voltage power supplies have been developed for the clinotron based oscillators. These supplies demonstrate a high efficiency and reliability. The possibility of the development of clinotron-based synthesized oscillators with the relative frequency stability of about 10^{-7} has also been demonstrated [15].

The theoretical study [16] indicates that the clinotron has a large potential for further increase of both the operating frequency and the output power. In order to implement this potential, clinotron tubes with a denser and more intensive electron beam should be manufactured. Tube sources suffer from metallic losses and the need for high magnetic and electric fields as well as high current densities. Therefore, it worth to be mentioned that the space charge and temperature effects do not impose serious limitations on an essential current density increase in comparison with the corresponding values in the present clinotron design. According to recent simulation results, the output power levels of about 2 and 70 W can be achieved at the frequencies around 1 THz in CW and pulsed operating modes, respectively. To embody such tubes, the beam current density should be increased to about 100 A/cm^2 for the CW mode and to about 1,000 A/cm^2 for the pulsed mode. The beam cross section should be 0.05 mm $\times$ 2.5 mm, which is the same as in the already produced tubes.

12.3 Clinotron Based Imaging System

A clinotron based imaging system has been recently designed and built to study potentials of clinotrons for such applications. A block-diagram of the imaging system is shown in Fig. 12.4. The imager is based on a simple single-pixel design that produces raster scene with a minimal loss of the radiated energy to utilize

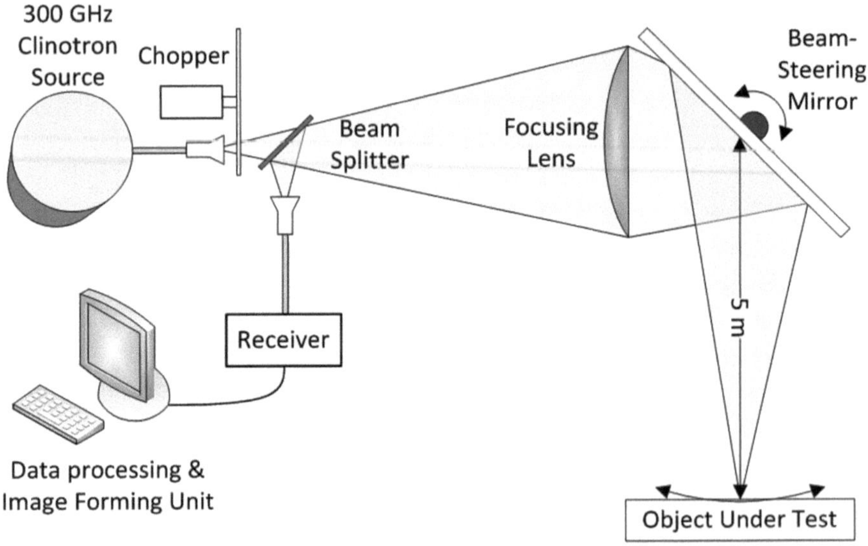

Fig. 12.4 Schematic of the single-pixel scanner setup

the high output power capability of the clinotron oscillator. A 2D single-pixel scanner setup is manufactured and tested so far. The imaging system is based on a clinotron oscillator that is capable to deliver about 200 mW of CW power in the 300 GHz frequency range. The clinotron waveguide output is directly connected to a diagonal horn, which forms the radiation pattern with a high level of the radial symmetry and the gaussicity. The output radiation is modulated at the frequency of 0.8 kHz by means of a mechanical chopper. A 3-dB quasi-optical beam-splitter is used to decouple the outgoing and incoming radiation while keeping the Tx and Rx radiation patterns oriented along the same scan axis. On the receiver side, the detector is equipped with the same horn which is pointed normally to the transmitter horn axis into the beam splitter.

The quasioptical system is comprised of a 34 mm thick Teflon focusing lens with the diameter of 220 mm that is used to focus the incident radiation. A flat scanning mirror is used to scan the investigated scene. The lens is illuminated by a transmitting horn with a −10 dB taper at the edge to provide both an optimal gain and a maximum Gaussian beam transformation efficiency. The manufactured lens gives the diffraction-limited beam waist, which is less than 24 mm at a 5 m distance. To perform the raster scanning with the beam, the flat flapping mirror is placed on a PC controlled positioner. The mirror has a rectangular shape of 250 mm × 450 mm. The mirror was made of a silver-plated glass plane. The achieved signal to noise ratio for this setup is about 15 dB that is mainly determined by a simple square-law detector used on the receiver side. Nevertheless, even such relatively simple setup as shown in Fig. 12.4 makes it possible to use the developed 300 GHz imaging system for various security and other applications. Our first imaging experiments indicate

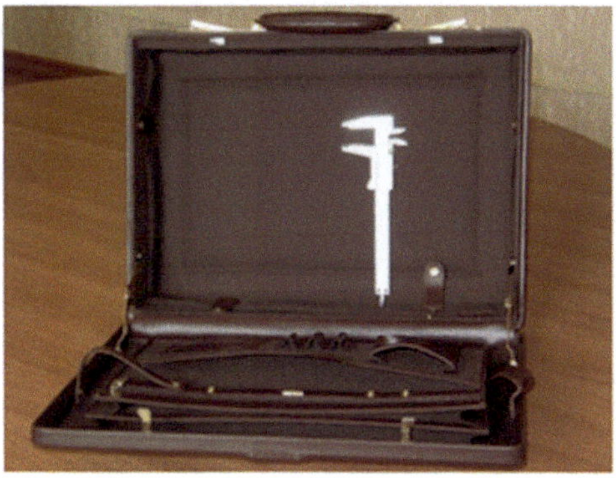
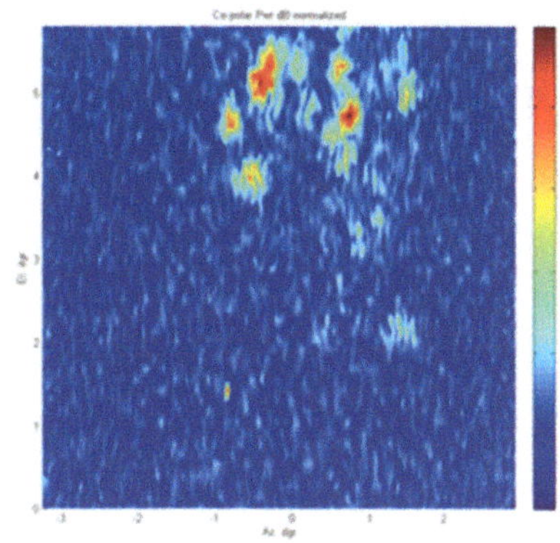

Fig. 12.5 Briefcase with a metallic calliper and THz image of the calliper obtained at the distance of 5 m through the briefcase wall

the ability of this scanner to reveal concealed metallic objects under the layer of dense but transparent for THz radiation materials. As an example, a THz image of a metallic calliper hidden inside a briefcase is shown in Fig. 12.5. This image was obtained from the distance of 5 m.

The future improvement of the image quality can be accomplished by increasing the optical system effective aperture that leads to a better resolution. In addition, it is planned to introduce an advanced heterodyne receiver resulting in a higher dynamic range of scanned images (up to 80 dB) what will make possible to reveal more details of investigated samples.

12.4 Prospects for Clinotron Based Imaging Systems

The potential of the clinotron oscillators looks very attractive from the point of view of their possible applications in various THz systems. Modern clinotrons provide enough power to do heterodyne imaging, at the same time combining other unbeatable properties, such as compact size, and wide frequency tuning range. However, for their successful application in perspective high-resolution 3D imaging systems, still some further investigations and developments have to be carried out. The problem is that any clinotron oscillator is a self-oscillating THz source that means inherent difficulties in achieving high frequency stability along with the frequency modulation capabilities. This problem restricts, for example, the application of the conventional FMCW technique for the realization of a fine range resolution. Nevertheless, our preliminary simulation results indicate that this obstacle can be overcame by using the above technique in combination with the coherent on receiver technique. A conceptual diagram of clinotron based 3D radar/imager is shown in Fig. 12.6.

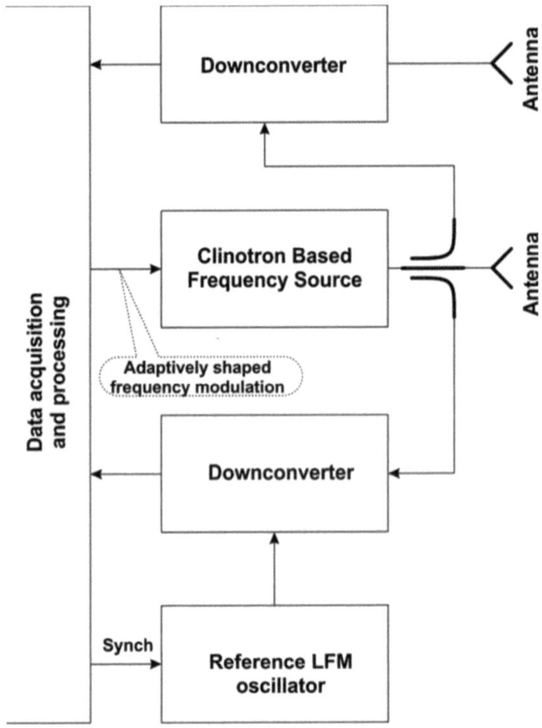

Fig. 12.6 Conceptual diagram of clinotron based 3D radar/imager

In this case, the frequency variations of each individual frequency sweep are digitized, recorded and utilized for the processing of the received backscattered signals. This approach does not require a high frequency reproducibility or ultimate stability of the source, and a high range resolution can be achieved even with self-running oscillators. The effectiveness of such approach has been already demonstrated during the development of high-resolution, magnetron-based, millimeter-wavelength radar systems [17]. Our estimations prove that, for example, the implementation of the FMCW in combination with the coherent on receiver technique in the described 300 GHz clinotron based imaging system would allow achieving the range resolution of about 2 cm. This opens up a way for the development of novel high-resolution, long range 3D THz imaging systems for a wide variety of emerging applications yet to be seen.

References

1. Siegel PH (2002) Terahertz technology. IEEE MTT 50(3):910–920
2. Linfield E (2007) Terahertz applications: a source of fresh hope. Nat Photonics 1:257–258
3. Dragoman D, Dragoman M (2004) Terahertz fields and applications. Prog Quantum Electron 28:1–66
4. Kurt H, Citrin DS (2005) Photonic crystals for biochemical sensing in the terahertz region. Appl Phys Lett 87:104–108

5. Phillips TG, Keene J (1992) Submillimeter astronomy. Proc IEEE 80:1662–1678
6. Golant MB et al (1965) Series of wide-range small power generators for submillimeter wave range. Pribory i Tekhnika Eksperimenta 4:136–139
7. Kantorowicz G, Palluel P (1979) Backward wave oscillators. In: Button K (ed) Infrared and millimeter waves, vol 4. Academic Press, New York, pp 185–211
8. Churilova SA et al (1992) The clinotron. Naukova Dumka Press, Kiev
9. Lysenko YY, Pishko OF, Chumak VG et al (2004) State of the development of CW clinotrons. Adv Mod Radio Electron, Foreign Radio Electron 8:3–12
10. Churilova SA, Pishko OF, Schünemann K et al (1999) Submillimeter-wave clinotrons with distributed energy output. In: Proceedings of 24th international conference on infrared and millimeter waves, Monterey, CA, USA, 6–10 Sept 1999
11. Karpowicz N, Zhong H, Zhang C et al (2005) Compact continuous-wave subterahertz system for inspection applications. Appl Phys Lett 86:54–105
12. Lettington AH, Dunn D, Alexander NE et al (2004) Design and development of a high performance passive mm-wave imager for aeronautical applications. Proc SPIE 5410:210–218
13. Dobroiu A, Yamashita M, Ohshima YN et al (2004) Terahertz imaging system based on a backward-wave oscillator. Appl Optics 43(30):5637–5646
14. Domey T, Symes WW, Baraniuk RG et al (2002) Terahertz multistatic reflection imaging. J Opt Soc Am A 19(7):1432–1442
15. Vavriv DM, Volkov VA, Bormotov VN et al (2002) Millimeter-wave radars for environmental studies. Radio Phys Radio Astron 7(2):121–138
16. Volkov VA, Vavriv DM, Chumak VG et al (2007) Clinotron-based synthesized oscillators for THz-regions. In: MSMW'07 symposium proceeding, Kharkov, Ukraine, 25–30 June 2007
17. Schünemann K, Vavriv DM (1999) Theory of the clinotron: a grating backward-wave oscillator with inclined electron beam. IEEE Trans Electron Device 46:5993–6006

Chapter 13
Emerging Thermal Infrared '*THz Torch*' Technology for Low-Cost Security and Defence Applications

Fangjing Hu and Stepan Lucyszyn

Abstract Terahertz (THz) systems are notoriously large and very expensive, from complete systems down to individual front-end active devices and passive components. This is a major reason why there are currently no ubiquitous applications in the terahertz frequency spectrum (from 0.3 to 10 THz). However, by moving into the high-THz part of the frequency spectrum (thermal infrared region, from 10 to 100 THz), for specific niche applications, it is possible to create affordable systems for commercial exploitation.

The thermal infrared frequency bands, typically from 20–40 THz and 60–100 THz, are best known for applications in thermography. This underused and unregulated part of the spectral range offers opportunities for the development of secure communications. To this end, the '*THz Torch*' concept was recently presented. This technology fundamentally exploits ultra-low cost engineered blackbody radiation, by partitioning thermally-generated spectral noise power into pre-defined frequency channels; the energy in each channel is then independently pulsed modulated and transmitted, creating a robust form of short-range secure communications in the far/mid infrared.

In this Chapter, the basic *THz Torch* architecture for short-range secure wireless communications is introduced. Single-channel working demonstrator from 25 to 50 THz is shown. This concept is further extended by multiplexing schemes with important benefits, including increased overall end-to-end data rates (with band-limited channels) and higher levels of security. Multi-channel '*THz Torch*' frequency division multiplexing (FDM) and frequency-hopping spread-spectrum (FHSS) schemes are introduced and the first proof-of-concept FDM working demonstrator is presented. Furthermore, fundamental limits and engineering solution are discussed to increase the data rate and transmission range.

F. Hu • S. Lucyszyn (✉)
Centre for Terahertz Science and Engineering, Department of Electrical and Electronic Engineering, Imperial College London, London, UK
e-mail: s.lucyszyn@imperial.ac.uk

C. Corsi and F. Sizov (eds.), *THz and Security Applications*, NATO Science for Peace and Security Series B: Physics and Biophysics, DOI 10.1007/978-94-017-8828-1_13,

By exploring a diverse range of methods, significant enhancements to both data rate and distance can be expected. Our thermodynamics-based approach represents a new paradigm in the sense that nineteenth century physics can be exploited with twentieth century multiplexing concepts for low-cost twenty-first century ubiquitous security and defence applications in the thermal infrared.

13.1 Introduction

The idea of performing "wireless" ship-to-ship communications, by combining optical signalling with Morse code, was first put into practice by the British Royal Navy in 1867. The signalling device, a lantern, was called the Aldis lamp (named after British inventor Arthur C. W. Aldis). It is also interesting to note that shortly after, in 1898, an English inventor, David Misell, filed Patent No. 617,592 for an electric device. One year later, American Electrical Novelty and Manufacturing Co. manufactured the flashlight. Today, the commonwealth navies and NATO forces still use Aldis lamps when secure radio communications need to be silent or electronic "spoofing" is likely. Signalling at night is usually conducted in the near-infrared part of the electromagnetic spectrum, using 300 W incandescent light bulbs and requiring night vision sights, making it less likely to be intercepted or jammed.

In addition to signalling, the infrared part of the electromagnetic spectrum has many other security and defence related applications. At one end of the cost spectrum, an amateur infrared security alarm system was reported in 1965, employing a domestic flashlight and an infrared filter made using exposed photographic film [1]. At the other end of the cost spectrum, BAE Systems incorporate their patented thermal/infrared/multi-spectral adaptive camouflage technology, known as ADAPTIV, on their CV90 armoured vehicles. Sheets of relatively large hexagonal 'pixels' can quickly change temperature. On-board cameras allow the vehicle to effectively display the background infrared image onto the vehicle, allowing even a moving tank to match its background infrared surroundings. This technology also facilitates identification friend or foe (IFF) capabilities. More ominously, in the recent armed conflict in Iraq, camouflaged passive infrared sensors, wired to the arming mechanisms of improvised explosive devices (IEDs) were discovered; which are triggered by the movement of personal and vehicles.

The Centre for Terahertz Science and Engineering, at Imperial College London, recently started to re-invent the nineteenth century ideas of signalling and the flashlight for twenty-first century applications, with the emerging *THz Torch* technology in the thermal infrared region. In the wider context of 'THz and security applications', this chapter will present the *THz Torch* technology; introducing the strengths and weaknesses of using uncooled infrared sources and detectors, intended for specific niche applications in security and defence. Unlike approaches normally associated with traditional THz systems, which bridge the gap between electronics and photonics, *THz Torch* technologies can exploit ultra-low cost engineered blackbody radiation, by partitioning thermally-generated spectral noise power into pre-defined frequency channels. For example, simple incandescent light bulbs are used as high-THz thermal noise power source generators; while pyroelectric infrared sensors are used as thermal noise power detectors.

13.1.1 Background

The terahertz (THz) spectral region is receiving increasing interest, due to its unique properties. For example, THz radiation is generally considered safe for humans because it is: (1) non-ionizing, due to the low photon energy; (2) sources of power are normally limited to micro/milliwatts; and (3) the depth of penetration does not extend beyond the thickness of human skin. Furthermore, electromagnetic waves having a frequency of up to *ca.* 0.6 THz are able to penetrate most non-metallic materials (e.g. dry clothing and plastics), making it useful for stand-off imaging systems. Moreover, THz waves can excite molecular and atomic resonances and intra-band electron transitions in materials; the resulting "fingerprints", in the frequency-domain, enable spectroscopy to be performed on material samples of interest. THz radiation also allows non-invasive and non-destructive testing and has already found its way into a number of civilian applications (e.g. packaging failure analysis, production quality control, food spoilage detection, historical art forensics, pollution monitoring, Tbit/s communications for backbone telecommunications networks and 'instant' high-definition video-on-demand optical links, etc.).

13.1.2 Spectral Range

In general, for commercial applications, the infrared regions (which extend from 0.3 to 400 THz) remain the least explored parts of the electromagnetic spectrum. However, the technological '*THz gap*', between conventional electronics and photonics, offers the real promise of creating new scientific discoveries and commercial exploitation. Depending on the particular community of origin, several definitions exist for defining the 'terahertz' (also known as "*T-Ray*") part of the frequency spectrum. This is important, as several enabling technologists and/or applications may exists within the range specified by one definition but not necessarily within another. The most loose definitions have their two-decade spectral ranges from 0.1–10 THz (corresponding to free space wavelengths of 3 mm–30 μm) [2, 3] or 0.3–30 THz (corresponding to free space wavelengths of 1 mm–10 μm) [3], and these encompass the vast majority of enabling technologies and applications. For example, at the lower end of the former, various electronics-based source and detector technologies can be employed, often in conjunction with a wide selection of guided-wave structures. However, with the latter, photonics-based source and detector technologies are generally employed, often in conjunction with free-space (quasi-)optics. A more pragmatic view, adopted by the *IEEE Transactions on*

Terahertz Science and Technology journal, has its range of interest from 0.3 and 10 THz [4]. This definition excludes the 0.1–0.3 THz frequency range that is freely open for commercial exploitation, including ultra-high speed telecommunications and security body scanners for the screening of illegal substances and concealed weapons (e.g. capable of detecting hidden metals, plastic, liquids, gels, ceramics and narcotics concealed beneath a person's clothing) [5]. However, a more exclusive, single-decade, definition of the THz spectral range exists from 0.3 to 3 THz; widely cited in the open literature and having the least amount of commercial exploitation. This definition, previously referred to as the submillimeter wave region before *ca.* 1990, excludes the 3–30 THz part of the frequency spectrum that has a long history of commercial applications associated with the generation and detection of thermal radiation within the far-infrared region.

There are still no ubiquitous applications on the market operating from 0.3 to 10 THz, due to the lack of affordable enabling technologies (e.g. sources, detectors or guided-wave interconnects). Conventional THz systems are notoriously large and very expensive; from complete systems down to individual front-end active devices and passive components. Notable exceptions are the relatively basic ultra-low cost sensors for motion (with applications ranging from security to energy saving lighting systems) and fire detection systems, which operate between *ca.* 15 and 50 THz – i.e. at the upper half of the far-infrared band (0.3–30 THz) and lower part of the mid-infrared band (30–120 THz).

While the near-infrared band (120–400 THz) is well above the THz region, there are some interesting applications that can be adopted for longer wavelength applications. For example, two decades ago, the Infrared Data Association (IrDA) published a set of protocols for wireless near-infrared (NIR) communications that provide physically secure data transfer; having very low bit error rates (e.g. typically 10^{-9}), in environments where interference would make conventional radio-based wireless technologies unusable. Operating in the NIR, between 333 and 353 THz (i.e. 850–900 nm), the best operational range for IrDA communications is only between 5 and 60 cm. The slowest, serial infrared (SIR) physical layer protocol has specifications that include a 9.6–115.2 kbit/s data rate with asynchronous return-to-zero-inverted (RZI) pulses. For low-power mode to low-power mode of operation, the maximum range is just 20 cm. When compared to radio-based wireless technologies (e.g. Bluetooth), IrDA hardware is still a great deal less expensive and does not have the same security problems, which makes it ideal for line-of-sight data transfer between mobile phone handsets. However, IrDA links operate in the slow half-duplex mode, because when the transmitter is sending NIR pulses its receiver is effectively 'blinded' by them; making full-duplex unfeasible.

Other NIR applications of interest are ubiquitous remote controls and bi-directional voice and data communications systems used in secure environments; neither of which have specifications that conform to any international standards. Remote controls have been around since the early 1980s and today rely on ultra-low cost LED sources operating at approximately 319 THz (i.e. 940 nm); using 33–60 kHz carriers, 100 % amplitude-shift keying (ASK) modulation pulses transfer signals at data rates of between 4 and 120 bit/s. With secure communications, a mobile user's voice is digitally encoded, having a bandwidth of only 3.4 kHz, and transmitted using Class 1 NIR LEDs across an area of 100 m^2 (indoors) or 25 m^2 (in direct sunlight) [6] or across a range of 3.2 km with a point-to-point link and with higher data rates [7]; offering a low probability of interception or jamming. For all these NIR applications, little R&D has been reported in the open literature on similar systems operating at longer wavelengths in the high-THz part of the frequency spectrum. To this end, the *THz Torch* concept was conceived.

13.1.3 Cost Driver

In general, wireless systems find countless applications in communications and remote sensing applications. Their front-end hardware can perform relatively complex signal processing functions directly at microwave (*ca.* 1–30 GHz) and optical (400–789 THz) frequencies. When combining ubiquitous wireless applications with advances in enabling technologies, a positive spiral can be created that continually drives down manufacturing costs while still allowing performance and functionality to grow.

In contrast, there are two inherent reasons why millimetre-wave (30–300 GHz) and terahertz systems have yet to find ubiquitous applications; high data rate 60 GHz wireless communications systems and vehicular 76.5 GHz autonomous or adaptive cruise control radar systems are still considered luxury (non-ubiquitous) products on the domestic market. The first reason for this lack of progress is the increase in power loss (or its equivalent noise temperature) with frequency, within metal-based passive components and circuits. With coherent systems operating at ever higher frequencies, this problem is exacerbated by the increasing inability to generate sufficient spectrally-clean carrier power and to maintain power gain within an amplifying stage as frequency increases. As a result, the overall end-to-end systems performance degrades as frequency increases, requiring ever-more expensive solutions (e.g. more exotic transistor technologies and/or the use of larger passive components – both of which may even have to be cryogenically cooled) that only high-end users (e.g. commercial, scientific or military) can afford and/or physically accommodate. The second reason is that wavelength is inversely proportional to frequency; thus, structures defined by a specific electrical length (e.g. resonators) generally become smaller in size as frequency increases. Therefore, fabrication tolerances become ever more important, requiring more expensive manufacturing technologies to be used for shorter wavelength applications. For example, to define a simple microstrip transmission line within a hybrid microwave integrated circuit, UV lithography can be used with simple organic-based printed circuit board (PCB) or low-temperature co-fired ceramic (LTCC) technologies. For millimetre-wave integrated circuits, a more expensive deep-UV lithographic system may be needed with the use of a microfabrication facility; while for terahertz monolithic integrated circuits it may be necessary to use very expensive E-beam/ X-ray lithography, with more advanced microfabrication processing techniques.

For the above reasons, in terms of consumer-based systems intended for mass production, the cost of implementing a conventional radio-based wireless system at upper-millimetre-wave frequencies (*ca.* 0.1–0.3 THz) becomes prohibitive for ubiquitous applications. As a result, manufacturers may be reluctant to invest in the kinds of research and development (R&D) needed to make breakthroughs for the various enabling technologies that can bring prices down to affordable levels. This impasse is even more acute with operation at THz frequencies. The relatively recent emergence of THz technologies had been predominantly through the use of free-space (quasi-)optical techniques, in preference to those found at longer wavelengths.

Historically, apart from purely scientific experiments, there was relatively little interest in THz frequencies. The main reasons for the fall-off in activity with reducing wavelength was the lack of demand for commercial exploitation, limited scope for realizing affordable front-end subsystems, high manufacturing costs and the lack of commercial equipment for undertaking accurate metrology. Today, the upper-millimetre-wave and terahertz parts of the frequency spectrum are slowly opening up to commercial exploitation. This is because of the introduction of new concepts in the development of low-cost front-end hardware and their associated manufacturing techniques, and the emerging market in commercial turnkey measurement systems operating up to 1 THz when implemented in the frequency-domain [8] and 4 THz using time-domain techniques [9].

Today, there is a sustained drive within both the scientific and engineering communities to promote the development of THz enabling technologies and their applications. With the former, new devices for THz generation and detection are been investigated, with ever increasing performances [10–16]. With the latter, there is real motivation for finding ubiquitous applications to commercially exploit the '*THz gap*'. When ubiquitous THz applications finally emerge, the costs associated with active devices and passive components (including their manufacture and metrology) will fall, creating a positive spiral of growth in all areas. To achieve and maintain this advancement, new engineering solutions are needed. Examples of such solutions, associated with: (1) material characterisation [17–20]; (2) analytical modelling of passive structures [21–25]; (3) stress testing of commercial numerical CAD software [26, 27]; and (4) design of passive components and circuits [28–37], can be found online. More recently, over the past 3 years, the *THz Torch* technology has emerged [38–42], in an attempt to dramatically cut the cost for implementing specific niche applications in security and defence.

13.1.4 Emerging Thermal Infrared THz Torch Technology

Unlike most conventional THz technologies, which are based on expensive coherent signal generation and detection techniques, the *THz Torch* technology is based on incoherent thermal radiation in the high-THz (thermal infrared) part of the frequency spectrum, which can exploit ultra-low cost thermodynamic approaches. It is this work that will be discussed in the remainder of this chapter. In 2011, a proof-of-concept ultra-low cost short-range *THz Torch* wireless link was demonstrated for the first time; having a maximum bit rate of only 5 bit/s and across a range of just 0.5 cm [42]. Here, digital modulation was implemented by the simple ON-OFF keying (OOK) of band-limited (25–50 THz) thermal noise. This performance soon improved to 40 bit/s, using four-channel frequency division multiplexing (FDM) scheme, operating across the 20–90 THz spectrum range [41] and then 380 bit/s over a single 25–50 THz channel [39]. Today, bit rates of more than 1 kbit/s have been achieved over the single 25–50 THz channel, without the use of any advanced technique.

This chapter is organized as follows: in the next section, the basic *THz Torch* prototype is presented in details. Physical modelling and proof-of-concept experiments are explained. The basic concept is then extended to include multiplexing schemes and a four-channel frequency division multiplexing prototype is presented. Fundamental limitations of this technology are investigated; with a view to improving the maximum bit rate and transmission range. Engineering solutions are proposed and verified experimentally.

13.2 Basic Single-Channel Architecture

13.2.1 Introduction to Basic Components

13.2.1.1 High-THz Thermal Noise Sources

Various approaches, based on either electronic or photonic techniques, have been used to generate coherent THz radiation. However, for ultra-low cost applications, blackbody radiation is considered to be the best option, due to its simplicity, ease of tuning and affordability. The concept of using thermodynamic approaches to generate high-THz (10–100 THz) thermal noise power is straightforward, as radiation is naturally emitted by all objects – as their temperature must be above zero degree Kelvin (as described by Planck's Law). Although thermally-generated THz radiation offers many benefits, its main drawback is that there is no coherency, as with all unmodulated noise sources. Therefore, when compared to conventional THz systems that employ coherent sources, only the intensity of the output noise power can be measured, even with ultra-low cost thermal sensors. Other information (e.g. phasor representations and polarization) cannot be determined for the noise "carrier" and so for any system coherency a known modulating signal would have to be introduced.

In principle, many types of thermal radiators can be employed to generate and radiate electromagnetic energy at high-THz frequencies. As a simple proof-of-concept demonstrator, miniature incandescent light bulbs were used. At high-THz frequencies, such commercial-off-the-shelf (COTS) bulbs do not contain perfect blackbody radiators (since their tungsten filaments have low emissivity), their windows have relatively poor transmission characteristics (due to the high absorption and reflectance of the glass envelopes) and without an additional collimating lens and back reflector the spreading loss is high. Nevertheless, with COTS bulbs, one can still find niche applications in security and defence.

13.2.1.2 High-THz Radiation Detectors

The specific detectivity is normally used to characterize the performance of a detector and this is defined as:

$$\textit{Specific Detectivity}, D^* = \frac{\sqrt{\textit{Area}, A\,[\mathrm{m}^2]}}{NEP} = \Re\frac{\sqrt{A}}{S_n}\left[\mathrm{m}\sqrt{Hz}/\mathrm{W}\right] \tag{13.1}$$

Input Noise Equivalent Power, NEP

$$= \frac{\textit{Output Noise Voltage Spectral Density}, S_n\left[V/\sqrt{Hz}\right]}{\Re}\left[W/\sqrt{Hz}\right] \tag{13.2}$$

Detector Responsivity, $\Re$

$$= \frac{\textit{Output Voltage}, u}{\textit{Input Power or Incident Radiation Flux}, \ \Phi s}[V/W] \tag{13.3}$$

For the detection of thermally-generated noise power radiation at high-THz frequencies, two classes of detector can be employed; high performance, high cost photon-based detectors or low performance, ultra-low cost thermal-based detectors. The former includes photoconductive, PC, and photovoltaic, PV, semiconductor devices. Photon detectors have a high specific detectivity of $\sim 10^{10}$ cm · $\mathrm{Hz}^{1/2}$/W and fast response times of <1 μs, but may require cryogenic cooling to minimise thermal noise. Furthermore, this type of detector has a relatively narrow instantaneous bandwidth, depending on the energy band gap of the specific semiconductor used. The latter sensors include thermopile, thermistor bolometer and pyroelectric infrared (PIR) sensors. Thermopiles, comprising several radiation thermocouples connected in series or, less commonly, in parallel, generates an output voltage proportional to a local temperature difference or temperature gradient. Typical specific detectivity and response times for thermopiles are 10^8 cm · $\mathrm{Hz}^{1/2}$/W and ~ 10 ms, respectively. Thermistor bolometers measure the power of incident electromagnetic radiation via the heating of a material with a temperature-dependent electrical resistance, and have typical values of specific detectivity and response times of 2×10^8 cm · $\mathrm{Hz}^{1/2}$/W and ~ 1 ms, respectively. Pyroelectric sensors exploit pyroelectric materials (e.g. crystalline triglycine sulphate, TGS, or lithium tantalate, LiTaO3; or the piezoceramic lead zirconate titanate, PZT) to generate a change in voltage due to a corresponding change in material temperature. Typical specific detectivity and response times are $>10^8$ cm $\cdot$ Hz$^{1/2}$/W and $\sim$10 ms, respectively. Therefore, when compared to photon detectors, thermal detectors typically have a specific detectivity that is two orders of magnitude lower, while the response times are typically four orders of magnitude lower. However, important advantages of thermal-based sensors include operation across an ultra-wide instantaneous bandwidth and at room temperature. For most applications, these advantages may not be enough to justify employing thermal detectors. However, with cost-sensitive applications, there may well be a good argument for investigating engineering solutions with the use of thermal detectors.

The Golay cell should also be mentioned, as it is the most sensitive amongst the thermal-based detectors that can operate at room temperature. It consists of a gas-filled cavity. On one face of the cavity is an infrared window and on the other a flexible membrane; embedded within the cavity is an infrared absorbing film. Infrared radiation passes through the window and its energy is absorbed by the film. The absorbed energy is dissipated as heat and the surrounding gas expands, causing the membrane to flex outwards. On the other side of the membrane, an optical beam from a laser or LED is focused onto it, such that the reflected light can be measured by a photodiode. Therefore, when the membrane flexes, the amount of light measured by the photodiode will decrease and its calibrated readout will give an accurate measure of the incident infrared radiation flux. Golay cells typically have a high specific detectivity of $>10^9$ cm $\cdot$ Hz$^{1/2}$/W, over an ultra-wide instantaneous bandwidth. The main disadvantages of Golay cells are that they have linearity issues, relatively slow response times of $\sim$10 ms, are relatively large (i.e. not compatible with monolithic integration), have an extremely fragile membrane and are relatively expensive.

For completeness, a comparison of specific detectivity, across the 7.5 and 300 THz spectral range, for different commercially-available detectors is shown in Fig. 13.1 [43]. Each detector is assumed to have a hemispherical field of view (FOV), at a temperature of 300 K. It is interesting to note that in Fig. 13.1 the background-limiting case for an ideal thermal detector has a specific detectivity of almost $2 \cdot 10^{10}$ cm $\cdot$ Hz$^{1/2}$/W, which is comparable to that of a commercial photon detector. The pyroelectric sensor has been selected for the *THz Torch* technology, due to its ultra-wide instantaneous bandwidth, room temperature of operation and most importantly its ultra-low cost; but this effectively represents the lowest benchmark performance. The pyroelectric sensor used in the first two experiments [41, 42] was the Murata IRA-E710ST1 (for use in the 15–300 THz range; the value of D* is unknown) [44], while later experiments [39, 40] employed the InfraTec LME-553 sensor, having a specific detectivity of 1.2 10^8 cm $\cdot$ Hz$^{1/2}$/W [45].

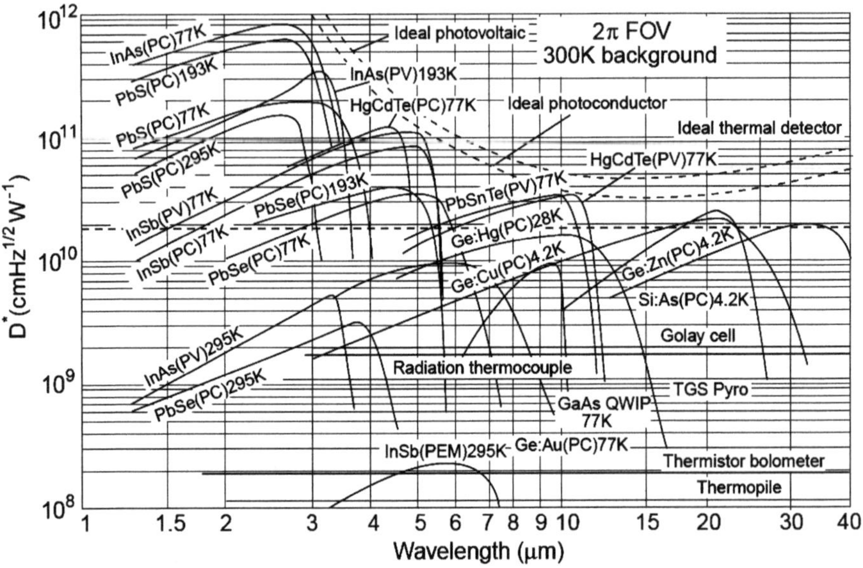

Fig. 13.1 Comparison of specific detectivity at room temperature for commercially-available detectors (chopping frequency is 1 kHz for photon detectors and 10 Hz for thermal detectors). Theoretical curves for the background-limiting case are shown by *dashed lines* (Courtesy of Progress in Quantum Electronics [43])

13.2.1.3 High-THz Filters

Thermal radiation has a continuous frequency spectrum. Therefore, filters have to be employed in order to ensure that communications channels operate within their prescribed spectral ranges. For example, in the earliest [42] and later [39] single-channel experiments, identical commercially-available 5 μm long-pass filters [44] were employed for both transmitter and receiver front ends. The spectral transmittance is >70.7 % across the 25–50 THz octave range and the mean insertion loss is ~2 dB, as shown Fig. 13.2. Many other COTS high-THz optical coating filters are readily available for multi-channel applications, by defining non-overlapping frequency bands within the working spectral range. Multi-channel multiplexing schemes will be introduced in Sect. 13.3.

13.2.2 Introduction to Basic Subsystems

This chapter considers the application of low bit rate signalling for secure communications over short distances. The basic architecture for a single-channel *THz Torch* wireless communications link is shown in Fig. 13.3, employing a simple bulb, PIR sensor and filter technologies introduced in Sect. 13.2.1.

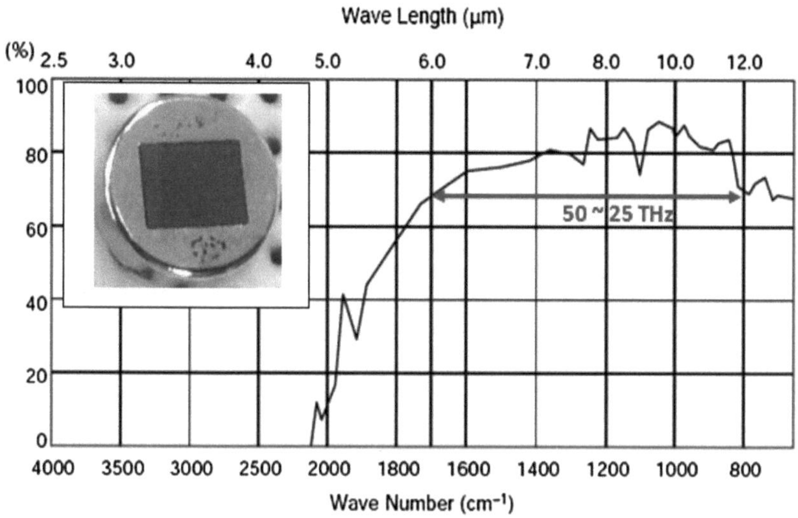

Fig. 13.2 Spectral transmittance for the 5 μm long-pass filter [44] (Courtesy of Murata Manufacturing Co.)

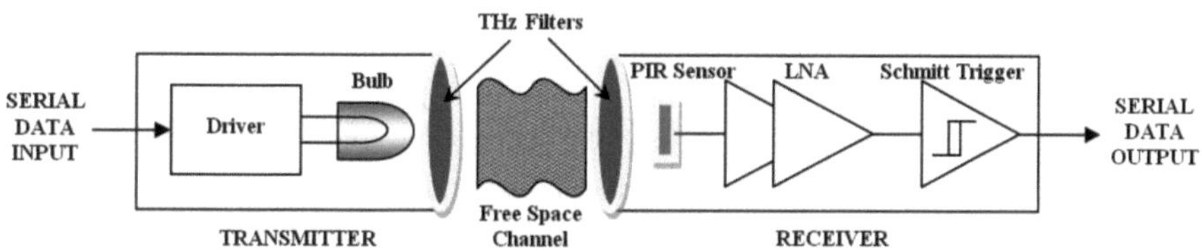

Fig. 13.3 Basic architecture for ultra-low cost ON-OFF keying *THz Torch* wireless links [42]

Simple digital OOK is performed by either:

1. direct (internal, electronic) modulation – used in the first experiment [42].
2. indirect (external, spatial) modulation – used in later experiments [39–41].

13.2.2.1 Basic Transmitter Subsystem

The transmitter consists of five Eiko 8666-40984 miniature incandescent light bulbs, having a length of 6.3 mm and diameter 2.6 mm. These bulbs are connected in series and assembled into a compact cylindrical package having an outer diameter of 8.2 mm, as shown in Fig. 13.4.

Fig. 13.4 Incandescent light bulb and filter assembly [42]

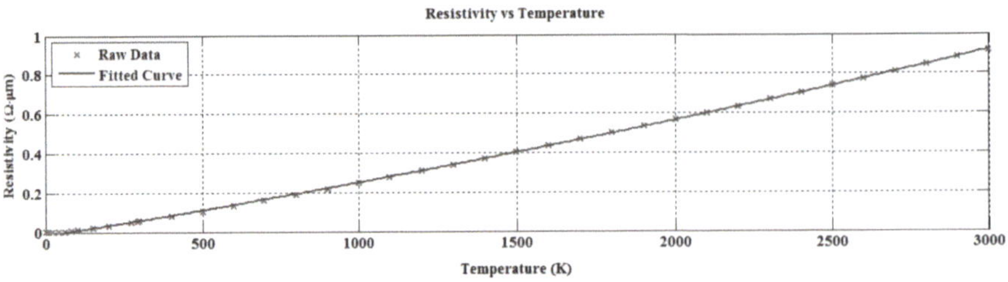

Fig. 13.5 Measured resistivity of tungsten against temperature and curve fitting

Filament Working Temperature Estimation

One of the most important parameters of the transmitter is the working temperature of the bulb's filament radiator. The ideal spectral radiance, as a function of wavelength, can be calculated using Planck's law if the temperature of the radiator is known. Moreover, knowing the spectral radiance and the instantaneous −3 dB bandwidth of the filter, the integrated output noise power from transmission can be estimated. The working temperature of the filament can be measured directly, with the use of a calibrated thermal camera. An alternative, widely used indirect approach is to measure the operating resistance of the filament, since this is a function of its working temperature. It is not possible to ascertain the exact geometry filaments within commercial bulbs. However, to a good degree of approximation, it can be assumed to be perfectly cylindrical, having a uniform cross-sectional area CSA [cm^2] and length l [cm]. The electrical resistivity of tungsten $\rho(T)$, as a function of absolute temperature T, can be expressed in terms of the filament resistance $R(T)$:

$$\rho(T) = R(T) \cdot \left(\frac{CSA}{l}\right)_{eff} \; [\Omega \cdot \text{cm}] \qquad (13.4)$$

An accurate curve fit can be applied to published measured data for the resistivity of tungsten $\rho(T)$ [46], shown in Fig. 13.5, as given by the following empirical quadratic expression:

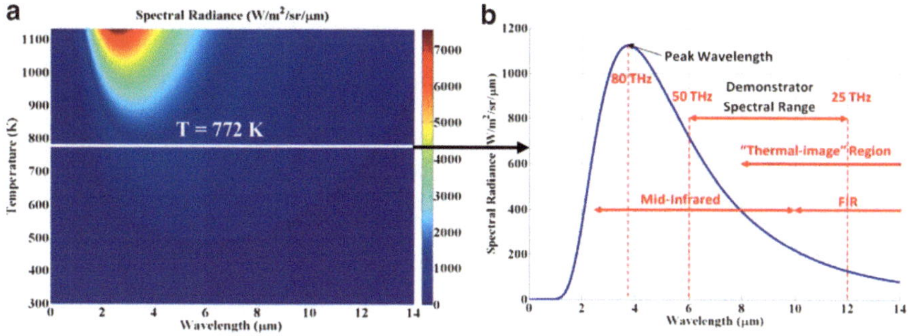

Fig. 13.6 Calculated ideal spectral radiance versus wavelength at: (**a**) various temperatures; (**b**) a working temperature of 772 K

$$\rho(T) = 2.228 \times 10^{-8} \cdot T^2 + 2.472 \times 10^{-4} \cdot T - 1.859 \times 10^{-2}\,[\Omega \cdot \mu\text{m}] \quad \text{for } 200 < T\,[\text{K}] < 3{,}000 \tag{13.5}$$

At room temperature, $\rho(300\text{ K}) = 5.14 \times 10^{-6}\ \Omega.\text{cm}$. The combined resistance of five Eiko 8666-40984 bulbs connected in series is directly measured to be 23.6 Ω at room temperature. Therefore, if the parasitic resistances of the short leads are considered negligible (which is a good assumption), this gives an average value of $R(300\text{ K}) = 4.72\ \Omega$ for each bulb filament. As a result, the effective ratio of cross-sectional area to length can be extracted using (13.4) and (13.5) to give a value of $(CSA/l)_{eff} = 1.09 \times 10^{-6}$cm, which is assumed to be temperature independent. By indirectly measuring the resistance of a bulb, at a specific bias level, the working temperature can be estimated to an acceptable degree of accuracy.

Spectral Radiance Estimation

After estimating the working temperature of the bulb filament radiator, the ideal (i.e. blackbody) spectral radiance $I(\lambda,T)$ can be obtained by applying Planck's law:

$$I(\lambda, T) = \frac{2hc^2}{\lambda^5} \cdot \frac{1}{e^{hc}/\lambda kT - 1}\,[\text{W/m}^2/\text{sr}/\mu\text{m}]$$

$$\text{having a spectral peak at } \lambda_{peak} = \frac{b}{T}\,[\text{m}] \tag{13.6}$$

where $I(\lambda,T)$ is the power radiated per unit area of emitting surface in the normal direction per unit solid angle per unit wavelength at absolute temperature T; λ is the free space wavelength; T is the absolute temperature of the filament; h is the Planck constant; c is the speed of light in a vacuum; k is the Boltzmann constant; and b is Wien's displacement constant. Figure 13.6a shows the ideal spectral radiance against wavelength at different temperatures.

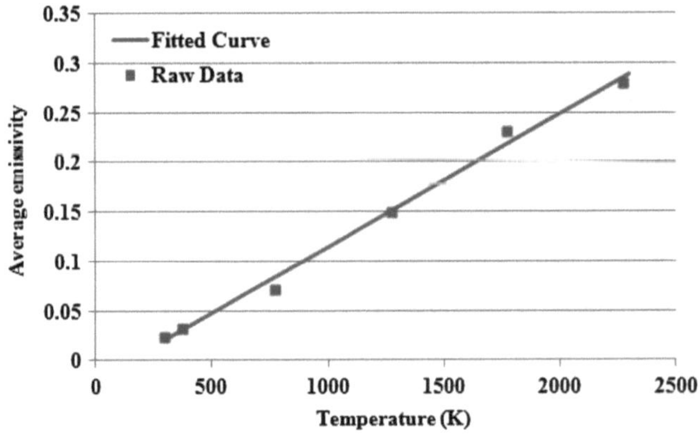

Fig. 13.7 Measured spectrum-average emissivity of tungsten against temperature and curve fitting

With the earliest experiments, the Eiko 8666-40984 bulbs had a DC biasing current of 44 mA, which gives an estimated filament working temperature of 772 K, $R(772\ \mathrm{K}) = 17.02\ \Omega$ and a corresponding spectral radiance peak at 80 THz, as shown in Fig. 13.6b. This peak frequency can be easily adjusted by changing the DC biasing current. At a larger biasing current, one can obtain higher spectral radiance levels, yielding an increased integrated output noise power for transmission. However, the penalty for this is a decrease in the band-limited output noise power to input DC power efficiency of the source; which may be an issue where available DC supply power is at a premium (e.g. security key fob applications).

Bulb Filament Emissivity Estimation

Most radiator materials are far from being perfect blackbody radiators. As a result, they are not efficient at emitting blackbody radiation. Emissivity $\varepsilon(\lambda, T)$, defined as the ratio of energy radiated by a material to that radiated by an ideal blackbody at the same wavelength and temperature, is a parameter that characterises this efficiency. In practice, if the spectral range is not too large, the emissivity at the surface of a material is only a function of temperature (to a good approximation). An accurate curve fit can be applied to measured data for the spectrum-average or total emissivity of tungsten $\overline{\varepsilon(T)}$ [46], shown in Fig. 13.7, as given by the following linear expression:

$$\overline{\varepsilon(T)} \approx 1.343 \times 10^{-4} \cdot T - 2.019 \times 10^{-2} \tag{13.7}$$

Band-Limited Output Noise Power and DC-to-THz Power Conversion Efficiency Estimation

Having previously obtained the -3 dB bandwidth (i.e. 25–50 THz) of the filters and estimated the working temperature of the filament radiator, spectral radiance and radiator emissivity, the band-limited output noise power P_{OUT} from each bulb can be estimated using the following expression:

$$P_{OUT} = \overline{\varepsilon(T)} \cdot \pi \cdot A_{eff} \int_{\lambda_1}^{\lambda_2} I(\lambda, T)\, d\lambda \tag{13.8}$$

where A_{eff} is the effective radiating area of the filament, and λ_1 and λ_2 are lower and upper wavelengths of the band-limited spectral range, respectively.

The filament is assumed to be a uniform cylinder having a cross-sectional area *CSA* and length l. Therefore, if one of these parameters can be physically measured then the other can be extracted using the previously estimated value of $(CSA/l)_{eff} = 1.09 \cdot 10^{-6}$ cm. It then follows that the effective radiating area can be estimated from $A_{eff} = 2l_{eff}\sqrt{\pi CSA_{eff}}$. Using a scanning-electron microscope (SEM), the average value for the diameter of the Eiko 8666-40984 bulb's filament was measured to be 22.40 μm, as shown in Fig. 13.8, yielding an effective radiating area of $2.54 \cdot 10^{-6}$ m^2. From (13.8), the band-limited output noise power from each bulb-filament is calculated to be 1.31 mW for a quiescent DC biasing current of 44 mA. Moreover, having previously derived a value for the filament resistance at its working temperature of $R(772\text{ K}) = 17.02\ \Omega$, the input DC power can be calculated to be $P_{DC} = 33$ mW for each bulb and, thus, the band-limited output noise power to input DC power efficiency of the source can be estimated to be ~4 %. It should be noted that the non-ideal transmission characteristics of the glass envelope, spreading loss and ~2 dB insertion loss of the filter are not taken into account. Nevertheless, an estimated DC-to-THz power conversion efficiency of ~4 % is orders of magnitude greater than those associated with coherent sources. For specific niche applications, this may be an important attribute for the *THz Torch* technology.

It is interesting to note that the Eiko 8666-40984 incandescent light bulb is intended for general purpose applications, with a recommended DC operating power of $P_{DC} = 100\text{ mA} \times 4\text{ V} = 400$ mW. For this quiescent bias condition the measured working temperature is 1,610 K, which is almost half the 3,695 K melting point of tungsten. This temperature corresponds to a spectral radiance peak at 167 THz (i.e. 1.8 μm), in the NIR spectral range. As a result, since it is far away from the visible part of the spectral range, the DC-to-Visible power conversion efficiency under recommended DC operating power conditions is very poor indeed and orders of magnitude lower than the previously estimated DC-to-THz power conversion efficiency. Furthermore, by reducing the quiescent bias current from 100 mA down to 44 mA, the peak value of spectral radiance (where DC-to-THz power conversion efficiency is at its highest) is easily tuned by more than an octave in frequency; from

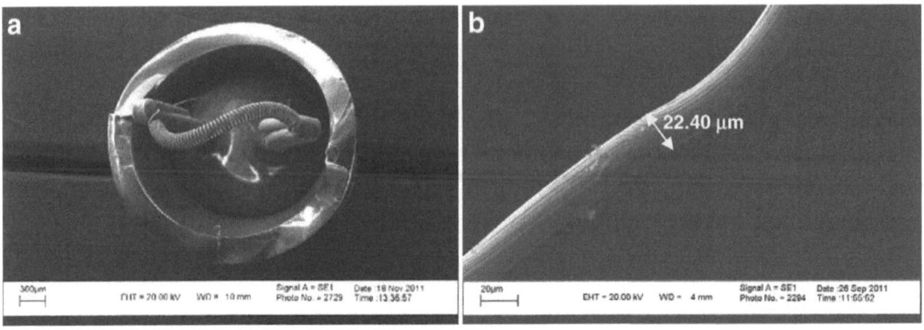

Fig. 13.8 SEM images of the Eiko 8666-40984 bulb's tungsten filament: (**a**) showing the complete filament within the *circular* perimeter of the broken glass envelope; (**b**) close-in view of the filament

167 THz down to 80 THz. This can be compared to the alternative low-cost approach of employing infrared LEDs and photodiodes, operating at shorter wavelengths; switching between devices is necessary, as no continuous spectral tuneability is possible. Finally, for *THz Torch* applications (having $P_{DC} < 100$ mW, which is well below the bulb's intended DC operating power level), a long lifetime of operation can be expected.

Filament Response Time Estimation

The response times associated with incandescent light bulbs are not of importance if the source radiation is to be modulated indirectly (e.g. via spatial modulators) or if applications do not require fast switching times when modulated directly (e.g. via electronic modulation). However, for applications that require fast switching speeds and that cannot employ indirect modulated, the physical limitations associated with direct modulation response times need investigating. In practice, there are two types of response times; electrical and thermal. With the former, the low frequency inductance of the Eiko 8666-40984 bulb's filament was measured to be $L_{LF} = 5.18$ μH, which is expected to be relatively temperature independent. Therefore, for the worst-case room temperature condition, the electrical time constant is only $L_{LF}/R(300\,\text{K}) \sim 1.1$ μs. It will be found that this is many orders of magnitude lower that those associated with the thermal time constants. For this reason, as one would expect with technologies that rely on thermodynamics, it is the thermal time constants that impose the limitations on response times.

There are two thermal time constant that need to be considered, corresponding to the heating and cooling processes. The heating thermal time constant τ_H is defined as the time it takes to go from 10 % to 90 % of the temperature difference between the initial and final steady-state temperatures. The cooling thermal time constant τ_C is defined in a similar way, as the time it takes to decrease from 90 % to 10 % of the temperature difference between the final and initial steady-state temperatures. Figure 13.9 shows the calculated instantaneous filament temperature,

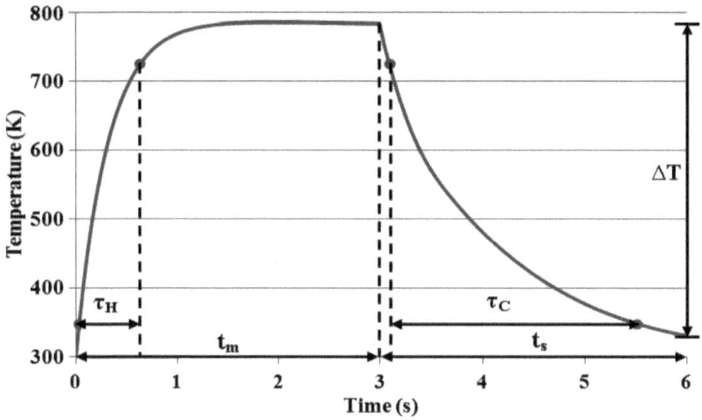

Fig. 13.9 Calculated instantaneous filament temperature for a *square wave* having M:S = 1:1 and modulating frequency of 0.16667 Hz ($t_m > \tau_H$ and $t_s > \tau_C$)

for a square wave having a mark(M):space(S) ratio of 1:1 and modulating frequency of 0.16667 Hz (representing a very slow bit rate).

When the bulb is electrically turned ON with a step response function, at its initial temperature $T(0)$, there is a large injection of current and the temperature of the filament increases due to Joule heating. Since the resistivity of tungsten has a positive temperature coefficient, the instantaneous bulb resistance also increases from its initial value of R(T(0)) until a steady-state value is reached; for example, from R(300 K) = 4.72 Ω to R(772 K) = 17.02 Ω. The steady-state temperature is reached when the input DC power is exactly balanced by all of the dissipative (i.e. ohmic and radiative) loss mechanisms. Figure 13.10a shows the measured transient turn-ON current profile for a quiescent DC biasing current of 44 mA, corresponding to a final steady-state working temperature of 772 K. The transient temperature of the filament can also be extracted, according to the measured bulb resistance, as shown Fig. 13.10b. It is worth mentioning that tungsten has a specific heat capacity $Cp = 0.134$ J/g.K, which is normally considered to be low. However, this aspect of the thermodynamic system will need to be taken into account as the bit rate increase. Fortunately, since the instantaneous bulb resistance is extracted from instantaneous I-V measurements, the specific heat capacity for this thermodynamic system is already taken into account.

A simple empirical curve fit can be applied to the instantaneous turn-ON filament temperature, shown in Fig. 13.10b, as given by the following expression with less than 1.5 % error:

$$T(t) = T(\infty) - \Delta T_{MAX} \cdot e^{-1.9995t/\tau_H}, \tag{13.9}$$

where t is the instantaneous heating time from the initial working temperature $T(0)$, $\Delta T_{MAX} = [T(\infty) - T(0)]$ is the maximum change in filament temperature

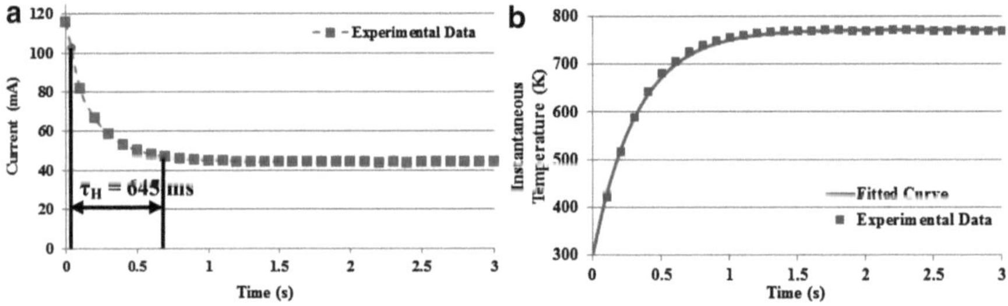

Fig. 13.10 Turn-ON responses for instantaneous: (**a**) current; (**b**) extracted filament temperature

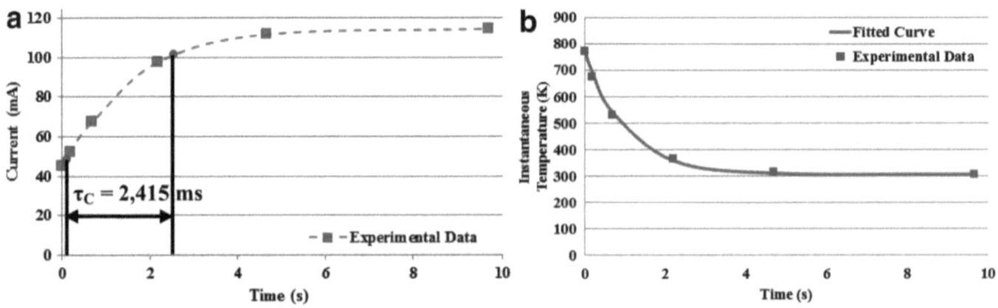

Fig. 13.11 Turn-OFF responses for instantaneous: (**a**) current; (**b**) extracted filament temperature

and $T(\infty)$ is the final working temperature; with this example, $T(0) = 300$ K and $T(\infty) = 772$ K. The heating or turn-ON thermal time constant was found to be $\tau_H = 645$ ms.

When the bulb is electrically turned OFF with a step response function, at its initial temperature $T(0)$, the filament starts cooling down and its instantaneous resistance decreases with time from its initial steady-state value of $R(772\text{ K})$ to its final steady-state value of $R(300\text{ K})$. However, the current will instantly drop to zero and so a similar techniques used to determine the turn-ON thermal time constant cannot be used. In order to measure the turn-OFF thermal time constant, a new technique is proposed. The instantaneous turn-ON currents are measured at different delayed intervals of cooling after the turn-OFF from the initial steady-state ON condition. Figure 13.11a shows six turn-ON current measurements; the interpolated profile represents the transient cooling response for the Eiko 8666-40984 bulb having a quiescent DC biasing current of 44 mA. The corresponding filament temperature can also be extracted, according to the transient bulb resistance, as shown Fig. 13.11b.

Cooling of the hot tungsten filament is a relaxation process [47]. However, a simple empirical curve fit can be applied to the instantaneous turn-OFF filament temperature, shown in Fig. 13.11b, as given by the following expression with less than 3 % error

$$T(t) = T(\infty) + \Delta T_{MAX} \cdot e^{-1.9995\,t/\tau_H}, \qquad (13.10)$$

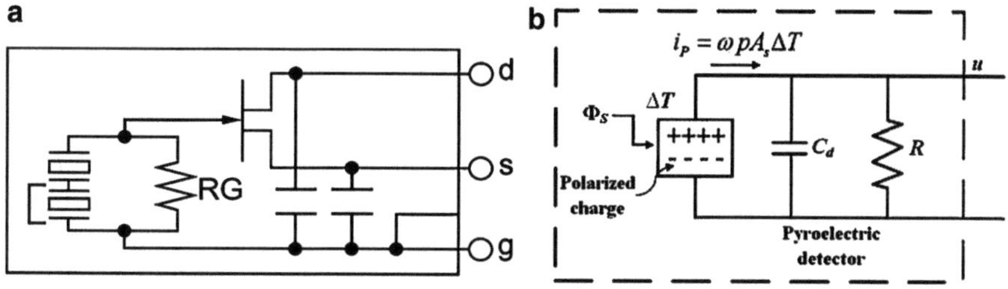

Fig. 13.12 Equivalent circuit models for PIR sensors: (**a**) Murata IRA-E710ST1 sensor (Courtesy of Murata) [44]; and (**b**) front-end of the InfraTec LME-553 sensor (Courtesy of InfraTec) [48]

where t is the instantaneous cooling time from the initial working temperature $T(0)$, $\Delta T_{MAX} = [T(0) - T(\infty)]$ is the maximum change in filament temperature and $T(\infty)$ is the final working temperature; with this example, $T(0) = 772$ K and $T(\infty) = 300$ K. The cooling or turn-OFF thermal time constant was found to be $\tau_C = 2{,}415$ ms. This could severely limit the turn-OFF speed for the bulb and set a practical limit on the bit rate for OOK digital modulation in a communications channel – although this may not be an issue with other applications, where fast switching is not required. In practice, as will be seen later, the bit rate can greatly exceed the reciprocal of the turn-OFF thermal time constant when the change in working filament temperature $\Delta T \ll \Delta T_{MAX}$, but this is at the expense of a degraded end-to-end systems performance for the wireless link and assumes that the thermal time constant for the PIR sensor is much shorter, i.e. $\tau_{sensor} \ll \tau_C$.

13.2.2.2 Basic Receiver Subsystem

The pyroelectric sensor is able to detect objects by measuring the difference in temperature between a target and its background. Their pyroelectric materials create a spontaneous polarization charge, with a magnitude depending on the material's change of temperature. The equivalent circuit models for the PIR sensors used are shown in Fig. 13.12.

When incident radiation flux Φ_s [W] is absorbed by the sensing element's material there is a corresponding change in its physical temperature. This temperature change then can be expressed as [48]:

$$\Delta T = \alpha \frac{\Phi_S}{G_T} \frac{1}{\sqrt{1 + (\omega \tau_T)^2}} \tag{13.11}$$

where α is the absorption efficiency, $\tau_T = H_p/G_T$ [s] is the thermal time constant of the detector (the value for the Murata IRA-E710ST1 sensor is not known; while that specified for the InfraTec LME-553 sensor is 200 ms [45] – since no distinction is made between heating and cooling, it is assumed that this is the worst-case value of the two), H_p [W.s/K] and G_T [W/K] are the heat capacity and thermal conductance of the sensing element's material, respectively. For the same reason given previously (for the filament turn-OFF thermal time constant), in practice, the bit rate can greatly exceed the reciprocal of the sensor's thermal time constant when the physical temperature change of the sensing element's material $\Delta T \lll \Delta T_{MAX}$, but again this is at the expense of a degraded end-to-end systems performance for the wireless link (i.e. reduced output signal-to-noise ratio and, therefore, a reduction in the maximum transmission distance), as will be seen in Sect. 13.4.3.

A change in temperature will generate a small polarization current, i_p, due to the pyroelectric effect, which is proportional to the temperature difference and the sensing element's surface area:

$$i_P = \omega p A_s \cdot \Delta T \tag{13.12}$$

where ω is the angular modulation frequency of the incident radiation, p is the pyroelectric coefficient [C/cm^2/K] and A_s [cm^2] is the cross-sectional area of the sensing element. Therefore, the magnitude of the output voltage is given by:

$$u = \omega p A_s \cdot \left[\alpha \frac{\Phi_S}{G_T} \frac{1}{\sqrt{1 + (\omega \tau_T)^2}} \right] \cdot \frac{R}{\sqrt{1 + (\omega \tau_E)^2}} \tag{13.13}$$

where $\tau_E = R\,C_d$ is the electrical time constant of the detector circuit (the value for the Murata IRA-E710ST1 sensor is not known; while that specified for the InfraTec LME-553 sensor is 0.4 ms [45]), R is the internal resistance of the voltage-follower (RG seen in Fig. 13.12a, used to convert current to voltage) and C_d is the output capacitance of the PIR sensor.

The Murata IRA-E710ST1 sensor used in the first two experiments [41, 42] has two 2×1 mm^2 sensing elements. Such dual-element sensors are normally used for detecting objects in motion, as the background environment can be subtracted from the moving target. As a result, it was necessary to remove this function by introducing a short circuit to one of the elements.

Back-End Electronics

Commercial pyroelectric sensors normally have an integrated preamplifier [44, 45]. However, the output voltages are normally small, in the mV range. As a result, back-end electronics is needed. Analogue circuitry is introduced, consists of a two-stage high gain low-noise amplifier (LNA), employed for signal amplification and DC blocking; followed by a Schmitt trigger for simple analogue-to-digital conversion. This simple back-end electronics represents an ultra-low cost solution for OOK digital modulation applications.

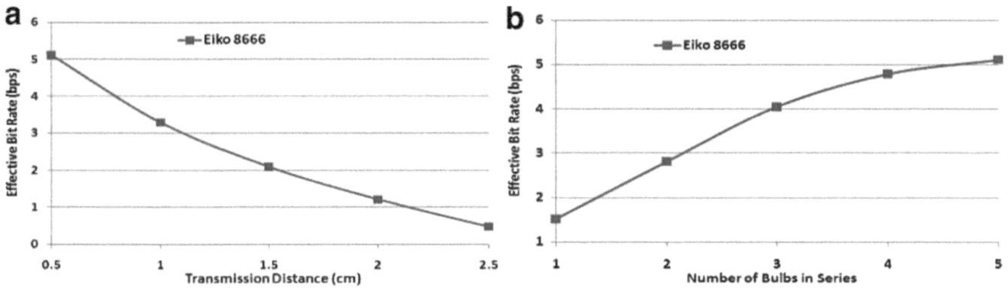

Fig. 13.13 Maximum effective bit rates: (**a**) against transmission distance; (**b**) against number of bulbs

13.2.3 First Proof-of-Concept Single-Channel THz Torch *Demonstrator*

The basic single-channel architecture, shown in Fig. 13.3, was implemented. The transmitter and receiver were positioned 0.5 cm apart, to create a very short line-of-sight wireless communications link. A rectangular pulse generator was used to digitally modulate the bulb, via an intermediate driver circuit. The quiescent DC bias current through five bulbs connected in series was set to 44 mA, giving the desired peak spectral radiance at ~80 THz. Experimental result shows a maximum bit rate of 5 bit/s over the 0.5 cm transmission distance, as shown in Fig. 13.13 [42]. From Fig. 13.13a, it can be seen that the bit rate decreases dramatically as the transmission distance increases. This is mainly due to losses from beam spreading (since no back reflector or collimating lenses were employed) and atmospheric absorption. These contributions will be discussed further in Sect. 13.4. While this attenuation characteristic would otherwise be viewed as undesirable, for specific niche security and defence applications this offers a low probability of interception or jamming. It can also be seen, in Fig. 13.13b, that as more bulbs are introduced the maximum effective bit rate increases for a fixed transmission distance. However, the law of diminishing returns applied, where a saturation effect can be seen as the number of bulbs increases beyond three, for this particular experimental setup.

13.3 Multi-channel Multiplexing Scheme Architectures

13.3.1 Introduction to Multiplexing Schemes

In communications systems, multiplexing schemes can offer important benefits; including the increased robustness to interference and jamming (both natural and manmade) and enhanced protection from interception, which are essential physical layer attributes for secure communications.

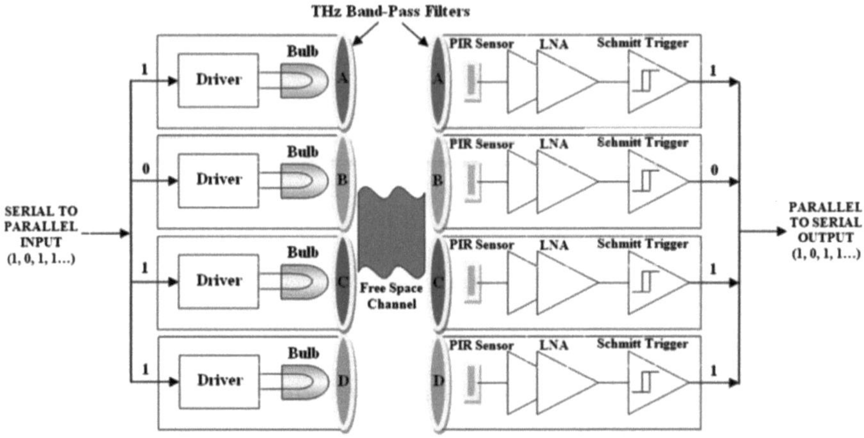

Fig. 13.14 Example of an ultra-low cost *THz Torch* FDM system architecture [41]

13.3.1.1 Frequency Division Multiplexing

For example, to increase the end-to-end bit rate beyond that of a single band-limited channel, the *THz Torch* concept can be enhanced by implementing frequency division multiplexing (FDM), while still maintaining its low cost advantage. Here, an input data stream is de-multiplexed and data is simultaneously transmitted in a number of channels, with each channel operating within different non-overlapping frequency bands, which are defined by COTS filters, within the 10–100 THz region. A simple *THz Torch* FDM scheme can be implemented as illustrated in Fig. 13.14, for a four-channel system [41]. The bandwidth, selectivity and transmittance of the associated channel filters should be carefully chosen, so as to allow roughly equal levels of energy transfer through each channel. The recovered bit streams from the single-channel receivers are then multiplexed to reproduce the original transmitted bit stream.

13.3.1.2 Frequency-Hopping Spread-Spectrum

In addition to FDM, a frequency-hopping spread-spectrum (FHSS) system can also be implemented to further increase the level of security. Here, the input bit stream is periodically switched from one channel to another, with each channel operating within different frequency bands, within the 10–100 THz region, so that only one channel is used at any moment in time. A simple *THz Torch* FHSS scheme can be implemented as illustrated in Fig. 13.15, for a four-channel system [41]; the pseudo-random channel allocation, synchronization and locking subsystems are not shown. While there is no advantage in the overall end-to-end bit rate, beyond that

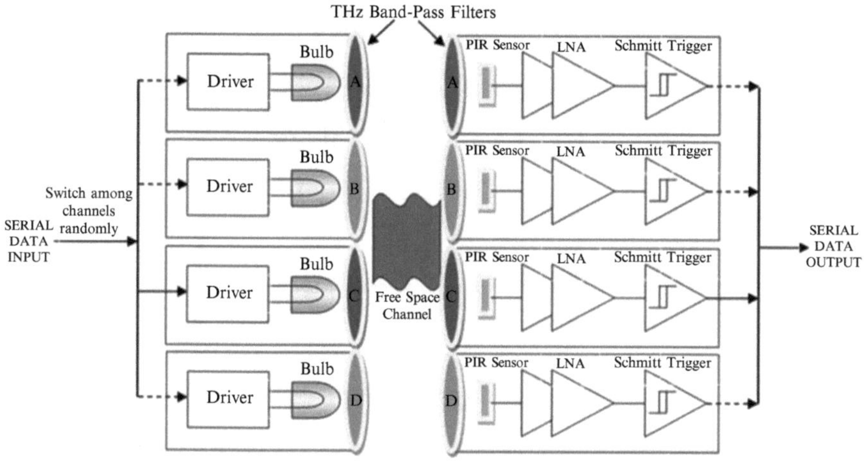

Fig. 13.15 Example of ultra-low cost *THz Torch* FHSS system architecture [41]

Table 13.1 Optical properties COTS filters

	Specifications		
Northumbria Optical Coatings Ltd. stock codes	50 % cut off [THz] (μm)	50 % cut on [THz] (μm)	Average transmittance (%)
SLWP-8506-000240	N/A	34 (8.801)	~79.6
SWBP-6177-000111	42 (7.059)	57 (5.295)	~84.2
SWBP-4596-000070	60 (5.004)	72 (4.188)	~75.7
SWBP-3685-000091	75 (4.001)	89 (3.372)	~72.2

of a single band-limited channel, there can be a significant enhancement in physical layer security.

13.3.2 First Proof-of-Concept THz Torch *FDM Demonstrator*

The first proof-of-concept *THz Torch* FDM system demonstrator has already been reported [41]. Here, four channels were defined within the high-THz thermal infrared region from 20 to 90 THz. The four COTS filters selected for this application were sourced from Northumbria Optical Coatings Ltd [49]; the associated specifications for each of the 1 mm thick filters are listed in Table 13.1 and their measured spectral transmittances are shown in Fig. 13.16.

The first *THz Torch* FDM system employed transmitters and receivers similar to those used for the single-channel *THz Torch* demonstrator. However, within the 8.2 mm diameter packaging can, a conformal metallic reflective film was also included, to reduce the spreading loss (i.e. by increasing the level of radiation in the forward direction). Four channel transmitter-receiver pairs were mounted face-to-face. Unlike the single-channel *THz Torch* demonstrator, where the bulbs were

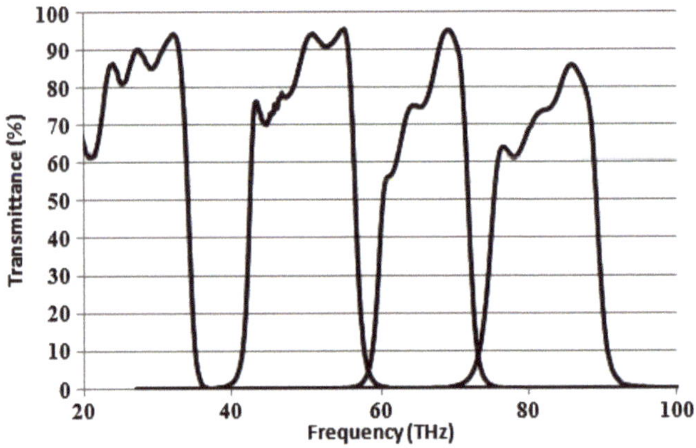

Fig. 13.16 Measured spectral transmittance responses for COTS filters given in Table 13.1 [49]

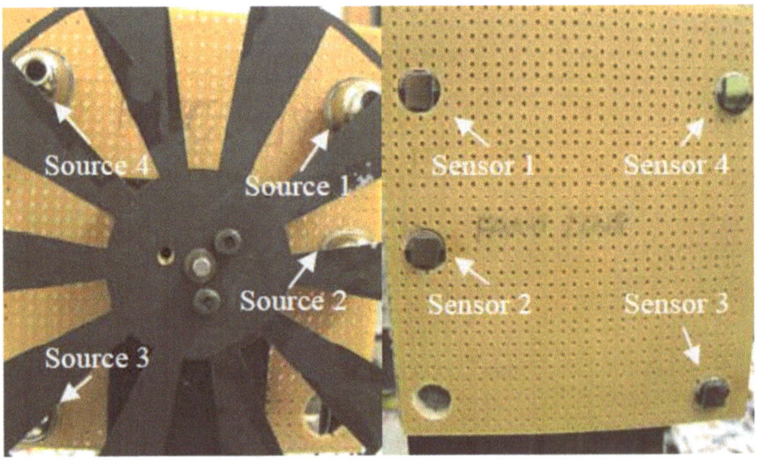

Fig. 13.17 Assembled channel transmitter array behind the optical chopper (*left*) and receiver array (*right*) [41]

directly modulated electronically, indirect modulation using an optical chopper was adopted for simplicity, as seen in Fig. 13.17. The quiescent DC bias current through all the bulbs was 44 mA, to give the same peak spectral radiance at ~80 THz before channel filtering. The separation distance between the channel transmitter-receiver pairs was fixed at a distance of 1.0 cm (i.e. double the distance of the first experiment).

It is worth mentioning that the proximity of adjacent channel transmitter-receiver pairs may have to be considered; for example, in order to avoid adjacent channel interference. This is not important in this particular application, as there are a number of simple solutions that can be considered (e.g. employing collimating lenses, ensuring band separation and increasing filter selectivity). However, for this

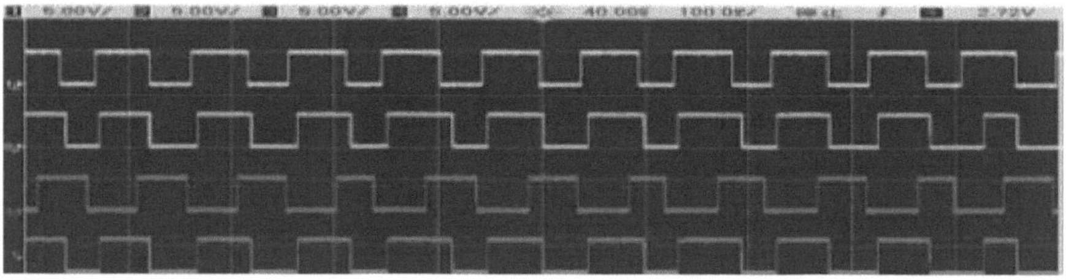

Fig. 13.18 Output bit steams from the four channel transmitter-receiver pairs, each having a bit rate of 10.3 bit/s

particular experimental setup, the adjacent channel transmitter-receiver pairs were spatially separated in order to allow them to be completely shielded by the blades of the 12 cm diameter optical chopper, as shown in Fig. 13.17 [41].

The maximum bit rate for each individual channel was measured to be 10.3 bit/s, as shown in Fig. 13.18. It was found that the recovered signal from the channel transmitter-receiver pair using filter SWBP-3685-000091 was distorted if the chopping frequency is increased further. This is because the bandwidth and/or transmittance for this particular channel filter was smaller than desirable. The maximum effective bit rate for this *THz Torch* FDM system demonstrator was achieved after combining the parallel outputs from the four channel transmitter-receiver pairs into one serial output bit stream. In this experiment, as expected, the maximum end-to-end bit rate was measured to be 41.2 bit/s. This represents an eight-fold increase in bit rate, and over twice the distance, when compared to first experiment. Recent results have demonstrated a much faster 1,280 bps FDM implementation. In addition, an experimental FHSS scheme has also been demonstrated for the first time, having a 320 bps data rate. With both 4-channel multiplexing schemes, measured bit error rates (BERs) of $< 10^{-6}$ are achieved over a distance of 2.5 cm.

Future *THz Torch* multiplexing schemes may involve having more channels, requiring more filters with bespoke bandwidths. Low-cost enabling technologies for realising such filters are already available. For narrower fractional bandwidths of operation (e.g. between 10 and 30 %), metal mesh filters can be employed, as demonstrated by a 14 % fractional bandwidth filter at 10 THz, patterned by a two-dimensional array of half-wavelength sized crosses (15 μm in size) [50]. At 40 THz, the half-wavelength size of 3.75 μm is very close to the 1 μm minimum feature size limit for low-cost shadow mask manufacturing.

With both FDM and FHSS applications, a number of standard COTS filters and windows can be employed with sufficient bandwidth, selectivity and transmittance to create filter banks for multi-channel operation. Figure 13.19 shows COTS metal mesh filters, ranging in centre frequency from 0.5 to 30 THz, having Gaussian-shaped pass band characteristics, 90 % spectral transmittance and full-width half-maximum (FWHM) bandwidths between 7 and 25 % [51].

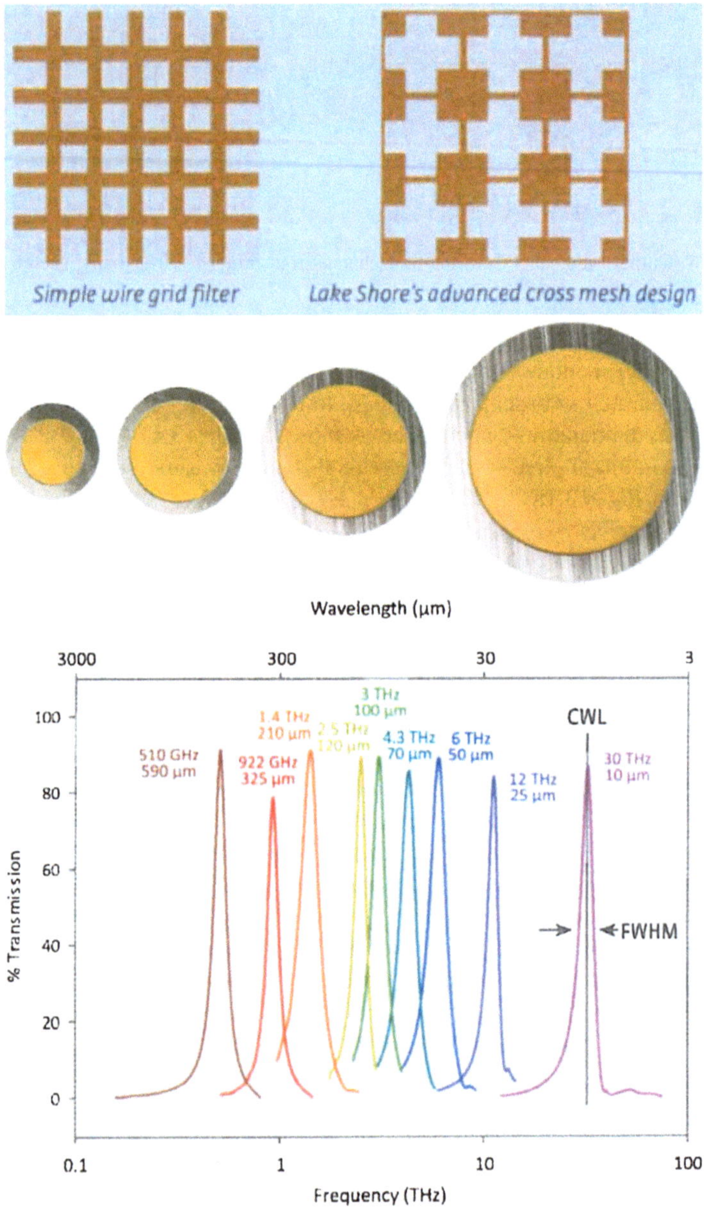

Fig. 13.19 COTS band-pass filters from 0.5 to 30 THz (Courtesy of Lake Shore Cryotronics, Inc. [51])

13.4 Fundamental Limits and Engineering Solutions

Ultra-low cost short-range wireless communications links based on the *THz Torch* technology have been successfully demonstrated experimentally, with single-channel and multi-channel frequency division multiplexing architectures. While the measured performances of these first proof-of-concept demonstrators have been limited, in terms of bit rate and transmission range, it is important to recognise that this emerging technology is still in its infancy. By understanding the physical limits of the individual technologies, depending on the particular application, the design engineer will know whether it is worth pushing the performance of an individual component to its limit, replace it with a more expensive higher performance component or adopt an alternative solution.

13.4.1 Operating Spectrum

The recommended operating spectrum for the *THz Torch* technology has been identified as lying in the high-THz region of 10–100 THz; well above the frequencies normally associated with terahertz technologies. This recommendation is based on the levels of spectral radiance, which are relatively low below 10 THz; lying in the long wavelength tail creating by applying Planck's law. Increasing the bulb's quiescent DC bias current will increase the working temperature of the filament radiator and, therefore, increase both the amplitude and frequency of the spectral radiance peak. However, the levels of spectral radiance in the long-wavelength tail will not increase by anywhere near the same amount as that at the peak. As a result, of the high levels of spectral noise power generated near the peak will be wasted (since the peak will be above 100 THz) and the DC-to-THz power conversion efficiency will reduce dramatically. An example of this was previously discussed for the Eiko 8666-40984 bulb's recommended DC operating power level for general purpose applications.

The operating spectrum is also limited by the background thermal noise associated with the transmitter hardware (excluding the filament radiators) or target, ambient channel environment and receiver hardware. At a room temperature of 300 K, the equivalent photon energy $E = kT = 26$ meV, where k is the Boltzmann constant, and by using the Planck relation $E = h\nu$, where h is the Planck constant, this corresponds to a frequency $\nu \sim 6$ THz. Therefore, for room-temperature operation, the *THz Torch* concept has a lower frequency limit of ~6 THz. The thermodynamics of the *THz Torch* concept relies on temperature changes; temporal (in the case of secure RFID, smart key fobs, secure data transfer, covert communications, night signalling and IED trigger applications) or spatial (in the case of IED triggers, high 'pixel' resolution FIR counter-measures for adaptive thermal camouflage and sentry decoys and identification friend or foe applications). Introducing an active cooling subsystem (e.g. Peltier effect cooler, PEC) to the transmitter/target can increase its temperature change ΔT, improving the end-to-end

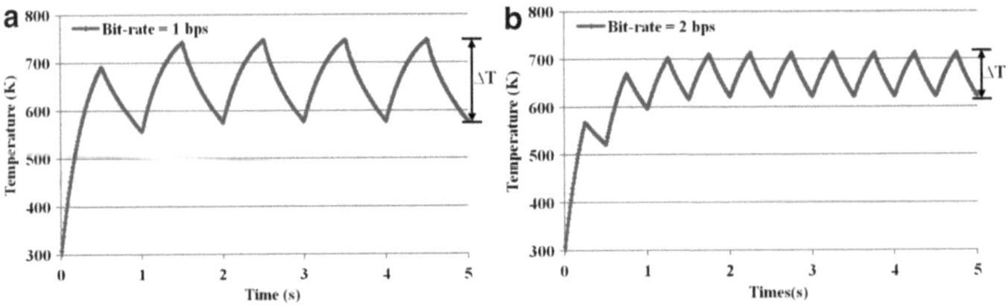

Fig. 13.20 Measured instantaneous filament temperature, showing the ΔT of the source as bit rate increases: (**a**) 1 bit/s and *larger* source ΔT; (**b**) 2 bit/s and *smaller* source ΔT

systems performance, but this is only when the transmitter/target dominates the receiver's FOV. Introducing a PEC to the receiver's back-end electronics only will help to improve its output signal-to-noise ratio. When employing a PEC, for either the transmitter or receiver, there are the inevitable increases in associated DC power consumption, complexity, size, mass and cost.

13.4.2 Filament and Sensor Thermal Time Constants

If the incandescent light bulbs employ direct modulated, both the heating and cooling thermal time constants impose fundamental limitations on their signalling speed; this was observed in the first experiment, where the maximum bit rate was only 5 bit/s. Since pyroelectric sensors can only detect the change in temperature ΔT of the source, increasing the bit rate will result in a smaller ΔT, as shown in Fig. 13.20; giving a lower output signal-to-noise ratio.

One solution is to employ indirect modulation, whereby a constant level of spectral noise power (generated from a bulb having a fixed quiescent DC bias current) is externally pulsed using an optical switch. As a result, this avoids any issues of thermal time constant limitations associated with the transmitter. If a simple optical chopper is employed, as used in the FDM experiment [41], the thermal time constant limitation associated with the PIR sensor can be investigated.

An experiment was conducted with the use of the Murata IRA-E710ST1 sensor, across a single channel covering the 25–50 THz spectral range. Here, the optical chopper was positioned 1.0 cm from the transmitter and three values of quiescent DC bias current were considered: with $I = 44$, 50 and 60 mA, having estimated spectral radiance peaks at 80, 93 and 108 THz, respectively. The performances with the lowest quiescent DC bias current of 44 mA can be compared with the very first experiment having direct modulation. As shown in Fig. 13.21a, with a fixed transmission distance of 1.0 cm, the maximum bit rate was 50 bit/s. This represents a ten-fold increase in bit rate with double the distance, when compared with the

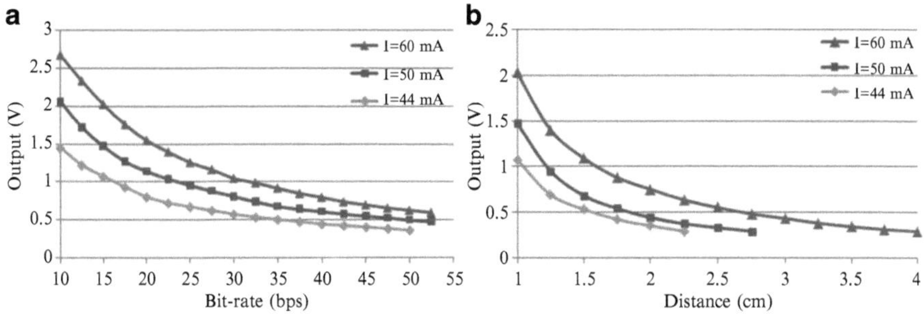

Fig. 13.21 Experimental results for different quiescent DC bias currents, with the use of an optical chopper, showing the output voltage from the LNA against: (**a**) bit rate for a fixed transmission distance of 1.0 cm; (**b**) transmission distance with a fixed bit rate of 15 bit/s

same conditions using direct modulation. Moreover, as shown in Fig. 13.21b, with a fixed bit rate of 15 bit/s, the maximum transmission distance was 2.25 cm. This represents almost a five-fold increase in the transmission distance with three times the bit rate.

As expected, Fig. 13.21b shows that when the quiescent DC bias current is increased there is an increase in transmission distance. However, the same characteristic cannot be seen in Fig. 13.21a. This is because the Murata IRA-E710ST1 sensor cannot respond fast enough to a bit rate that exceeds 52.5 bit/s. The reason for this is due to the detector's responsivity, having a 70.7 % cut-off modulation frequency at ~1.3 Hz; this will be discussed further in the next subsection.

For most practical applications, an optical chopper is of little use. However, there are other indirect modulation technologies that can be employed. For example, depending on the speed of operation, mechanical shutters (realises using miniature mechanical or even microelectromechanical systems, MEMS [52], technologies) can provide perfect transmission and extinction, although this may present an expensive solution. High-contrast terahertz modulators, based on extraordinary transmission through ring aperture arrays, can be ultra-fast (with expected switching speeds >10 Gbit/s) and potentially low cost [53], although being based on resonant arrays they will have a narrow instantaneous bandwidth.

13.4.3 *Detector Responsivity and Microphonics*

The pyroelectric sensor has a detector responsivity that is given by the following expression:

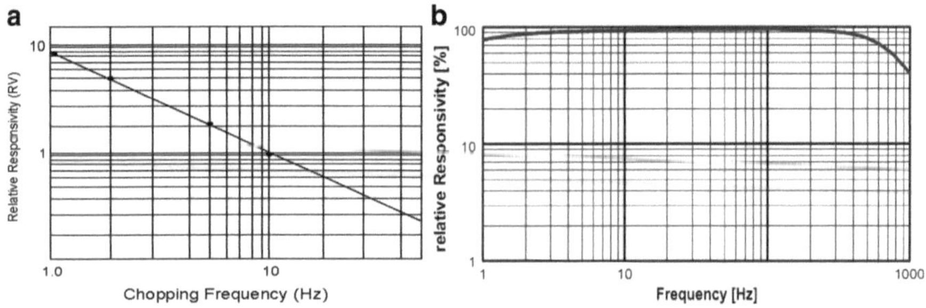

Fig. 13.22 Modulation frequency dependency of PIR sensor responsivity: (**a**) Murata IRA-E710ST1, indicating a 70.7 % cut-off frequency at ~1.3 Hz (Courtesy of Murata Manufacturing Co. [44]); and (**b**) InfraTec LME-553, indicating a 70.7 % cut-off frequency at ~600 Hz (Courtesy of InfraTec [45])

$$\Re = \frac{u}{\Phi s} = \omega p A_s \cdot \left[\frac{\alpha}{G_T} \frac{1}{\sqrt{1 + (\omega \tau_T)^2}} \right] \cdot \frac{R}{\sqrt{1 + (\omega \tau_E)^2}} \tag{13.14}$$

It can be see that its modulation frequency dependant characteristics are dictated by the thermal τ_T and electrical τ_E time constants. The modulation frequency dependency of detector responsivity for the Murata IRA-E710ST1 and InfraTec LME-553 devices can be seen in Fig. 13.22a, b, respectively. The former has a very low 70.7 % cut-off modulation frequency at ~1.3 Hz. Indeed, the Murata IRA-E710ST1 sensor was originally developed for detecting relatively slow movements of warm bodies (i.e. for human motion sensing). As a result, the low modulation frequencies of operation (normally between 1 and 10 Hz) can be realised with a basic technology that is very cheap to manufacture in large volumes. The latter detector has a much larger 70.7 % cut-off modulation frequency at ~600 Hz. This is because this more expensive InfraTec LME-553 sensor has an internal preamplifier with a low input impedance. Therefore, with $\tau_T >> \tau_E$ (since $\tau_T = 200$ ms and $\tau_E = 0.4$ ms), it is the thermal time constant that dictates the lower cut-off modulating frequency of the InfraTec LME-553 sensor and, thus, the maximum bit rate for an indirectly modulated *THz Torch* wireless link.

The InfraTec LME-553 sensor element is made from the crystalline pyroelectric material lithium-tantalate, which also exhibits piezoelectric properties. As a result, unless corrected for, the performance of this sensor will be severely degraded by microphonic effects, from external mechanical vibrations and acoustic pressure waves. The simple engineering solution is to connect two sensors (one blind) directly to the differential inputs of a high gain, low noise, operational amplifier. The output from this LNA is then band-pass filtered before the analogue signal reaches the Schmitt Trigger.

Dual-sensor experiments were conducted with a single channel, covering the 25–50 THz spectral range. The optical chopper was positioned 1.0 cm from the transmitter and three values of quiescent DC bias current were again considered:

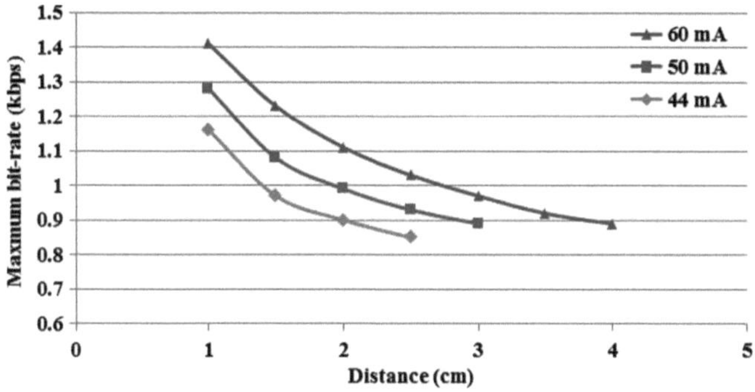

Fig. 13.23 Measured maximum bit rate against transmission distance for transmitter bulbs having different quiescent DC bias currents; for the single-channel architecture, covering the 25–50 THz spectral range, using indirect modulation and InfraTec LME-553 sensors

with $I = 44$, 50 and 60 mA. As can be seen in Fig. 13.23, both the maximum bit rate and the maximum transmission distance increase with the quiescent DC bias current. In other words, the InfraTec LME-553 was able to respond with a recorded bit rate of 1.4 kbit/s (with $I = 60$ mA, over this single channel), even though its 70.7 % cut-off modulation frequency is only ~600 Hz. Indeed, much better wireless link performances can be achieved with higher levels of quiescent DC bias current.

13.4.4 Bulb Glass Envelope Absorption

COTS incandescent light bulbs have a hermetically-sealed envelope, to prevent the hot tungsten filament from oxidizing in air. The envelope material (e.g. fused silica or soda lime glass) has a high spectral transmittance in the optical and NIR spectral ranges, but this is far from true at longer wavelengths. Figure 13.24a gives the measured refractive index (n) and extinction coefficient (k) of the complex refractive index for a typical soda line window glass from 3 to 100 μm (i.e. 3–100 THz) [54]. The Eiko 8666-40984 bulb has a glass envelope thickness of ~300 μm at its sides. However, it is thicker at the end section (~400 μm), from where the majority of the energy radiates in the experimental setup shown in Fig. 13.4. Assume the thickness of the glass envelope is 350 μm, corresponding power transmittance, reflectance and absorptance can be calculated [22], as shown in Fig. 13.24b. The calculated results show a low transmittance for frequency range from 1 to 65 THz. Therefore, the bulb's glass envelope is expected to contribute significantly to the overall transmission path loss. In this range, a secondary radiation mechanism dominates from heating of the glass envelope.

One method of minimizing the glass envelope loss is to selectively remove its end section and replace it with a more transparent window material. For example

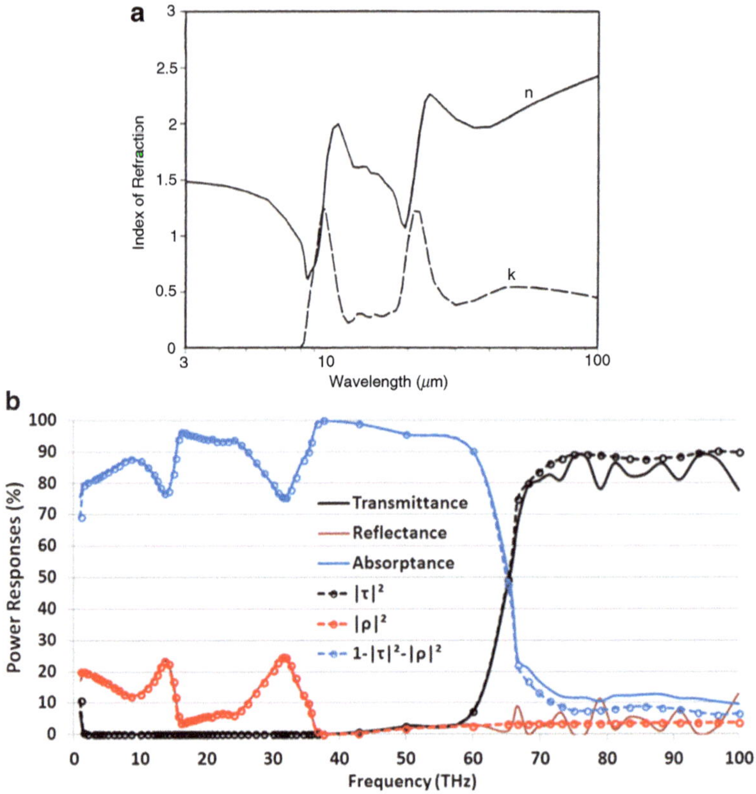

Fig. 13.24 (**a**) Measured refractive index and extinction coefficient for typical soda lime window glass (Courtesy of Solar Energy Materials [54]); (**b**) Calculated power transmittance, reflectance and absorptance for 350 μm thick glass envelope sing the measured dated in (**a**)

potassium bromide (KBr) has excellent transmission properties, with a minimum transmission level of 80 % from 15–750 THz (i.e. 0.4–20 μm), but is hygroscopic. With the use of KBr collimating lenses, faster PIR sensors and improved back-end electronics, recent results have demonstrated 1,000 bps single-channel links having a range of >10 cm and no detectable errors. Zinc selenide (ZnSe) has a relatively low absorption coefficient across a wide spectral range, from 16 to 600 THz (i.e. 0.5–19 μm); while also being non-hygroscopic and stable for practically all user environments. As a result, it is used in optical component (e.g. windows, lenses and beamsplitters) for pyrometry applications. The Russian company TYDEX® supply such materials with broadband anti-reflection (BBAR) coatings. Figure 13.25 shows the spectral transmittance of a ZnSe window material with BBAR coating for the 25–100 THz range. Synthetic (CVD grown) diamond has a high transmittance (~70 %) from UV to far-IR.

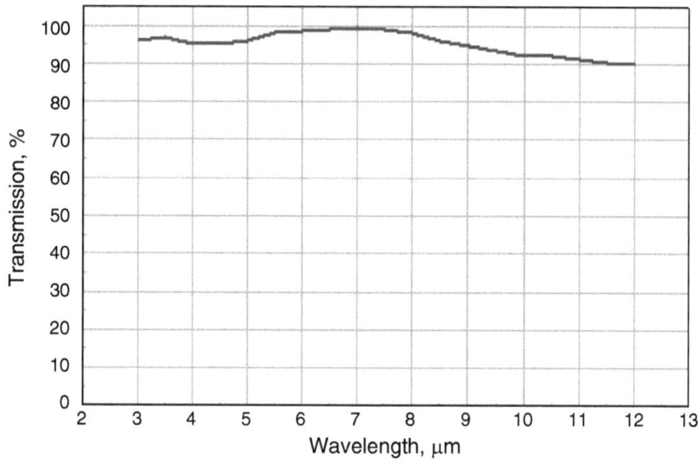

Fig. 13.25 Spectral transmittance of a ZnSe window material with BBAR coating for the 25–100 THz range (Courtesy of TYDEX® [55])

13.4.5 Free Space Attenuation and Spreading Loss

The *THz Torch* concept has been demonstrated to work over short transmission distances; to date, of the order of centimetres (e.g. a 4 cm range, with a quiescent DC bias current of 60 mA into five miniature bulbs connected in series). One of the inherent limitations of operating the high-THz spectral range of 10–100 THz is the high free space attenuation in air. Figure 13.26 shows various spectral windows, including a low attenuation band from 21–40 THz (i.e. 7.6–14 μm), followed by a high attenuation band from 40–56 THz (i.e. 5.4–7.6 μm), followed by mixed low and high attenuation bands to 100 THz. In principle, it may be possible implement a multi-channel *THz Torch* architecture that can exploit the spectral locations of the low attenuation windows.

Since the transmission distances of the early proof-of-concept experimental demonstrators are of the order of centimetres, the dominant loss mechanism is beam spreading. For a compact design (whereby five miniature bulbs are assembled into a compact cylindrical package, having an outer diameter of only 8.2 mm), it is not appropriate to employ a much large parabolic mirror. A simple engineering solution is to use a conformal metal reflecting film (e.g. foil), placed behind the five-bulb array, to reduce the spreading loss. Moreover, by including collimating lenses made from low absorption materials, located at both the transmitter and receiver, multi-kbit/s links having transmission distances of the order of metres can be expected.

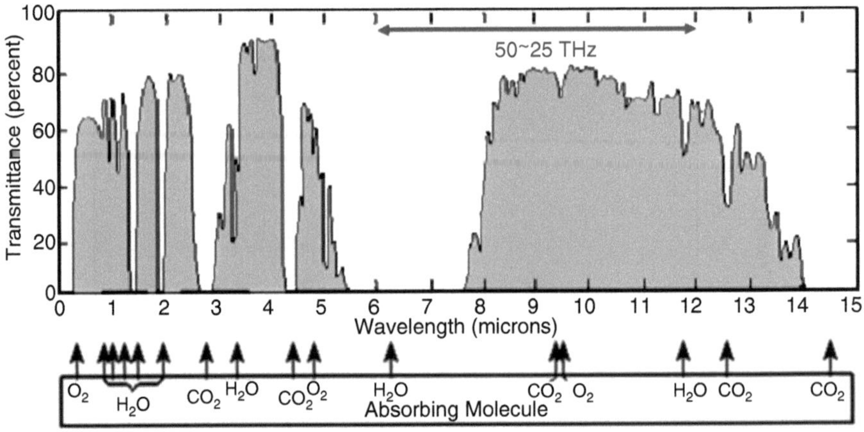

Fig. 13.26 Atmospheric transmittance from UV to far infrared [56]

13.5 Conclusions

This chapter has examined the emerging *THz Torch* technology, with a view to implementing low cost solutions for specific niche security and defence applications. More specifically, the focus in this chapter has been on its application to short range, low bit rate wireless communications links.

After introducing the basic concepts, results from the first proof-of-concept experimental demonstrators are given, with a view to highlighting some of the inherent benefits and challenges. The physical limitations of various enabling technologies, along with simple engineering solutions are discussed, with a view to increasing the bit rate and transmission distance for a wireless communications link, while also maintaining the very high DC-to-THz power conversion efficiency and limited form factor.

This technology is still in its infancy and so it should not be of any surprise that the initial performances fall well short when compared to mature, coherent, radio-based wireless solutions. However, there are strong reasons to suggest that performance levels are set to increase dramatically (e.g. much faster speeds of operation and across greater transmission distances), by exploiting a raft of ultra-low cost techniques and, where necessary, by selectively introducing higher performance specification components (e.g. sources, detectors, etc.) at extra cost.

The motivation for this research should always be kept in mind, as it offers the following benefits that may be critical with specific niche applications: (1) only low precision components are needed; (2) easy to manufacture in large volumes; (3) inherently ultra-low cost; (4) contactless data transfer that is electrically/mechanically robust; (5) tuneable performance at no extra cost; (6) diverse enabling technology solutions; (7) milliwatt output power levels without the need for cryogenic cooling; (8) very high DC-to-THz power conversion efficiency;

and (9) operation is in a virtually unused part of the electromagnetic spectrum. This last point lends itself to covert operation. Indeed, with such large amounts of freely available spectrum and high atmospheric attenuation, there is an extremely low probability of intercept and code grabbing; making this technology ideal for security applications.

The *THz Torch* technology is expected to find niche security and defence applications that do not require high data rates but must be ultra-low cost (e.g. secure RFID, smart key fobs, secure data transfer, covert communications and night signalling, IED triggers, high 'pixel' resolution FIR counter-measures for adaptive thermal camouflage and sentry decoys, identification friend or foe (IFF), and potentially even absorption spectroscopy for explosives detection).

Acknowledgments The initial work was started using funding from the UK Engineering and Physical Science Research Council (EPSRC), under Platform Grant EP/E063500/1. The authors would like to thank Hanchao Lu. The final proof-reading of this chapter by William J. Otter and Stephen M. Hanham is gratefully appreciated. In addition, Fangjing Hu would like to thank the China Scholarship Council (CSC) for funding his PhD.

References

1. King EV (1965) Infra-red ray alarm system. The Radio Constructor, October, pp 173–176
2. Nezih P, Abbas AN (2012) Terahertz technology for nano applications. In: Bhushan B (ed) Encyclopedia of nanotechnology. Springer, New York, pp 2653–2788
3. Lin H, Fischer BM, Mickan SP, Abbott D (2006) Review of THz near-field methods. Smart materials, nano-and micro-smart systems. Proc SPIE 6414:64140L
4. http://ieeexplore.ieee.org/xpl/aboutJournal.jsp?punumber=5503871#AimsScope
5. http://www.digitalbarriers.com/inside-thruvision/
6. http://www.azdec.com/
7. http://tplogic.net/wp-content/uploads/2013/07/product_brochure.pdf
8. http://cp.literature.agilent.com/litweb/pdf/5989-7620EN.pdf
9. http://www.teraview.com/products/terahertz-pulsed-spectra-3000/index.html
10. Williams BS (2007) Terahertz quantum-cascade lasers. Nat Photonics 1:517–525
11. Preu S, Dohler GH, Malzer S, Wang LJ, Gossard AC (2011) Tunable, continuous-wave terahertz photomixer sources and applications. J Appl Phys 109:1–56
12. Nagatsuma T (2009) Generating millimeter and terahertz waves. IEEE Microw Mag 10:64–74
13. Knap W, Kachorovskii V, Deng Y, Rumyantsev S, Lü JQ, Gaska R, Shur MS, Simin G, Hu X, Khan MA, Saylor CA, Brunel LC (2002) Nonresonant detection of terahertz radiation in field effect transistors. J Appl Phys 91:9346–9353
14. Meziani YM, Garcмa-Garcмa E, Velázquez-Pérez JE, Coquillat D, Dyakonova N, Knap W, Grigelionis I, Fobelets K (2013) Terahertz imaging using strained-Si MODFETs as sensors. Solid State Electron 83:113–117
15. Rauter P, Fromherz T, Winnerl S, Zier M, Kolitsch A, Helm M, Bauer G (2008) Terahertz Si: B blocked-impurity-band detectors defined by nonepitaxial methods. Appl Phys Lett 93:261104-1–261104-3
16. Komiyama S, Astafiev O, Antonov V, Kutsuwa T, Hirai H (2000) A single-photon detector in the far-infrared range. Nature 403(4768):405–407
17. Lucyszyn S (2007) Evaluating surface impedance models for terahertz frequencies at room temperature. PIERS Online J 3:554–559

18. Lucyszyn S (2005) Investigation of Wang's model for room temperature conduction losses in normal metals at terahertz frequencies. IEEE Trans Microw Theory Tech 53:1398–1403
19. Lucyszyn S (2004) Investigation of anomalous room temperature conduction losses in normal metals at terahertz frequencies. IEE Proc Microw Antenna Propag 151:321–329
20. Lucyszyn S (2001) Comment on terahertz time-domain spectroscopy of films fabricated from SU-8. IEE Electron Lett 37:1267
21. Lucyszyn S, Zhou Y (2011) Engineering approach to modelling metal THz structures. Online J Terahertz Sci Technol 4:1–8
22. Lucyszyn S, Zhou Y (2010) Characterising room temperature THz metal shielding using the engineering approach. PIER J 103:17–31
23. Lucyszyn S, Zhou Y (2010) THz applications for the engineering approach to modelling frequency dispersion within normal metals at room temperature. PIERS Online J 6:293–299
24. Lucyszyn S, Zhou Y (2010) Engineering approach to modelling frequency dispersion within normal metals at room temperature for THz applications. PIER J 101:257–275
25. Zhou Y, Lucyszyn S (2009) HFSS™ modelling anomalies with THz metal-pipe rectangular waveguide structures at room temperature. PIERS Online J 5:201–211
26. Episkopou E, Papantonis S, Otter WJ, Lucyszyn S (2012) Defining material parameters in commercial EM solvers for arbitrary metal-based THz structures. IEEE Trans Terahertz Sci Technol 2:513–524
27. Episkopou E, Papantonis S, Otter WJ, Lucyszyn S (2011) Demystifying material parameters for terahertz electromagnetic simulation. In: 4th UK/Europe-China conference on millimetre waves and terahertz technologies, Glasgow, pp 80–81
28. Otter WJ, Hanham SM, Episkopou E, Zhou Y, Klein N, Holmes AS, Lucyszyn S (2013) Photoconductive photonic crystal switch. In: 38th international conference on Infrared, Millimeter and Terahertz Waves (IRMMW-THz 2013), Mainz, Germany
29. Episkopou E, Papantonis S, Holmes AS, Lucyszyn S (2012) Optically-controlled plasma switch for integrated terahertz applications. In: 39th IEEE International Conference on Plasma Science (ICOPS2012), Edinburgh
30. Lucyszyn S, Zhou Y (2012) Reconfigurable terahertz integrated architecture (RETINA) – a paradigm shift in SIW technology. IEEE International Microwave Symposium (IMS2012) workshop proceedings, WFA: integration and technologies for mm-wave sub-systems, Montreal, Canada
31. Zhou Y, Lucyszyn S (2010) Modelling of reconfigurable terahertz integrated architecture (RETINA) SIW structures. PIER J 105:71–92
32. McPherson DS, Soe HC, Jung YL, Lucyszyn S (2001) 110GHz vector modulator for adaptive software-controlled transmitters. IEEE Microw Wirel Compon Lett 11:16–18
33. Lucyszyn S, Silva SRP, Robertson ID, Collier RJ, Jastrzebski AK, Thayne IG, Beaumont SP (1998) Terahertz multi-chip module (T-MCM) technology for the 21st century? IEE colloquium digest on multi-chip modules and RFICs, London, pp 6/1–8
34. Lucyszyn S (1997) The future of on-chip terahertz metal-pipe rectangular waveguides implemented using micromachining and multilayer technologies. IEE colloquium digest on terahertz technology and its applications, London, pp 10/1–10
35. Lucyszyn S, Budimir D, Wang QH, Robertson ID (1996) Design of compact monolithic dielectric-filled metal-pipe rectangular waveguides for millimetre-wave applications. IEE Proc Microw Antenna Propag 143(5):451–453
36. Sanchez-Hernandez D, Lucyszyn S, Robertson ID (1996) A study of integrated antennas for terahertz circuits. COST-245 workshop on applications of MMICs in active antenna systems 1996. ERA, Leatherhead
37. Lucyszyn S, Wang QH, Robertson ID (1995) 0.1 THz rectangular waveguide on GaAssemi-insulating substrate. IEE Electron Lett 31:721–722
38. Lucyszyn S, Hu F (2013) *THz torch* technology for low-cost security applications. NATO conference on THz and security applications 2013, Kiev, Ukraine, May 2013
39. Hu F, Lucyszyn S (2013) Improved '*THz Torch*' technology for short-range wireless data transfer. IEEE International Wireless Symposium (IWS2013), April, Beijing, China

40. Hu F, Lucyszyn S (2012) *THz torch* technologies for 21st century applications. IoP Photon 12 2012, September, Durham
41. Hu F, Lucyszyn S 2011 Ultra-low cost ubiquitous THz security systems. In: Proceedings of the 25th Asia-Pacific Microwave Conference (APMC2011) 2011, Melbourne, Australia, pp 60–62
42. Lucyszyn S, Lu H, Hu F (2011) Ultra-low cost THz short-range wireless link. IEEE international microwave workshop series on millimeter wave integrated technologies 2011, Sitges, Spain, pp 49–52
43. Rogalski A (2003) Infrared detectors: status and trends. Progress Quant Electron 27:59–210 http://www.murata.com/products/catalog/pdf/s21e.pdf
44. Murata Manufacturing Co., Pyroelectric infrared sensor & sensor module. Catalogue No S21E-2
45. http://www.infratec-infrared.com/Data/LME-553.pdf
46. Lide DR (ed) (1996) CRC handbook of chemistry and physics, 77th edn. CRC Press, Boca Raton
47. Durakiewicz T, Hala S (1999) Thermal relaxation of hot filaments. J Vacuum Sci Technol A: Vacuum Surf Films 17:1071–1074
48. InfraTec 'Detector basics'. http://www.infratec.de/fileadmin/media/Sensorik/pdf/Application_Detector_Basics.pdf
49. http://www.noc-ltd.com/catalogue
50. Melo AM, Kornberg MA, Kaufmann P, Piazzetta MH, Bortolucci EC, Zakia MB, Bauer OH, Poglitsch A, Alves da Silva AMP (2008) Metal mesh resonant filters for terahertz frequencies. Appl Opt 47:6064–6069
51. http://www.lakeshore.com/products/optical-filters/THz-and-IR-Band-Pass-Filters/Pages/Overview.aspx
52. Lucyszyn S (ed) (2010) Advanced RF MEMS 2010. Cambridge University Press, Cambridge
53. Shu J, Qiu C, Astley V, Nickel D, Mittleman DM, Xu Q (2011) High-contrast terahertz modulator based on extraordinary transmission through a ring aperture. Opt Express 19: 26666–26671
54. Rubin M (1985) Optical properties of soda lime silica glasses. Solar Energy Mater 12: 275–288
55. http://www.tydexoptics.com/en/products/pyrometry/cvd_znse/
56. http://en.wikipedia.org/wiki/Absorption_(electromagneticradiation)

Chapter 14
New THz Technologies and Applications in Applications in Support of Safety and Security

Ashok Vaseashta

Abstract Recent incidents have prompted changes to the methods employed for security screening at airports and border security check-points. At cargo screening facilities and major border check-points, where thousands of containers need to be screened rapidly, it is a challenging task to effectively screen each container. As a result, there is an increasing focus on new technologies that can be applied for security screening in a stand-off mode, either to simplify or speed up the screening process, or to provide additional functionality. Terahertz (THz) technology is a promising and emerging technology and has been considered in various forms. Additionally in the battlefield, one of the major threat vectors is improvised explosive devices (IEDs) used in different forms such as vehicle borne IEDs (VBIEDs) or strapped to humans at inconspicuous locations. THz imaging systems can be used to image such threat vectors, since such materials have characteristic THz spectra. The use of THz illumination of sufficient power levels and fast image detection and processing, has shown that non-metallic weaponry can be imaged when concealed beneath clothing. Some of the barrier and potential confusing materials have smooth spectra with relatively low attenuation. However, use of the THz frequencies, initially aimed at narrow-band at 830 GHz along with optical mixing can be used to identify metal and dielectric objects. The ultimate possibility to identify the chemical compositions of explosive materials and mixed chemical compositions needs the wide-band antennas. Potential use of THz imaging in active and passive imaging systems for detection of chemical and biological agents and remote monitoring of signals is described. Selected applications of THz for stand-off

A. Vaseashta (✉)
Institute for Advanced Sciences Convergence, NUARI, 13873 Park Center Rd. Suite 500, Herndon, VA 20171, USA

International Clean Water Institute, NUARI, 13873 Park Center Rd. Suite 500, Herndon, VA 20171, USA
e-mail: avaseash@norwich.edu

C. Corsi and F. Sizov (eds.), *THz and Security Applications*, NATO Science for Peace and Security Series B: Physics and Biophysics, DOI 10.1007/978-94-017-8828-1_14,

detection are described by using nanomaterials to generate and detect response signal, and also to demonstrate that the THz spectra of several common chemicals, explosive compounds, and pharmaceuticals are distinct for ease of identification.

14.1 Introduction – Threat Vectors: Emerging, Persistent, Dual Use and Avant Garde

The current geopolitical landscape is exceedingly complex, dynamic, and unpredictable. Many threat vectors have become highly asymmetric, kinetic, and non-linear. Traditional rules-of-engagement do not apply or have different interpretations. Figure 14.1 shows many ways by which adversaries can launch overt/covert attacks, including psychological operations (PSYOPS), Chemical, biological, radiological, nuclear (CBRN), improvised explosive devices (IEDs),

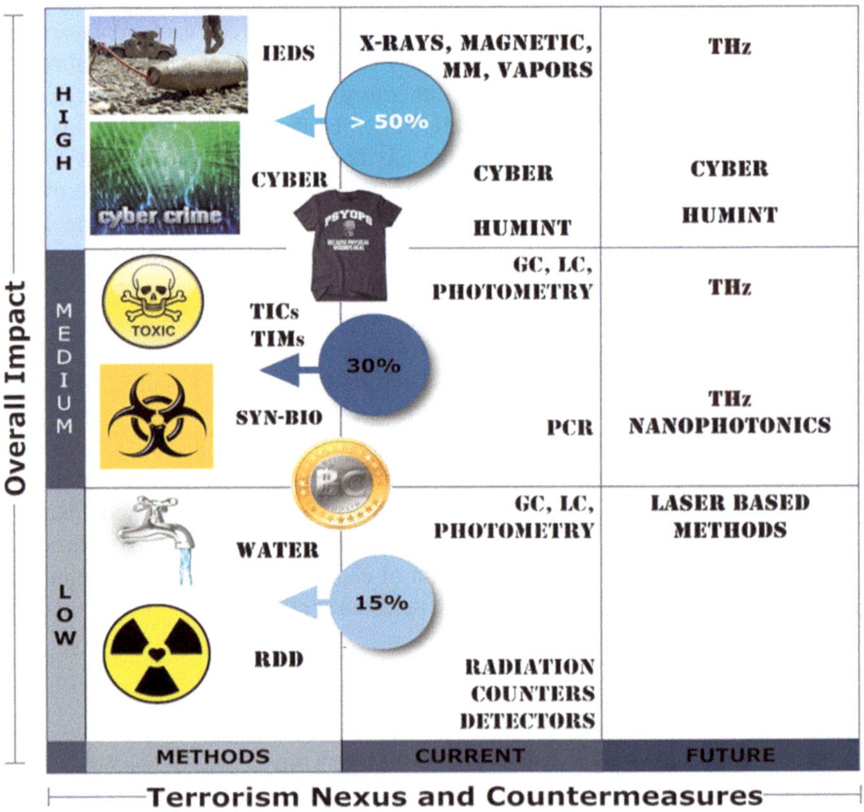

Fig. 14.1 Methods of terrorism including CBRNE, cyber, water, and Bitcoin with current and future countermeasure platforms

non-traditional agents (NTA), cyber-attacks, and raise funds using non-traditional methods (viz. *hawala*, Bitcoins) to support such operations, along with current and futuristic countermeasure platforms to defeat terrorism nexus. It is worth noting that THz is currently considered as a potential technological platform for many applications. STANDEX[1] is one such platform and is jointly considered by many countries.

Notwithstanding the unparalleled level of technological advances, the asymmetric threat from terrorist groups continues to evolve. S&T advances coupled with universal access to the Internet provides the same means by which state and non-state sponsored actors develop warfare agents with a certain level of sophistication. Effective countermeasures also use essentially the same tools, hence understanding of transformational emerging sciences, concepts and theories, and their potential applications in support of defense and security is exceedingly critical as effective countermeasures. Numerous technological advances arise from the potential of nanoscale materials to exhibit unique properties that are attributable to their reduced dimensions [1]. Furthermore, advances in material synthesis, device fabrication and characterization have provided the means to study, understand, control, or even manipulate the transitional characteristics between isolated atoms and molecules, and bulk materials. Consequently, various new "materials by design" capable of producing devices and systems with remarkable, tunable, and specific properties have recently been fabricated. Such advances coupled with information technology, cognitive sciences, biotechnology, artificial intelligence, and genetics offer an ecosystem of innovations and potential pathways to counter threat vectors in ways never imagined possible earlier. A nexus of technological innovations to include deployment of systems with enhanced and remote maneuverability, remote interrogation and mitigation, enhanced information gathering, and thwarting threats at the point-of-origin (PO2) is described elsewhere [2].

This report describes use of THz spectrum to detect explosives and harmful vapors, image hidden metal objects that can serve as weapons, and receive electromagnetic signals from devices emitting such signals from a distance. Examples of innovative countermeasures using advanced and nano materials exploiting THz region of the electro-magnetic spectrum to provide various functionalities are also discussed. Tactically, the unsophisticated nature of threats poses significantly greater technical challenges in both point and stand-off detection. This threat is growing due to increased globalization and mobility within society, the explosion in chemical and biotech expertise and the relative ease with which chemical weapons can be prepared at off-sites, and the means by which pathogens can be covertly engineered, transported and released or dispersed.

Understanding the emerging security challenges (ESC) and nature and potential of threats may prevent or minimize a potentially catastrophic occurrence. Most potential threats have been characterized by a severity of hazard (SH) ranking based on toxicity, flammability, and/or reactivity. Notwithstanding many conventional and well characterized toxic industrial chemicals (TICs)/toxic industrial materials

---

[1] http://factsindia.wordpress.com/category/standex/

(TIMs), there are many other emerging, persistent, dual-use, and unconventional threat vectors that must be identified and mitigated using advanced technological platforms. Despite its many useful applications, research on genetically modified organisms (GMO) coupled with recent advances in nucleic acid synthesis, lack of in-situ detection, reference materials and standards have caused public and scientific concerns. Synthetic biology (syn-bio), likewise, is modifying the DNA of an organism to alter its coded information, such as selecting desired mutations, inducing mutagenesis, altering genes or regulatory elements, cloning options, designing biological components and systems that otherwise do not naturally exist. Syn-bio provides the tools to "redesign" the genomes of existing microbes to increase their efficacy or offer brand new functionalities. As an example, the successful "redesign" of the bacterium Mycoplasma genitalium, which has the smallest known bacterial genome, yet possesses all of the bio-chemical machinery needed to metabolize, grow, and reproduce, has been readily available in open literature and on the Internet [citation withheld]. Syn-bio is subject to potential misuse, in terms of its enhanced virulence, resistance to control agents, altering host defense, and increasing environmental stability and/or dispersal. Similar to GMO, limited detection methods exist for emerging syn-bio threats requiring countermeasures using advanced technology innovations platforms. Dual-use is a term often used in the political and diplomatic context to express a technology that can be used for both peaceful and military applications. Although used primarily in context of nuclear proliferation, the use of GMO and (bio) weaponized syn-bio agents pose a major concern. Hence, THz systems require imaging that use complex algorithms for processing of information and strategic assessments and modeling of mixed and complex hazardous environments to delineate signal from noise and background interactions.

14.1.1 Sampled vs. Remote Detection

The two most important metrics for the sensing and transduction functions are time to detect material under investigation to enable appropriate response and the resulting response function consistent with the species detected. Given the complex environment described above, the challenges at subsystem levels occur to evaluate overall effectiveness and efficacy of the sensor/detector systems. In point/direct/sampled detection platforms, the analysis is accurate, real-time (in most cases), and conform to sensor/detector metrics of specificity, selectivity, and sensitivity. However, due to the extreme nature of some chemical-biological agents, it is not always feasible to have either a direct or close contact with such environments. In such cases, stand-off detection/imaging systems are required. Stand-Off detection/imaging systems consist of a set of methodologies adopted to detect CBRNE agents and contamination to provide fast, reliable, and real-time detection and differentiation of chemical, biological (e.g., bacteria, virus, pathogens), volatile organic compounds (VOCs), TICs/TIMs, returned/unused pharmaceuticals, and other contaminants at a distance. Generally, optical properties of nanomaterials are considered for stand-off detection/sensing/imaging applications when combined with biotechnology and quantum mechanics [2]. Several other approaches include use of nanoparticles, more specifically, quantum dots for catalysis or carriers to enable transduction [3]. A few examples are surface-enhanced Raman spectroscopy (SERS) [4] and localized surface plasmon resonance (SPR) [5]. Yet another application is in metamaterials or negative refractive index materials for potential applications in satellite imaging elements [2]. Potential challenges arise in that the source of illumination must have sufficient light intensity with minimum power consumption. Further, intensity based measurements are susceptible to intensity-based noise in signals and require appropriate signal extraction software. Yet another stand-off detection methodology is generation of high frequency electromagnetic waves using carbon nanotubes (CNTs) that identify signatures in reflected/scattered beam of potential chemical-biological agents [6]. Significant challenges still remain in the generation of high frequency using CNTs and also producing CNTs of same chirality [7]. Electromagnetic signals at THz frequencies have the advantage that their photon energies are in the low meV range and therefore not capable to break up organic molecules of living cells. Therefore they have been considered for many applications such as airport surveillance. Such scanners can detect metals such as weapons hidden under clothing from a distance of several meters, without exposing the persons to harmful radiation. There are many other applications, such as identifying the validity of documents by identifying THz metallic thin-film structures covered by only optically opaque paints. THz spectroscopy is under extensive investigation to be used to successfully characterize many materials, including packaging, explosives and drugs.

14.1.2 Signal Intelligence (SIGINT)

SIGINT[2] is remote intelligence information gathering from communications intelligence (COMINT), electronics intelligence (ELINT), and/or telemetry intelligence (TELINT). Intelligence is derived from instrumentation signals consisting of a category of devices either individually or in combination. The interception/receiving/deciphering of such transmissions can provide information on the type and location of even low power transmitters. Most military communications are restricted by encryption algorithms. To de-encrypt, complex processing is required in conjunction with additional layer of intelligence to analyze patterns and contents of transmissions over time. Generally speaking, electronic intelligence (ELINT) is analysis of non-communications electronic transmissions to include TELINT and/or radar transmitters (RADINT).

[2]http://factsindia.wordpress.com/category/standex/

Generating an Electronic order of battle (EOB) (covers both COMINT and ELINT) requires identifying SIGINT emitters in an area of interest, determining their geographic location or range of mobility, characterizing their signals, and, where possible, determining their role in the broader organizational order of battle. The Defense Intelligence Agency (DIA) maintains an EOB by location and the Joint Spectrum Center (JSC) of the Defense Information Systems Agency (DISA) supplements this location database with five more technical databases: Frequency Resource Record System (FRRS); Background Environment Information (BEI); Spectrum Certification System (SCS); Equipment Characteristics/Space (EC/S); and Tactical Database (TACDB) – i.e. platform list sorted by nomenclature. A comprehensive analysis is beyond the scope of this publication, however, it is emphasized that THz plays an incremental role in SIGINT and measurement and signature intelligence (MASINT). THz, although in its initial stage, has the potential to become a dominant feature of modern warfare. With shifting operational spectrum, the term electronic warfare (EW) is likely to be changed to electromagnetic warfare (EMW) – to effectively suppress the use of communication channels of adversaries, MASINT, optimizing its use by friendly forces, and remote plume analysis.

14.2 THz Imaging Technology – Basic Operation of Systems and Phenomenology

The following sections provide an overview of the latest information of THz technology, system concept implementation, and its implementation in terms of safety and security in the battlefield. THz region, as shown in Fig. 14.2, is a fundamental frequency range having origin in rotational and vibrational modes of molecules. Transition between rotational modes with angular momentum l–1 and l, where the symbols have their usual meaning,

$$\Delta E_l = E_{l-1} - E_l = \frac{l\,(l+1)\,\hbar^2}{2\mu r_0^2} - \frac{l\,(l-1)\,\hbar^2}{2\mu r_0^2} = \frac{l\hbar^2}{\mu r_0^2}$$

with,

$$m_1 r_1^2 + m_2 r_2^2 = \mu r_0^2$$

provides, $\Delta E_l \approx$ few meV, which overlaps with the intermolecular vibration modes of many known and potential explosives.

The electromagnetic spectrum from mm to THz can be used to create an image of an object and to gather information on its chemical composition by measuring the absorption (or refection) of electromagnetic energy simply by measuring the intensity of reflected or emitted energy. Generally, the imaging technique consists

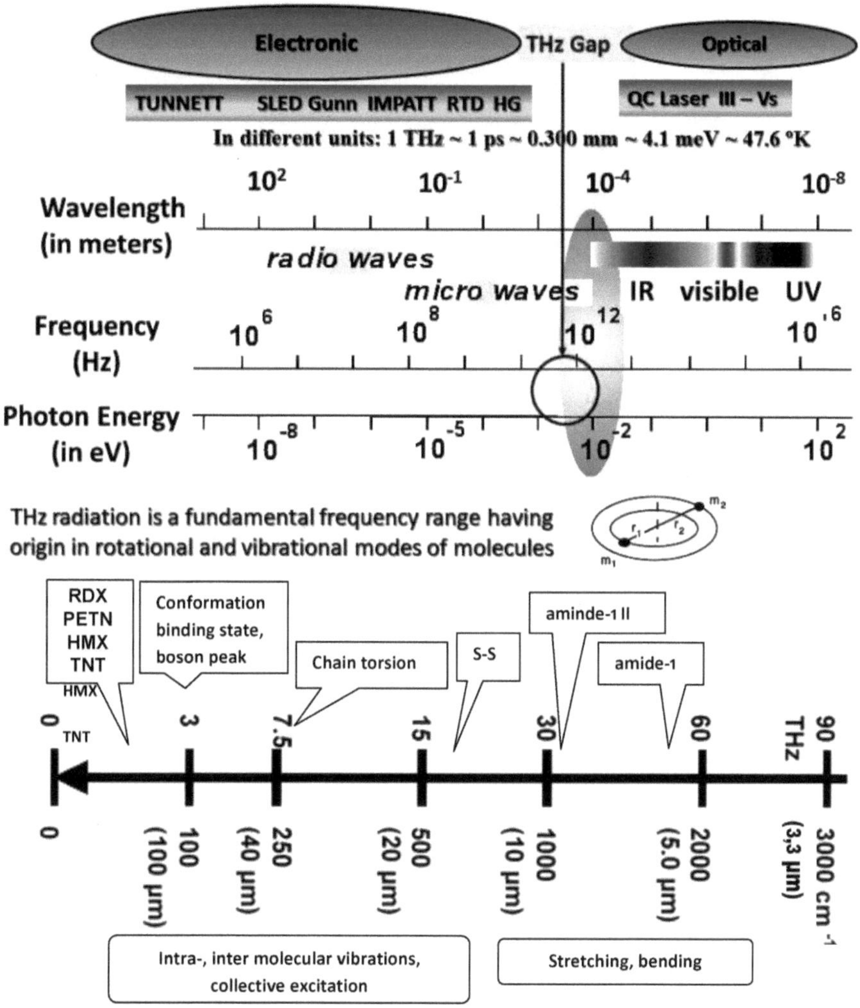

Fig. 14.2 *Upper panel*: Electromagnetic spectrum showing THz range occurring at the intersection of electronic and optical region. Tunnel injection transit-time (*TUNNETT*), super-lattice electron device (*SLED*), IMPact ionization Avalanche Transit-Time (*IMPATT*), resonant tunneling diode (*RTD*), HG, quantum cascade (*QC*) laser. *Lower panel*: THz range showing regions of intra, inter molecular vibration, stretching, bending and absorption bands of several materials of interest

of two classes – active and passive. Active imaging systems illuminate the detection space with a beam of THz power, either by illumination of the entire space or as a focused beam scanned over the project, with detectors specifically sensitive to the illuminating frequencies. Since the energy passes through most materials but only to a skin depth, hence potential adverse health impacts are significantly less than those from the competitive imaging technologies using x-rays. In passive imaging

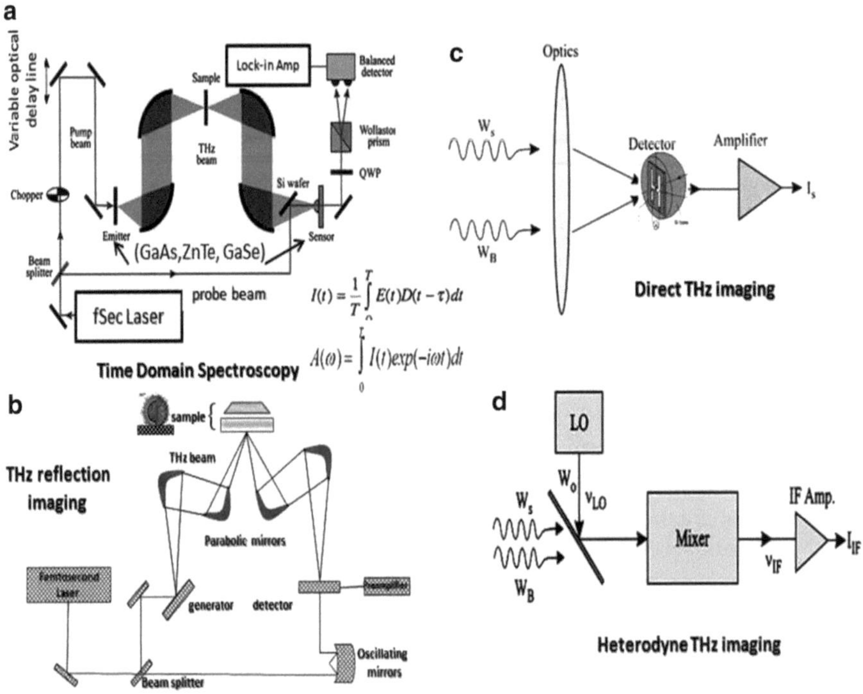

Fig. 14.3 Imaging methodologies of THz – (**a, b**) time domain spectroscopy (TDS), (**c**) direct (passive) imaging, and (**d**) heterodyne imaging

detection techniques, the methods rely on collecting naturally occurring radiation and using the emissivity contrast between objects.

Figure 14.3 shows basis of several active and passive THz imaging methods. Time domains spectroscopy (TDS) as shown in Fig. 14.3a, b is used to focus THz beam to an intermediate focus using a pair of lenses or parabolic reflectors, which are inserted into the region where the THz beam is collimated.

An object of interest (to be scanned) is placed at the focus and then the amplitude and delay of the wave that has traversed through the object is measured. By translating the object and measuring the transmitted THz waveform for each position of the object, a pixel by pixel image is formed. TDS in reflection mode is shown in Fig. 14.3b. Usually, materials with high absorption coefficients are better suited for reflection geometry and those with low absorption coefficients for transmission geometry.

The detector provides a current signal proportional to the electric field but not to the intensity, which allows the determination of the absorption coefficient and the refractive index of the sample. Heterodyning, the beating together of two closely spaced frequencies to yield the sum and difference of the original signals, has been in use since the early days of radio. The primary advantage is the acquisition of very weak, narrow band signals where direct detection and post amplification (or

Table 14.1 Object emissivity

Object	Emissivity %
Human skin	65–95
Plastics	30–70
Paper	30–70
Ceramics	30–70
Water	50
Metal	Approx. 0

even pre-amplification) followed by detection, adds electronic noise to a point that extracting signal needs long integration times. Heterodyne imaging, Fig. 14.2d, can be useful for both passive or active imaging, and in fact, in many instances can simultaneously employ the same system for both observing modes by simply turning off (or chopping) the coherent illumination source. Many different systems are used based on commercially available components and hence it is not within the scope of this publication to simply review all of them. Nevertheless, there is wide range of applications under consideration; however the scope here is limited primarily to safety and security.

Most imaging techniques rely on the contrast of temperature or emissivity. Passive systems use natural background radiation for the illumination of detection space. Every object generates EM emissions at all wavelengths with intensity proportional to the product of its physical temperature and its emissivity in accordance with Planck's radiation law. Passive imaging systems require that there be an apparent temperature difference, either positive or negative, between the body and its surroundings. While the surrounding environment is generally colder than the human body, some passive systems use non-coherent sources that surround the body to enhance contrast by making reflective objects appear warmer than the body. The passive detection systems require the ability to differentiate a temperature differential. Similar to a camera, a passive THz imager, as shown in Fig. 14.2c is able to image concealed weapons based on the implementation of detection hardware. For comparison, the Table 14.1 above lists objects emissivity. Several of such systems are in developmental stage and/or in the process of being deployed.

14.3 System Concepts and Implementation Strategies

For this report, the system requirements are limited in scope to the operational needs based on safety and security. The system configuration is further based on a narrow premise of explosive detection in a stand-off configuration in support of concealed-object identification, primarily for transport security; stand-off detection of explosives; and measurement and signature intelligence – albeit very broad, the premise includes picking up "a" signature from distance and analyzing the received information. Many different variations are envisaged and possible.

Primarily, the THz system-development activity is to primarily develop a technology to accomplish the following tasks:

(a) Stand-off detection of the presence, location, and identification of weapon, explosive compounds and devices, and other items of interest.
(b) Distant monitoring of a person carrying metallic objects, weapons, explosive materials and devices, other items of security interest, while maintaining privacy of all individuals.
(c) Stand-off detection of plume of TICs/TIMs, biologics, and other agents of interest. Collection and analysis of signatures of various compounds and capability to conduct real-time analysis. An extension of the project is to gather an electronic signature to analyze, jam/defeat if necessary, and other inferences that can be derived from SIGINT, in support of MASINT and HUMINT.

The principal system components are provided below:

(a) A detector array and/or scanning device
(b) Image acquisition – software/hardware
(c) Image interpretation/recognition/computation
(d) Database containing images/spectra for comparison
(e) Display hardware/devices
(f) Network interface and other elements of the layered system

The system development depends on the requirement, taking into account several trade-offs between design related issues and performance requirements. In fact, there exist many guidelines for various commercial systems presently under development. With the advancement of new technologies, such as nanotechnologies, many new configurations are possible. The discussion below is limited to system capabilities and how new, advanced, and nanotechnologies can improve the system performance.

The process of identification begins with detection followed by recognition and classification using image-recognition algorithm and matching it with items of interest. The recognition process goes through several hierarchical thresholds for identification and matching with high probability of occurrence and low probability of false recognition. Such steps are based on reflective properties of substances, as shown below in Table 14.2, showing reflective properties of basic explosives, and human flesh.

This data is further normalized with reflection of clothing material, which is also stored in the database. The final assessment of identification is difficult and requires creative and innovative algorithm design. Many active millimeter wave imaging systems are commercially available, such as Pacific Northwest National Laboratory[3], Safe-view by L3 communication[4], QinetiQ[5], Agilent technologies[6],

[3]http://www.technet.pnnl.gov/sensors/chemical/projects/ES4THzSpec.stm

[4]http://www.sds.l-3com.com/

[5]http://www.qinetiq.com/Pages/default.aspx

[6]http://thznetwork.net/index.php/archives/1813

Table 14.2 Reflective properties of basic explosives, and human flesh

Substance/name	Molecular weight	Density (g/cm^3)	Dielectric constant	Reflectivity: Reflectance (R)	Reflectivity: Decibel (−dB)
2,4,6 Trinitrotoluene (TNT)	227.13	1.65	2.7	−0.24	12.3
Hexahydro-1,3,5-trinitro-1,3,5-triazine (RDX)	222.26	1.83	3.14	−0.28	11.1
Cyclotetramethylene-tetranitramine (HMX)	296.16	1.96	3.08	−0.27	11.2
Pentaerythritol tetranitrate (PETN)	316.2	1.78	2.72	−0.25	12.2
2,4,6-Trinitrophenyl-N-methylnitramine (Tetryl)	287.15	1.73	2.9	−0.26	11.7
Nitroglycerin (NG)	227.09	1.59	19	−0.63	4.1
Ammonium nitrate (AN)	80.05	1.59	7.1	−0.45	6.9
RDX TNT (COMP B)			2.9	−0.26	11.7
RDX (COMP C-4)			3.14	−0.28	11.1
PETN (Detasheet)			2.72	−0.25	12.2
HMX TNT (Octol)			2.9	−0.26	11.7
RDX-PETN (Semtex-H)			3	−0.27	11.4
Human flesh (H_2O + Nacl)		0.93	88	−0.81	1.9

Millitech[7], Trex Enterprise Corporation[8], Brijot[9] and Millivision[10], as some of the primary vendors.

As the THz systems are currently under development, some of the latest technologies are considered with objective that such THz sources will ultimately provide;

(a) Nondestructive inspection through dielectrics using TTDS pulse techniques,
(b) Medical diagnostics through skin or thin tissue for non-intrusive purposes,
(c) Detection of undesired metals and contraband hidden underneath clothing,
(d) Emission of signal detected/identifies from a finite distance.

With the reduced dimensional materials, new and unique characteristics evolve and are considered for integrations from systems perspectives [2]. The first focus is on THz generation such that relatively large signal powers can be realized. One of the configurations is based on an electron resonance structure formed by

[7] http://www.millitech.com/

[8] http://www.trexenterprises.com/

[9] http://www.microsemi.com/products/screening-solutions (redirected)

[10] http://www.microsemi.com/products/screening-solutions (redirected)

a semiconductor hetero-junction structure [8]. The electrons in n-layer needs to be accelerated by V_e – applied alternating voltage, such that they reach (length $L \approx 150$ nm) the barrier of the high-mobility n-type wide bandgap semiconductor and are reflected back without a loss of kinetic energy. The electrons then travel towards the opposite barrier ballistically, where they are reflected again. When V_e changes the polarity, this process continues resulting in electron resonance producing THz signals. An extension of the ballistic device is the use of simultaneous D.C. biasing and the generation of mobile electrons in a short quantum well of smaller gap values by a pulsed optical signal. The optical pulses can be longer than half of the THz period due to space charge limited effect of electron bunch transfer.

Another configuration under consideration is a heterostructure equivalent to the well-known step recovery diode, arranged as a double structure. This provides a new nonlinear device configuration for highly efficient harmonic THz signal generation. In fact, two barriers opposing each other via narrow-gap semiconductors, such as AlGaAs/GaAs/AlGaAs or InAlAs/InGaAs/InAlAs with a proper n-doping can be considered as two step-recovery-diode junctions in opposition, which functions without any applied D.C. bias requirement. To generate THz radiation using a set of suitable materials as targets, the optical excitation can be provided [8] by typically a 12-fs mode-locked Ti:Sapphire laser of center frequency $\approx$790 nm and repetition rate $\approx$75 MHz. The emitted THz radiation is detected either using a pneumatic Golay cell (incoherent detection) or in a conventional TDS arrangement using electro-optic (coherent) detection.

New types of THz generators are based on field emission of electrons into vacuum using carbon nanotubes (CNTs) based structures that are capable of producing high current densities (based on assumptions of field amplification factor and estimation of electron work function) [9]. Industry requires a miniaturized source to enhance speed and resolution of the scanning process. A portable THz source for this frequency range uses the Dynatron oscillator concept based on travelling-wave structures [10]. This is composed of a triode tube with a grid voltage higher than the anode voltage. This configuration accelerates secondary electrons from the anode to the grid, which makes the dynatron to act as a negative resistance device. A serial or parallel oscillator circuit is linked between the anode and a working point potential source using a lower value than at the extraction grid.

14.4 Forward Thinking and Pathways for Future Implementation in Safety and Security

Due to advantages described above, the field of THz generation, detection, and imaging is still in the R&D infancy phase. Limited or no commercial or operationally deployed systems exist as of yet, for security applications.

Many approaches are pursued and can be divided in the following categories to achieve desired applications, as shown in Fig. 14.4.

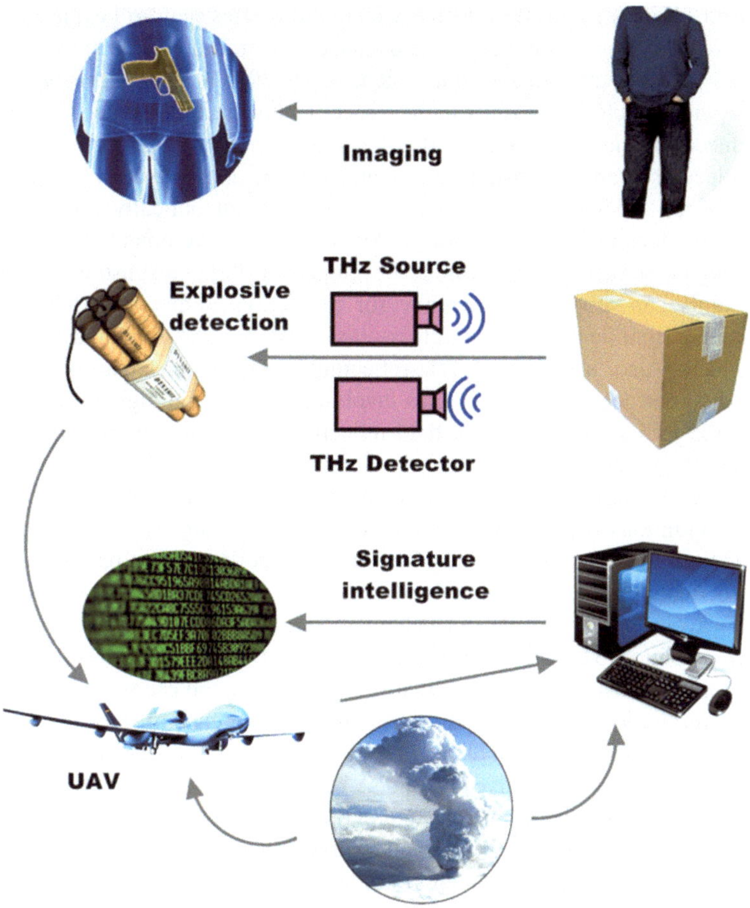

Fig. 14.4 Applications of THz imaging including imaging, detection, MASINT, and plume monitoring

(a) Sub millimeter wave electronic components and system for single frequency operation between 300 and 600 GHz.
(b) Broad-band imaging and spectroscopy
(c) Component and system development for systems above 600 GHz.

Sub millimeter wave systems are designed for security applications with electronic source at 600 GHz for targeting at approx. few 10s of meters. Receivers mainly concentrate on heterodyne techniques. Above 600 GHz, the work is focused primarily at component level. Microbolometers are used for broadband solution for detector arrays. In addition to being low cost solution, the devices can be used for wide range of frequency and temperature range – based on the sensitivity required.

Pulsed THz system, as described earlier demonstrates stand-off explosive detection using reflection spectroscopy at a distance of 1 m. Liu et al. [11] have shown a collimated THz beam with detection distance up to 30 m in air, however with optical system placed in close proximity to the target to gather reflection spectra accurately.

Reduced dimensional systems offer many additional options and are considered as "molecular scanners" using broadband THz at approximately 10 nm scale. Furthermore, "optical matrix codes" covered with an optically non-transparent coat can be deciphered by THz waves. Such systems have potential for inventory management, if such patterns can be developed on flexible substrates. Similarly, such paints can also cover currency for protection from counterfeit currency, deciphering confidential digital information, and reading digital authentication of products. Also, a matrix of THz-reflecting metal antennas can be employed, where the presence of such structures involving confidential digital information is deciphered by identifying the far-field reflection pattern by an array of receiving antennas.

Imaging techniques is a critical issue towards the realization of a THz camera. To obtain Continuous Wave (CW) operation the main challenge is power handling via suitable heat sinking strategies [12]. The image detection is then based on an array of photoconductive antenna heterodyne receivers illuminated with two phase-locked optical wavelengths obtained from an Optical Frequency Comb Generation (OFCG) for CW operation. This architecture allows for an efficient, phase controlled, Local Oscillator (LO) distribution with low losses intrinsic to the use of optical fibers, as well as amplitude and phase recovery due to the coherence of the LO distributed. Tunable operation is achieved using the different lines of the OFCG and broadband design for the photoconductive antenna. A recent development of the OFC source in combination with radiofrequency and photonic electronic technique has resulted in synthesizing very high quality signals, while reducing overall size[11].

THz radiation can pass through clothing and packaging, but they are strongly absorbed by metals and many other inorganic substances. THz sources use a number of basic techniques, namely either harmonic extraction from the mm-waves or using various methods from the optical signals. The possibility of deep-infrared lasers by quantum-cascading reaches the low THz frequencies of interest only by cooling to liquid nitrogen or below.

THz however presents some drawbacks thus limiting its use in every day medical use. These limitations cover a wide range from the low-performance of emitting sources to the low sensitivity or selectivity to pathological tissues. Nanotechnology-based techniques seems to be a crucial key tool in their efforts to improve these imaging modalities, by using nanoparticles as contrast agents. As mentioned earlier, CNT is a suitable candidate for compact THz source. Similarly, many nanomaterials such as quantum dots (QDs) based systems contain free electrons, thus due to their intrinsic discreet energy level, long carrier relaxation times and the ability to control these times, nanomaterials offer the way forward for QD-based THz optoelectronic

[11]http://portal.uc3m.es/portal/page/portal/actualidad_cientifica/noticias/terahertz_luz_wavelabs

devices [13]. Furthermore, due to existence of surface plasmons, non-linear optical phenomena are enhanced due to the strong interactions and the high field strengths due to SP excitation and multiphoton photoelectric effect.

14.5 Conclusion

After successful launch of millimeter wave systems, higher frequencies in the THz range are now beginning to become more widely used in this field. Higher frequencies offer compact systems, especially using materials in the reduced dimensions. Many components and systems are currently under development up to sub-millimeter to low THz frequency range. The current, THz system are still at the R&D laboratory scale, nevertheless the envisaged applications range from remote/stand-off weapon depletion, imaging concealed metal objects, and analysis plume from a distance. Many other applications exist such as counterfeit currency detection, inventory management, protecting classified information, and digital authentication of products to list a few. Before systems can be further developed, additional research is required, both in source and detector technology. With additional emphasis on stand-off detection systems, use of nanomaterial based systems offer plausible solutions as light weight THz generators and detectors. THz continues to demonstrate a strong promise as a technique for screening due to its potential for material specific detection. Finally, with lighter weight devices, it will be possible to mount such devices on a UAV for MASINT, plume monitoring, and also for digital signature recognition and localized monitoring for Command, Control, Communications, Computers, Intelligence, Surveillance and Reconnaissance (C4ISR).

References

1. Vaseashta A, Mihailescu I (2007) Functionalized nanoscale materials, devices, and systems. Springer, Dordrecht
2. Vaseashta A (2012) Ecosystem of innovations in nanomaterials based CBRNE sensors and threat mitigation. In: Vaseashta A, Khudaverdyan S (eds) Advanced sensors for safety and security 2012. Springer, Dordrecht
3. Denkbas E et al (2012) Nanoplatforms for detection, remediation, and protection against chem-bio-warfare. In: Vaseashta A, Braman E, Susmann P (eds) Technological innovations in sensing and detection of chemical, biological, radiological, nuclear threats and ecological terrorism. Springer, Dordrecht. ISBN 978-94-007-2488-4
4. Sharma B, Frontiera RR, Henry A-I, Ringe E, VanDuyne R (2012) SERS: materials, applications, and the future. Mater Today 15(1–2):16–25
5. Vaseashta A, Irudayaraj J (2005) Nanostructured and nanoscale devices and sensors. J Optoelectron Adv Mater 7(1):35–42
6. Hartnagel HL, Ong DS, Oprea I (2009) Ballistic Electron Wave Swing (BEWAS) to generate THz-signal power. Frequenz 63(3–4):60–62

7. Vaseashta A (2003) Field-emission characteristics of carbon nanotubes and their applications in photonic devices. J Mater Sci Mater Electron 14(10–12):653–656
8. Criado AR, De Dios C, Acedo P, Hartnagel HL (2012) New concepts for a photonic vector network analyzer based on THz heterodyne phase-coherent techniques. Proceedings of European Microwave Week, Nov 2012, Amsterdam, The Netherlands
9. Brodie I, Spindt CA (1992) Adv Electron Electron Phys 83:1–106
10. Koops HWP, Al–Daffaie S, Hartnagel HL (2012) Portable source with free electron beams for 0.1 to 10 THZ. In: Proceedings of WOCSDICE-EXMATEC 2012, CRHEA-CNRS in Porquerolles, France
11. Liu HB, Chen Y, Bastiaans GJ, Zhang XC (2006) Detection and identification of explosive RDX by THz diffuse reflection spectroscopy. Opt Express 14(1):415–423
12. Al-Daffaie S, Acedo P, Hartnagel HL (2012) Simulation of a CW THz camera scheme. In: Proceedings of WOCSDICE-EXMATEC 2012, CRHEA-CNRS in Porquerolles, France
13. Ferguson B, Zhang XC (2002) Materials for terahertz science and technology. Nat Mater 1(1):26–33

MIX
Papier aus verantwortungsvollen Quellen
Paper from responsible sources
FSC® C105338

If you have any concerns about our products,
you can contact us on
ProductSafety@springernature.com

In case Publisher is established outside the EU,
the EU authorized representative is:
Springer Nature Customer Service Center GmbH
Europaplatz 3, 69115 Heidelberg, Germany

Printed by Libri Plureos GmbH
in Hamburg, Germany